STATISTICAL METHODS for GEOGRAPHY

SAGE was founded in 1965 by Sara Miller McCune to support the dissemination of usable knowledge by publishing innovative and high-quality research and teaching content. Today, we publish more than 750 journals, including those of more than 300 learned societies, more than 800 new books per year, and a growing range of library products including archives, data, case studies, reports, conference highlights, and video. SAGE remains majority-owned by our founder, and on her passing will become owned by a charitable trust that secures our continued independence.

Los Angeles | London | Washington DC | New Delhi | Singapore

Peter A. Rogerson

STATISTICAL METHODS *for* GEOGRAPHY

FOURTH EDITION

A STUDENT'S GUIDE

$SAGE

Los Angeles | London | New Delhi
Singapore | Washington DC

Los Angeles | London | New Delhi
Singapore | Washington DC

SAGE Publications Ltd
1 Oliver's Yard
55 City Road
London EC1Y 1SP

SAGE Publications Inc.
2455 Teller Road
Thousand Oaks, California 91320

SAGE Publications India Pvt Ltd
B 1/I 1 Mohan Cooperative Industrial Area
Mathura Road
New Delhi 110 044

SAGE Publications Asia-Pacific Pte Ltd
3 Church Street
#10-04 Samsung Hub
Singapore 049483

Editor: Robert Rojek
Assistant editor: Keri Dickens
Production editor: Katherine Haw
Copyeditor: Catja Pafort
Proofreader: Neil Dowden
Marketing manager: Michael Ainsley
Cover design: Francis Kenney
Typeset by: C&M Digitals (P) Ltd, Chennai, India
Printed in India at Replika Press Pvt Ltd

This edition first published 2015
2nd edition: Reprinted 2008 and 2009, 3rd edition: Reprinted
2010, 2011 and 2013

Library of Congress Control Number: 2014931420

British Library Cataloguing in Publication data

A catalogue record for this book is available from
the British Library

MIX
Paper from
responsible sources
FSC® C016779

ISBN 978-1-4462-9572-4
ISBN 978-1-4462-9573-1 (pbk)

At SAGE we take sustainability seriously. Most of our products are printed in the UK using FSC papers and boards.
When we print overseas we ensure sustainable papers are used as measured by the Egmont grading system.
We undertake an annual audit to monitor our sustainability.

CONTENTS

ABOUT THE AUTHOR

Peter A. Rogerson is SUNY (State University of New York) Distinguished Professor in the Department of Geography at the University at Buffalo, Buffalo, New York, USA. He also holds an adjunct appointment in the Department of Biostatistics and is a member of the National Center for Geographic Information and Analysis.

PREFACE

The development of geographic information systems (GIS), an increasing availability of spatial data, and recent advances in methodological techniques have all combined to make this an exciting time to study geographic problems. During the late 1970s and throughout the 1980s, there had been among many an increasing disappointment in, and questioning of, the methods developed during the quantitative revolution of the 1950s and 1960s. Perhaps this reflected expectations that were initially too high – many had thought that sheer computing power coupled with sophisticated modeling would 'solve' many of the social problems faced by urban and rural regions. But the poor performance of spatial analysis that was perceived by many was at least partly attributable to a limited capability to access, display, and analyze geographic data. During the last decade, geographic information systems have been instrumental in not only providing us with the capability to store and display information, but also in encouraging the provision of spatial datasets and the development of appropriate methods of quantitative analysis. Indeed, the GIS revolution has served to make us aware of the critical importance of spatial analysis. Geographic information systems do not realize their full potential without the ability to carry out methods of statistical and spatial analysis, and an appreciation of this dependence has helped to bring about a renaissance in the field.

Significant advances in quantitative geography have been made during the past decade, and geographers now have both the tools and the methods to make valuable contributions to fields as diverse as medicine, criminal justice, and the environment. These capabilities have been recognized by those in other fields, and geographers are now routinely called upon as members of interdisciplinary teams studying complex problems. Improvements in computer technology and computation have led quantitative geography in new directions. For example, the new field of geocomputation (see, e.g., Longley et al. 1998) lies at the intersection of computer science, geography, information science, mathematics, and statistics. The recent book by Fotheringham et al. (2000) also summarizes many of the new research frontiers in quantitative geography.

The purpose of this book is to provide undergraduate and beginning graduate students with the background and foundation that is necessary to be prepared for spatial analysis in this new era. I have deliberately adopted a fairly traditional approach to statistical analysis, along with several notable differences. First, I have attempted to condense much of the material found in the beginning of introductory texts on the subject. This has been done so that there is an opportunity to progress further in important areas, such as regression analysis and the analysis of geographic patterns in one semester's time. Regression is by far the most common method used in geographic analysis, and it is unfortunate that it is often left to be covered hurriedly in the last week or two of a 'Statistics in Geography' course.

The level of the material is aimed at upper-level undergraduate and beginning graduate students. I have attempted to structure the book so that it may be used as either a first-semester or second-semester text. It may be used for a second-semester course by those students who already possess some background in introductory statistical concepts. The introductory material here would then serve as a review. However, the book is also meant to be fairly self-contained, and thus it should also be appropriate for those students learning about statistics in geography for the first time. First-semester students, after completing the introductory material in the first few chapters, will still be able to learn about the methods used most often by geographers by the end of a one-semester course; this is often not possible with many first-semester texts.

In writing this text, I had several goals. The first was to provide the basic material associated with the statistical methods most often used by geographers. Since a very large number of textbooks provide this basic information, I also sought to distinguish it in several ways. I have attempted to provide plenty of exercises. Some of these are to be done by hand (in the belief that it is always a good learning experience to carry out a few exercises by hand, despite what may sometimes be seen as drudgery!), and some require a computer. Although teaching the reader how to use computer software for statistical analysis is *not* one of the specific aims of this book, some guidance on the use of *SPSS for Windows* is provided. It is important that students become familiar with *some* software that is capable of statistical analysis. An important skill is the ability to sift through output and pick out what is important and what is not. Different software will produce output in different forms, and it is also important to be able to pick out relevant information whatever the arrangement of output.

In addition, I have tried to give students some appreciation for the special issues and problems raised by the use of geographic data. Straightforward application of the standard methods ignores the special nature of spatial data, and can lead to misleading results. Topics such as spatial autocorrelation and the modifiable areal unit problem are introduced to provide a good awareness of these issues, their consequences, and potential solutions. Because a full treatment of these topics would require a higher level of mathematical sophistication, they are not covered fully, but pointers to other, more advanced work and to examples are provided.

Another objective has been to provide some examples of statistical analysis that appear in the recent literature in geography. This should help to make clear the relevance and timeliness of the methods. Finally, I have attempted to point out some of the limitations of a confirmatory statistical perspective, and have directed the student to some of the newer literature on exploratory spatial data analysis. Despite the popularity and importance of exploratory methods, inferential statistical methods remain absolutely essential in the assessment of hypotheses. This text aims to provide a background in these statistical methods and to illustrate the special nature of geographic data.

A Guggenheim Fellowship afforded me the opportunity to finish the manuscript during a sabbatical leave in England. I would like to thank Paul Longley for his careful reading of an earlier draft of the book. His excellent suggestions for revision have led to a better final result. Yifei Sun and Ge Lin also provided comments that were very helpful in revising earlier drafts. Art Getis, Stewart Fotheringham, Chris Brunsdon, Martin Charlton, and Ikuho Yamada suggested changes in particular sections, and I am grateful for their assistance. Emil Boasson and my daughter, Bethany Rogerson, assisted with the production of the figures. I am thankful for the thorough job carried out by Richard Cook in editing the manuscript. Finally, I would like to thank Robert Rojek at SAGE Publications for his encouragement and guidance.

PREFACE TO THE SECOND EDITION

In the first edition of *Statistical Methods for Geography*, a primary goal was to ensure adequate introductory treatment of regression and the statistical analysis of geographic patterns. These topics are of paramount importance to geographers, and there often is not time in a first course on quantitative methods to give them the attention that they deserve. A primary goal in this second edition has been to give more space to introductory topics in probability and statistics; consequently, the presentation of this important foundational material is not as condensed as it was in the previous edition. As a result, I am hopeful that the book will now be more appealing as a choice for a first-semester, as well as a second-semester text.

In particular, more introductory material has been added to the first chapter. Much of the second chapter, on descriptive statistics, is entirely new. Material on additional probability distributions has been added to what are now Chapters 3 and 4. In addition, the number of exercises has been increased, and implementation of the methods is described for users of *Excel*, as well as for users of *SPSS*. Some of the added exercises are built around two new datasets – one consisting of measures of cell or mobile phone signal strength for over 200 geographic locations near Buffalo, New York, and the other consisting of the sales price for over 500 houses, along with associated characteristics of both the houses and their neighborhoods in Tyne and Wear.

Some of the more technical and peripheral material from the first edition has been removed, and greater emphasis has been placed upon interpretation. For example, operational details for statistical tests of the assumptions of homoscedasticity and normality have been dropped, and there is now a more extended discussion of the consequences of the failure to meet the assumptions of the various tests. In particular, further emphasis has been placed upon the assumption of statistical dependence.

The development of this second edition has benefited from the many constructive suggestions made by the anonymous reviewers who were solicited to provide feedback on the first edition and ideas for the second. Barry Solomon and others pointed out typos in the first edition, and I acknowledge gratefully their contribution. Jessie Poon pointed out a number of recent and relevant articles. Daejoing Kim assisted with the production of several figures, and Gyoungju Lee provided helpful comments. I thank Stewart Fotheringham for providing the Tyne and Wear housing data. I am grateful to Vanessa Harwood at Sage for her fine efforts in the production of the manuscript. I also thank Anjana Narayanan and Frances Morgan at Keyword Group for their work with the manuscript. Finally, I would once again like to thank Robert Rojek at SAGE for his efforts and advice. It is not uncommon for the life cycle of a book to experience several different editors (which says something about the time it takes to develop a book and/or the turnover rate of editors!); I have been fortunate to have the steady guidance of a top-notch editor for both editions.

PREFACE TO THE THIRD EDITION

Two of the primary goals in this third edition have been to (a) add illustrative, worked exercises at the end of each chapter, and (b) give additional attention to the implementation of spatial regression (through the use of the free and widely used *GeoDa* software; see Anselin et al. 2006). With regard to the former, the worked exercises are intended to supplement both the illustrations that already appear within the text and the exercises that appear at the end of each chapter. They are what might be regarded as 'mixed exercises'; unlike the examples that are introduced with each concept, these require the reader to think about the collection of ideas introduced in the chapter. With regard to the latter, it is essential that geographers be aware of the effects of spatial dependence on statistical analyses, and that they be able to use tools designed to uncover geographic patterns, and additions to this edition should help in that regard.

The third edition has benefited from constructive suggestions made by the anonymous reviewers who were solicited by SAGE to provide feedback on the second edition and ideas for the third. I would like to thank Jared Aldstadt, Greg Aleksandrovs, Richard Brandt, Jerry Davis, David Mark, Tom Panczykowski, Gregory Taff, and Enki Yoo for pointing out typos and inconsistencies in the second edition; and I would like to acknowledge the contribution to this edition made by Raymond Greene – Department of Geography, Western Illinois University – who prepared the chapter ancillaries on the book's companion website (www.sagepub.co.uk/rogerson). I am grateful to Sarah-Jayne Boyd at SAGE for her assistance with the manuscript. Finally, I would like to thank Katherine Haw, and Robert Rojek at SAGE a third time for all of his kind guidance.

PREFACE TO THE FOURTH EDITION

This new edition includes new sections on the measurement of distance (in Chapter 2) and Type II errors and the concept of statistical power (Chapter 5). Perhaps more importantly, a complete re-working of the presentations, explanations, and examples from the third edition should (I hope!) improve both the clarity and the reader's level of understanding.

Another new feature is the addition of 'Answers for Selected Exercises', which appear at the end of the book.

Like the previous editions, this one has also benefited from the helpful comments from many people. I am grateful to Jared Aldstadt, Richard Brandt, Sarah Glann, Eva Johansson, Paul Johnson, Xiaoxing Qin, Fernando Rios, and Enki Yoo for pointing out typos and for their suggestions. Rusty Weaver assisted by providing information that was useful in updating some of the material. I am also grateful for Peter Kedron's assistance with a number of the figures. I would especially like to thank Keri Dickens and Katherine Haw at SAGE for their help in putting this edition together. Finally, I am grateful for the support of Robert Rojek at SAGE – he has now provided both encouragement and sound advice on all four editions.

ABOUT THE COMPANION WEBSITE

PRELIMINARY MATERIAL: GUIDE TO HOW TO USE THIS WEBSITE

Specially developed for the fourth edition, be sure to visit the *Statistical Methods for Geography* companion website at **study.sagepub.com/rogerson4e** to find a range of interactive teaching and learning materials for both lecturers and students including:

- Datasets: Each dataset utilised in the text is included with a description of its format and fields
- Further Resources: Links to websites, videos, podcasts, and other media which supplements the information presented in the text
- Hyperlinked text that provides further information on the topics discussed
- Links to journal articles and book chapters mentioned in the text

1

INTRODUCTION TO STATISTICAL METHODS FOR GEOGRAPHY

1.1 INTRODUCTION

The study of geographic phenomena often requires the application of statistical methods to produce new insight. The following questions serve to illustrate the broad variety of areas in which statistical analysis has been applied to geographic problems:

1. How do blood lead levels in children vary over space? Are the levels randomly scattered throughout the city, or are there discernible geographic patterns? How are any patterns related to the characteristics of both housing and occupants? (Griffith et al. 1998).

2. Can the geographic diffusion of democracy that has occurred during the post-World War II era be described as a steady process over time, or has it occurred in waves, or have there been 'bursts' of diffusion that have taken place during short time periods? (O'Loughlin et al. 1998).

3. What are the effects of global warming on the geographic distribution of species? For example, how will the type and spatial distribution of tree species change in particular areas? (MacDonald et al. 1998).

4. What are the effects of different marketing strategies on product performance? For example, are mass-marketing strategies effective, despite the more distant location of their markets? (Cornish 1997).

These studies all make use of statistical analysis to arrive at their conclusions. Methods of statistical analysis play a central role in the study of geographic problems – in a survey of articles that had a geographic focus, Slocum (1990) found that 53% made use of at least one mainstream quantitative method. The role of statistical analysis in geography may be placed within a broader context through its connection to the 'scientific method', which provides a more general framework for the study of geographic problems.

1.2 THE SCIENTIFIC METHOD

Social scientists as well as physical scientists often make use of the *scientific method* in their attempts to learn about the world. Figure 1.1 illustrates this method, from the initial attempts to organize ideas about a subject, to the building of a theory.

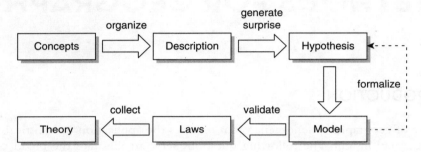

Figure 1.1   The scientific method

Suppose that we are interested in describing and explaining the spatial pattern of cancer cases in a metropolitan area. We might begin by plotting recent incidences on a map. Such descriptive exercises often lead to an unexpected result – in Figure 1.2, we perceive two fairly distinct clusters of cases. The surprising results generated through the process of description naturally lead us to the next step on the route to explanation by forcing us to generate hypotheses about the underlying process. A 'rigorous' definition of the term *hypothesis* is a proposition whose truth or falsity is capable of being tested. We can also think

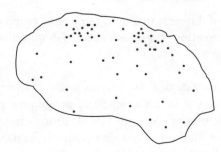

Figure 1.2   Distribution of cancer cases

of hypotheses as potential answers to our initial surprise. For example, one hypothesis in the present example might be that the pattern of cancer cases is related to the distance from local power plants.

To test the hypothesis, we need a *model*, which is a device for simplifying reality so that the relationship between variables may be more clearly studied. Whereas a hypothesis might suggest a relationship between two variables, a model is more detailed, in the sense that it suggests the nature of the relationship between the variables. In our example, we might speculate that the likelihood of cancer declines as the distance from a particular power plant increases. To test this model, we could plot cancer rates for a subarea versus the distance from the subarea centroid to the power plant. If we observe a downward sloping curve, we have gathered some support for our hypothesis (see Figure 1.3).

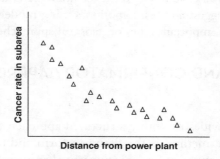

Figure 1.3   Cancer rates versus distance from power plant

Models are validated by comparing observed data with what is expected. If the model is a good representation of reality, there will be a close match between the two. If observations and expectations are far apart, we need to 'go back to the drawing board', and come up with a new hypothesis. It might be the case, for example, that the pattern in Figure 1.2 is due simply to the fact that the population itself is clustered. If this is true, the spatial pattern of cancer then becomes understandable; a similar rate throughout the population generates apparent cancer clusters because of the spatial distribution of the population.

Though models are often used to learn about particular situations, more often one also wishes to learn about the underlying process that led to it. We would like to be able to *generalize* from one study to statements about other situations. One reason for studying the spatial pattern of cancer cases is to determine whether there is a relationship between cancer rates and the distance to *specific* power plants; a more general objective is to learn about the relationship between cancer rates and the distance to *any* power plant. One way of making such generalizations is to accumulate a lot of evidence. If we were to repeat our analysis in many locations throughout a country, and if our findings were similar in all cases, we would have uncovered an empirical generalization. In a strict sense, *laws* are sometimes defined as universal statements of unrestricted range – that is, the statement may be applied to *any* place, during *any* time period. In our example, we might not expect our generalization to have an unrestricted range, and we might want, for example, to confine our generalization or empirical law to power plants and cancer cases in the country of interest.

Einstein called theories 'free creations of the human mind'. In the context of our diagram, we may think of theories as collections of generalizations or laws. The collection is greater than the sum of its parts in the sense that it gives greater insight than that produced by the individual generalizations or laws alone. If, for example, we generate other empirical laws that relate cancer rates to other factors, such as diet, we begin to build a theory of the spatial variation in cancer rates.

Statistical methods occupy a central role in the scientific method, as portrayed in Figure 1.1, because they allow us to suggest and test hypotheses using models. In the following section, we will review some of the important types of statistical approaches in geography.

1.3 EXPLORATORY AND CONFIRMATORY APPROACHES IN GEOGRAPHY

The scientific method provides us with a structured approach to answering questions of interest. At the core of the method is the desire to form and test *hypotheses*. As we have seen, hypotheses may be thought of loosely as potential answers to questions. For instance, a map of snowfall may suggest the hypothesis that the distance away from a nearby lake may play an important role in the distribution of snowfall amounts.

Geographers use spatial analysis within the context of the scientific method in at least two distinct ways. *Exploratory* methods of analysis are used to *suggest* hypotheses; *confirmatory* methods are, as the name suggests, used to help confirm hypotheses. A method of visualization or description that led to the discovery of clusters in Figure 1.2 would be an exploratory method, while a statistical method that confirmed that such an arrangement of points would have been unlikely to occur by chance would be a confirmatory method. In this book, we will focus primarily upon confirmatory methods.

We should note two important points here. First, confirmatory methods do not always confirm or refute hypotheses – the world is too complicated a place, and the methods often have important limitations that prevent such absolute confirmation and refutation. Nevertheless, they are important in structuring our thinking and in taking a rigorous and scientific approach to answering questions. Second, the use of exploratory methods over the past few years has been increasing rapidly. This has come about as a result of a combination of the availability of large databases and sophisticated software (including GIS), and a recognition that confirmatory statistical methods are appropriate in some situations and not others. Throughout the book, we will keep the reader aware of these points by pointing out some of the limitations of confirmatory analysis.

1.4 PROBABILITY AND STATISTICS

1.4.1 Probability

Probability may be thought of as a measure of uncertainty, with the measure taking on a value ranging from zero to one. Experiments and processes often have many possible outcomes, and the specific outcome is uncertain until it is observed. If we happen to know that a particular outcome will definitely not occur, that outcome has a probability of zero. At the other extreme, if we know that an outcome *will* occur, it is said to have a probability of one. At these two extremes, there is no uncertainty regarding whether the outcome will occur. When we are maximally uncertain about the outcome of an event, its probability is equal to 0.5 (i.e., 1/2); there is a '50–50' chance that it will occur. A major focus of the study of probability is the study of the likelihood of various outcomes. How likely or probable is it that a town will be struck by two hurricanes in one season? What is the probability that a resident in a community who is 4 km from a new grocery store will become a new customer?

Probabilities may be derived in a variety of ways, ranging from subjective beliefs, to the use of relative frequencies of past events. When guessing whether a coin will come up heads when tossed, you may choose to believe that the probability is 0.5, or you could actually toss the coin many times to determine the proportion of times that the result is heads. If you tossed it 1000 times, and it came up heads 623 times, an estimate of the probability of heads that relied on relative frequency would be 623/1000 = 0.623.

The study of probability has its origins, at least to some degree, in questions of gambling that arose in the 17th century. In particular, correspondence between Pascal and Fermat in 1654 concerned how to properly resolve a game of chance proposed by deMéré that had to be terminated before its conclusion. Suppose that the first player with three wins is declared the overall winner, and can lay claim to the prize of 64 pistoles (Pistole is the French word given at the time to a Spanish gold coin). How should the prize money be divided, given that the game had to be terminated, and given that one player had two wins, and the other had one win? DeMéré argued that the first player should receive 2/3 of the prize, because that person had two-thirds of the wins. Since 2/3 of 64 is 42.67, the first player would receive the remaining 21.33 pistoles.

Fermat wrote to Pascal with an alternative solution. Pascal agreed, and in a letter to Fermat suggested a more compact way of stating the solution. In particular, he argued that the two players should consider what could happen if they continued. With probability equal to ½, the player with only one win would win the next round, and they could then split the pot of money (each receiving 32 pistoles), since they would then have an equal chance of winning the contest. With probability also equal to ½, the player with two wins would win the next round, and consequently the entire prize of 64 pistoles. Pascal argued that this player's fair share was an average of these two, equally likely outcomes; his rightful share was therefore (32+64)/2 = 48 (and not 42.67, as DeMéré had suggested). Pascal's reasoning, which is based upon probabilities and possible outcomes, forms the basis of modern probability.

What is the difference between *probability* and *statistics*? The field of probability provides the mathematical foundation for statistical applications. Year-long courses in probability and statistics are often subdivided into a first-semester course on probability, and then a second-semester course in statistics. Probability is discussed in more detail in Chapters 3 and 4; in the next section, we describe in more detail the field of statistics.

1.4.2 Statistics

Historically, a 'statist' was a word for a politician, and statistics was 'the branch of political science concerned with the collection, classification, and discussion of facts bearing on the condition of a state or community' (Oxford English Dictionary, online at www.oed.com/view/Entry/189322?)). A good example of this usage that survives to the present is the term 'vital statistics' – used to describe the collection and tabulation of information on a region's rates and numbers of births and deaths.

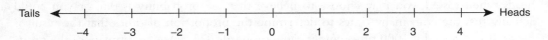

Figure 1.4 One-dimensional space for random walk

McGrew and Monroe (2000) define statistics as 'the collection, classification, presentation, and analysis of numerical data'. Note that this definition contains both the historical function of collection, classification, and presentation, but also the *analysis* of data. Modern definitions have in common the objective of inferring from a sample of data the nature of a larger population from which the sample was drawn. Statistics is often subdivided into two general areas – *descriptive statistics* are used to summarize and present information and this is in keeping with the more historical definition of the field; *inferential statistics*, as the name implies, allow inference about a larger population from a sample.

1.4.3 Probability Paradoxes

The following paradoxes are described for both mild amusement and to show that, although the use of probability to answer questions can often lead to intuitive outcomes, careful consideration is sometimes required to think through seemingly counterintuitive results.

1.4.3.1 A spatial paradox: random movement in several dimensions

This paradox is taken from Karlin and Taylor (1975). Consider a number line as in Figure 1.4, and suppose that our initial position is at the origin. We flip a single coin to govern our movement; if it is heads, we move one step to the right, and if it is tails, we move one step to the left. If we flip the coin many times, it is certain that we will, at some point, return to the origin (implying that, at that point in time, the total number of heads is equal to the total number of tails). This should not be surprising – it accords with our intuition that the number of heads and tails exhibited by a fair coin should be roughly equal.

Now consider generalizing the experiment to two dimensions (Figure 1.5), where the outcome of two coin flips governs the movement on the two-dimensional grid. One coin governs movement in the vertical direction and the other governs movement in the horizontal direction (for example, go one step up and one step to the right if both coins are heads, and one step down and one step to the left if both are tails). Again, it is possible to show that, although the path will certainly wander in the two-dimensional space, it is certain that there will be a return to the origin.

Finally, extend the procedure to three dimensions; each of three coins governs movement in one of the three dimensions. Movement starts at the origin and proceeds to lattice points within a cube. It now turns out that a return to the origin is no longer guaranteed! That is, there is a probability greater than zero that the random path will wander away from the origin and never return! This conclusion is also true for random walks in all dimensions greater than three. This is an example where the process of induction fails – what is true here in one and two dimensions cannot be generalized to higher dimensions. As the number of dimensions increases, the volume of space is increasing faster than the linear increase in the number of coins we have to govern our movement in the space. The example highlights the

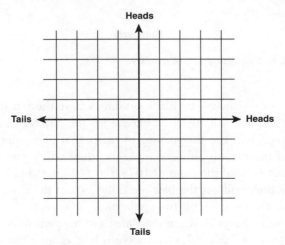

Figure 1.5 Two-dimensional space for random walk

fact that, while our intuition is often good, it is not perfect. We need to rely not only on our intuition about probability, but on a firmer foundation of the theory of probability.

1.4.3.2 A non-spatial paradox: quality pie

This probability paradox is taken from Martin Gardner's 'Mathematical Games' section of *Scientific American* (March 1976).

Consider an individual who goes into a diner each day to have a piece of pie. The diner always has apple and cherry pie, and it sometimes has blueberry pie. The quality of the pies is rated on a scale of one (lousy) to six (excellent), and the daily variability in the quality of each is summarized in Figure 1.6. For example, the cherry pie is either very good (it has a rating of five 49% of the time) or barely palatable (it has a rating of one 51% of the time). The person wishes to make a choice so as to maximize the proportion of time he or she ends up with the best pie. (With the exception of the blueberry pie, the person does not know the quality of the pie until after they have ordered!)

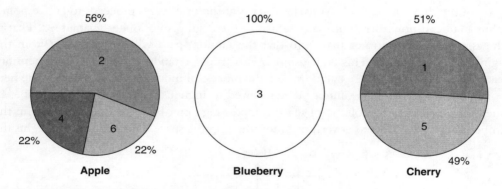

Figure 1.6 The relative frequency of pie quality

Consider first the decision that will be made on days where there is no blueberry pie. The possibilities are given in Table 1.1 (the best choice for the day is in bold).

The probabilities represent the proportion of times that particular combination of pie qualities will occur. If the person chooses apple pie, they will be getting the best pie the diner has to offer about 62% of the time (0.1078 + 0.1122 + 0.1122 + 0.2856 = 0.6178). If they choose cherry, they will get the best pie only 38% of the time (0.1078 + 0.2744 = 0.3822). The choice is clear – go with the apple pie.

Now let's examine what happens when the diner also happens to have blueberry pie. The possibilities are now given in Table 1.2. Here, apple is best about 33% of the time (0.1078 + 0.1122 + 0.1122 = 0.3322), cherry is best about 38% of the time (0.1078 + 0.2744 = 0.3822), and blueberry is best almost 29% of the time (it is only best on days when apple has a rating of two and cherry a rating of one – which occurs 28.56% of the time).

Table 1.1 Pie qualities and probabilities: apple and cherry

Apple	Cherry	Probability
6	5	.22 × .49 = .1078
6	1	.22 × .51 = .1122
4	5	.22 × .49 = .1078
4	1	.22 × .51 = .1122
2	5	.56 × .49 = .2744
2	1	.56 × .51 = .2856

Table 1.2 Pie qualities and probabilities: apple, blueberry, and cherry

Apple	Blueberry	Cherry	Probability
6	3	5	.22 × .49 = .1078
6	3	1	.22 × .51 = .1122
4	3	5	.22 × .49 = .1078
4	3	1	.22 × .51 = .1122
2	3	5	.56 × .49 = .2744
2	3	1	.56 × .51 = .2856

The best choice is now to go with the cherry pie. Thus, we have a rather bizarre scenario. The optimal strategy should be for the individual to ask the waiter or waitress if they happen to have blueberry pie; if they don't, the person should choose apple, and if they do, the person should choose cherry!

Recall that the objective here was to maximize the proportion of times that one would choose the best pie. A more common objective, employed in economic theory, is to maximize expected utility, which in this case would mean making a choice to maximize average pie quality. Apple has an average quality of $(6 \times 0.22) + (4 \times 0.22) + (2 \times 0.56) = 3.32$. Cherry has an average quality of $(5 \times 0.49) + (1 \times 0.51) = 2.96$, and blueberry has an average quality of 3. Using this objective, one should choose apple if they don't have blueberry (as before); if they have blueberry, one should still choose apple, since it has the best average quality. The economist's objective of maximizing expected utility leads to consistent results; other objectives can possibly lead to counterintuitive results.

As is pointed out in the original article, the example with pie is entertaining, but the example takes on more significance if one now imagines the information in Figure 1.6 representing the effectiveness of three alternative drugs in treating an illness. Imagine that there are two drugs, A and C, and their effectiveness is summarized by the 'pie' charts on the left and right sides in Figure 1.6 (where for example the '6' implies that drug A is found to be highly effective for 22% of all patients it is administered to). To obtain the best outcome the most frequently, one should choose drug A. Now suppose that drug B enters the market (the middle pie chart in Figure 1.6); the result is that in a three-way comparison of the available drugs, most frequently it is drug C that is best.

1.4.4 Geographical Applications of Probability and Statistics

This section provides examples of geographical applications of probability and statistics. The first two are what may be described as traditional, common applications, of the type we will address later in the book. The second two are illustrative of the unique and novel ways in which probability and statistics can be used to address geographical questions.

1.4.4.1 Are housing prices lower near airports?

An important objective in urban geography is to understand the spatial variation in housing prices. Housing characteristics such as lot size, the number of bedrooms, and the age of the house have a clear influence on selling prices. Characteristics of the neighborhood can also influence prices; whether a house is situated next to an industrial park or a recreational park is likely to have a clear effect on the price!

A nearby airport could potentially have a positive impact on prices, since accessibility is generally desirable. However, owning a home in the flight path of an airport may not necessarily be positive when the noise level is taken into consideration. We could take a sample of homes near the airport in question; we could also find a sample of homes that are not near the airport with similar characteristics (e.g., similar number of bedrooms, floor space, lot size, etc.). Suppose we find that the homes near the airport had a mean selling price that was lower than the homes not located near the airport. We need to decide whether (a) the sample reflects a 'true' difference in housing prices, based upon location with respect to the airport, or (b) the difference between the two locations is not significant, and the observed sample difference in prices is the result of sampling fluctuations (keep in mind that our samples represent a small fraction of the homes that could potentially be sold; if we went out and collected more data, the mean difference in selling price would likely be different). This is a problem in inferential statistics, based upon a desire to make an inference from a sample. We will return to this problem later in the book, and we will discuss how a critical difference threshold may be set: if the observed difference is below this threshold, we settle on conclusion (b); if the difference is above the threshold, we decide on option (a) above.

1.4.4.2 Do two places differ in terms of air quality?

Suppose we are interested in comparing the particulate matter in two cities. We collect daily data on PM10 (particles of 10 micrometers or less in size). Further suppose that we collect five daily samples in city A and five in city B, and these are designed to estimate the 'true' mean in each city. Table 1.3 presents the results.

The sample mean in city B is clearly higher than the mean in city A. But keep in mind that we have only taken a sample; there is certainly fluctuation from day to day, and so the 'true' means could possibly be the same (that is, if we took a very large sample over a very large number of days, the means could be equal).

We should not necessarily immediately conclude that city B has a higher 'true' mean particulate count; our results could be due to sampling fluctuations. Instead, we need to weigh the observed difference in the sample means against the difference in sample means that we might expect from sampling variation alone (when the true means are equal). If the observed difference in sample means is small relative to the difference that might be expected even when the true means are equal, we will accept the possibility that the true means are equal. On the other hand, if the observed difference in sample means is larger than the differences we would expect from such sampling fluctuations, we will conclude that the two cities have different levels of particulate matter. Details of problems like this (including the difference thresholds that must be set to distinguish between accepting and rejecting the idea that the means are equal) are covered in Chapter 5, which deals with questions of statistical inference.

Table 1.3 Hypothetical PM10 readings (units are micrograms per cubic meter)

City A	City B
40	45
38	41
52	59
35	34
26	25
Sample Mean:	Sample Mean:
38.2	40.8

1.4.4.3 Buffon's needle and migration distances

In the United States very little data are collected on the distances people move when they change their residential address. Yet this is a very basic measurement pertaining to a very important geographic process. Information is collected on the proportion of people changing their county of residence, and this may be used, together with concepts of probability, to estimate migration distances.

We begin with the work of Buffon, a 17th-century naturalist. Buffon was interested in many topics, ranging from subjects in botany to the strength of ships at sea. He was also interested in probability, and embedded in a supplement to the 4th volume of his 24–volume treatise on natural history is the following question.

Suppose we have a set of many parallel lines, separated by a constant distance, *s*. Now toss a needle of length *L* onto the set of parallel lines (see Figure 1.7). What is the probability that the needle will cross a line? Clearly, this probability will be higher as the length of the needle increases, and as the distance between the parallel lines decreases. Buffon found that the probability (*p*) of a randomly tossed needle crossing the lines was $p = 2L/(\pi s)$. Buffon's needle was actually used at the time to estimate π; if a needle of known length is tossed many times onto a set of parallel lines separated by a known distance, one can calculate *p* as the ratio of the number of crossings to the number of tosses. The only remaining unknown in the equation is π. Beckmann (1971), for example, refers to a Captain Fox, who passed at least some of his time this way while recovering from wounds received in the US Civil War. Unfortunately,

Figure 1.7 Buffon's needle on a set of parallel lines

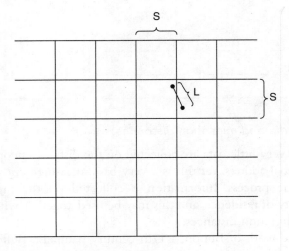

Figure 1.8 Buffon's needle on a square grid

the needle must be tossed a very large number of times to estimate π with any reasonable level of accuracy, and so such needle-tossing never developed into a popular pastime.

Laplace generalized this to the case of a square grid (see Figure 1.8). When the side of a square is equal to s (and $L<s$), the probability of crossing a line is now equal to $p = (4Ls - L^2/(\pi s^2))$.

Let us now turn to the connection to migration distance estimation. Define the ends of the needle as the origin and destination of a migrant; we wish to estimate this unknown needle length (L), which corresponds to migration distance. We will make the assumption that counties are approximately square, and that they are all the same size (i.e., we assume that a county map will look roughly like a square grid). We can estimate the side of a square, s, as the square root of the average area of a county. We may also estimate p if we have data on the proportion of all migrants who change their county of residence when they relocate. Finally, we of course now know that the value of π is 3.14.......We can solve Laplace's equation for the unknown migration distance:

$$L = 2s - s\sqrt{4 - p\pi}$$

Using data from the United States, $p = 0.35$ and $s = 33$ miles, and therefore we estimate the average migration distance as approximately $L = 10$ miles. Despite the perception that long-distance moving is perhaps the norm, the majority of individuals move a short distance when they relocate.

Although the assumption of square counties of equal size is of course unreasonable, a primary objective of a model is to simplify reality. We do not expect or contend that counties are equal-sized squares. We could be more exact by performing an experiment where we toss needles of a given size down onto a map of US counties; by trying different needle sizes, we will eventually find one that gives us county-crossing probabilities equal to about $p = 0.35$. We do not evaluate this assumption in greater detail here, but it turns out that the assumption of square counties of equal size is relatively *robust* – the conclusion doesn't change much when the assumption does not quite hold. Instead, such an assumption allows us to get a reasonable estimate of average migration distances.

1.4.4.4 Why is the traffic moving faster in the other lane?

Almost all would agree that traffic seems to always move faster in the other lane. There have been several statistical explanations for this. These explanations include:

(a) Redelmeier and Tibshirani (2000) created a simulation where two lanes had identical characteristics, in terms of the number of vehicles and their average speed. The only difference in the two lanes was the initial spacing between vehicles. In the simulation, the hypothetical vehicles would accelerate when traveling slowly, and would decelerate when approaching too closely to the vehicle in front of them. Not surprisingly, while moving quickly, vehicles were relatively far from one another; while moving slowly, they were closer together. Since the average speed in each lane was the same, and the number of cars in each lane was identical, each vehicle was passed by the same number of vehicles it had passed. However, the number of one-second time intervals during which a vehicle was passed was greater than the number of one-second intervals during which the vehicle was passing another vehicle. Thus, more time is spent being passed by other vehicles than is spent in passing vehicles (fast cars are spread out, and they are the ones overtaking you ... you are passing the slow cars, which are bunched up, so that it doesn't take long to pass them).

(b) Bostrom (2001) has, on the surface, a simpler answer to the question – cars in the other lane *are* moving faster! If cars in the fast lane are more spread out, the density of cars will be greatest in the slow lane. Now if you randomly choose a car at any time, there is a relatively high probability it will be from the slow lane, since that is where the density of cars is highest. So, at any given time, most drivers are in fact in the slow lane, and cars in the other lane *are* in fact moving faster.

(c) Dawson and Riggs (2004) note that if you are traveling at just under or just over the speed limit, and if you accurately observe the speeds of the vehicles passing you as well as the speed of the vehicles you are passing, there will be a misperception of the true average speed. In particular, drivers traveling at just under the average speed will perceive traffic to be going faster than it really is, while drivers traveling at just over the average speed will perceive traffic to be slower than it really is. The reason has to do with the selection of vehicles whose speeds are being observed – this sample will be biased in the sense that it will include many of the very fast and very slow vehicles, but not many of the vehicles going at your own speed. Although Dawson and Riggs do not mention this, if the distribution of speeds is skewed such that more than half of the vehicles are going slower than the mean speed (a likely assumption), then more than half of the vehicles will perceive traffic to be faster than it really is.

1.5 DESCRIPTIVE AND INFERENTIAL METHODS

A key characteristic of geographic data that brings about the need for statistical analysis is that they may often be regarded as a sample from a larger population. *Descriptive* statistical analysis refers to the use of particular methods that are used to describe and summarize the characteristics of the sample, while *inferential* statistical analysis refers to the methods that are used to infer something about the population from the sample. Descriptive methods fall within the class of exploratory techniques, while inferential statistics lie within the class of confirmatory methods. Descriptive summaries of data may be either visual (e.g., in the form of graphs and maps), or numerical; the mean and median are examples of the latter.

To begin to better understand the nature of inferential statistics, suppose you are handed a coin, and you are asked to determine whether it is a 'fair' one (that is, the likelihood of a 'head' is the same as the likelihood of a 'tail'). One natural way to gather some information would be to flip the coin a number of times. Suppose you flip the coin ten times, and you observe heads eight times. An example of a descriptive statistic is the observed proportion of heads − in this case 8/10 = 0.8. We enter the realm of inferential statistics when we attempt to pass judgment on whether the coin is 'fair'. We do this by *inferring* whether the coin is fair, on the basis of our sample results. Eight heads is more than the four, five, or six that might have made us more comfortable in a declaration that the coin is fair, but is eight heads really enough to say that the coin is *not* a fair one?

Table 1.4   Hypothetical outcome of 100 experiments of ten coin tosses each

No. of heads	Frequency of occurrence
0	0
1	1
2	4
3	8
4	15
5	22
6	30
7	8
8	8
9	3
10	1

There are at least two ways to go about answering the question of whether the coin is a fair one. One is to ask what *would* happen if the coin *was* fair, and to simulate a series of experiments identical to the one just carried out. That is, if we could repeatedly flip a known fair coin ten times, each time recording the number of heads, we would learn just how unusual a total of eight heads actually was. If eight heads comes up quite frequently with the fair coin, we will judge our original coin to be fair. On the other hand, if eight heads is an extremely rare event for a fair coin, we will conclude that our original coin is not fair.

To pursue this idea, suppose you arrange to carry out such an experiment 100 times. For example, one might have 100 students in a large class each flip coins that are known to be fair ten times. Upon pooling together the results, suppose you find the results shown in Table 1.4. We see that eight heads occurred 8% of the time.

We still need a guideline to tell us whether our observed outcome of eight heads should lead us to the conclusion that the coin is (or is not) fair. The usual guideline is to ask how likely a result equal to or more extreme than the observed one is, *if* our initial, baseline hypothesis that we possess a fair coin (called the *null* hypothesis) is true. A common practice is to not rule out the null hypothesis if the likelihood of a result at least as extreme as the one we observed is more than 5%. Here we would not rule out the null hypothesis of a fair coin if our experiment showed that eight or more heads occurred more than 5% of the time.

Alternatively, we wish to reject the null hypothesis that our original coin is a fair one if the results of our experiment indicate that eight or more heads out of ten is an uncommon event for fair coins. If fair coins give rise to eight or more heads less than 5% of the time, we decide to reject the null hypothesis and conclude that our coin is not fair.

In the example above, eight or more heads occurred 12 times out of 100, when a fair coin was flipped ten times. The fact that events as extreme, or more extreme than the one we observed, will happen 12% of the time with a *fair* coin leads us to accept the inference that our original coin is a fair one. Had we observed nine heads with our original coin, we would have judged it to be unfair, since events as rare or more rare than this (namely where the number of heads is equal to 9 or 10) occurred only four times in the 100 trials of a fair coin. Note, too, that our observed result does not prove that the coin *is* unbiased. It still *could* be unfair; there is, however, insufficient evidence to support the allegation.

The approach just described is an example of the *Monte Carlo method*, and several examples of its use are given in Chapter 10. A second way to answer the inferential problem is to make use of the fact that this is a *binomial* experiment; in Chapter 3, we will learn how to use this approach.

1.6 THE NATURE OF STATISTICAL THINKING

The American Statistical Association (1993, cited in Mallows 1998) notes that statistical thinking is:

(a) the appreciation of uncertainty and data variability, and their impact on decision making, and

(b) the use of the scientific method in approaching issues and problems.

Mallows (1998), in his Presidential Address to the American Statistical Association, argues that statistical thinking is not simply common sense, nor is it simply the scientific method. Rather, he suggests that statisticians give more attention to questions that arise in the beginning of the study of a problem or issue. In particular, Mallows argues that statisticians should: (a) consider what data are relevant to the problem; (b) consider how relevant data can be obtained; (c) explain the basis for all assumptions; (d) lay out the arguments on all sides of the issue; and only then (e) formulate questions that can be addressed by statistical methods. He feels that too often statisticians rely too heavily on (e), as well as the actual use of the methods that follow. His ideas serve to remind us that statistical analysis is a comprehensive exercise – it does not consist of simply 'plugging numbers into a formula' and reporting a result. Instead, it requires a comprehensive assessment of questions, alternative perspectives, data, assumptions, analysis, and interpretation.

Mallows defines statistical thinking as that which 'concerns the relation of quantitative data to a real-world problem, often in the presence of uncertainty and variability. It attempts to make precise and explicit what the data has to say about the problem of interest.' Throughout the remainder of this book, we will learn how various methods are used and implemented, but we will also learn how to interpret the results and understand their limitations. Too often, students working on geographic problems have only a sense that they 'need statistics', and their response is to seek out an expert on statistics for advice on how to get started. The statistician's first reply should be in the form of questions: (1) What is the problem? (2) What data do you have, and what are their limitations? (3) Is statistical analysis relevant, or is some other method of analysis more appropriate? It is important for the student to think first about these questions. Perhaps a simple description will suffice to achieve the objective. Perhaps some sophisticated inferential analysis will be necessary. But the subsequent course of events should be driven by the substantive problems and questions of interest, as constrained by data availability and quality. It should not be driven by a feeling that one needs to use statistical analysis simply for the sake of doing so.

1.7 SPECIAL CONSIDERATIONS FOR SPATIAL DATA

Fotheringham and Rogerson (1993) categorize and discuss a number of general issues and characteristics associated with problems in spatial analysis. It is essential that those working with spatial data have an awareness of these issues. Although all of their categories are relevant to spatial *statistical* analysis, among those that are most pertinent are:

(a) the modifiable areal unit problem;

(b) boundary problems;

(c) spatial sampling procedures;

(d) spatial autocorrelation or spatial dependence.

1.7.1 The Modifiable Areal Unit Problem

The modifiable areal unit problem refers to the fact that results of statistical analyses are sensitive to the zoning system used to report aggregated data. Many spatial datasets are aggregated into zones, and the nature of the zonal configuration can influence interpretation quite strongly. Panel (a) of Figure 1.9 shows one zoning system and panel (b) another, where the solid lines represent regional boundaries. The arrows represent the migration flows of individuals, and they are identical in each panel. The tail of an arrow represents a migrant's origin and the head represents a migrant's destination. In panel (a), no interzonal migration is reported, while a regional interpretation of panel (b) would lead to the conclusion that there was a strong southward movement, since five migrations from one zone to another would be reported. More generally, many of the statistical tools described in the following chapters would produce different results had different zoning systems been in effect.

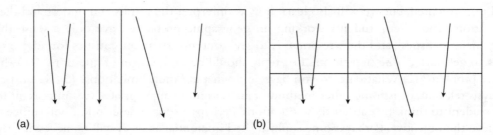

Figure 1.9 Two alternative zoning systems for migration data (note: arrows show origins and destinations of migrants)

The modifiable areal unit problem has two different aspects that should be appreciated. The first is related to the placement of zonal boundaries, for zones or subregions of a given size. If we were measuring mobility rates, we could overlay a grid of square cells on the study area. There are many different ways that the grid could be placed, rotated, and oriented on the study area. The second aspect has to do with geographic scale. If we replace the grid with another grid of larger square cells, the results of the analysis would be different. Migrants, for example, are less likely to cross cell boundaries in the larger grid than they are in the smaller grid.

As Fotheringham and Rogerson (1993) note, GIS technology now facilitates the analysis of data using alternative zoning systems, and it should become more routine to examine the sensitivity of results to modifiable areal units.

1.7.2 Boundary Problems

Study areas are bounded, and it is important to recognize that events just outside the study area can affect those inside of it. If we are investigating the market areas of shopping malls in a county, it would be a mistake to neglect the influence of a large mall located just outside the county boundary. One solution is to create a buffer zone around the area of study to include features that affect analysis within the primary area of interest. An example of the use of buffer zones in point pattern analysis is given in Chapter 10.

Both the size and shape of areas can affect measurement and interpretation. There are a lot of migrants leaving Rhode Island each year, but this is partially due to the state's small size – almost any move will be a move out of the state! Similarly, Tennessee experiences more out-migration than other states with the same land area in part because of its narrow rectangular shape. This is because individuals in Tennessee live, on average, closer to the border than do individuals in other states with the same area. A move of given length in some random direction is therefore more likely to take the Tennessean outside of the state.

1.7.3 Spatial Sampling Procedures

Statistical analysis is based upon sample data. Usually, one assumes that sample observations are taken randomly from some larger population of interest. If we are interested in sampling point locations to collect data on vegetation or soil, for example, there are many ways to do this. One could choose *x*- and *y*-coordinates randomly; this is known as a *simple random sample*. Another alternative would be to choose a *stratified* spatial sample, making sure that we chose a predetermined number of observations from each of several subregions, with simple random sampling within subregions. Alternative methods of sampling are discussed in more detail in Section 5.7.

1.7.4 Spatial Autocorrelation

Spatial autocorrelation refers to the fact that the value of a variable at one point in space is related to the value of that same variable in a nearby location. The travel behavior of residents in a household is likely to be related to the travel behavior of residents in nearby households, because both households have similar accessibility to other locations. Hence, observations of the two households are not likely to be independent, despite the requirement of statistical independence for standard statistical analysis. Spatial autocorrelation (or spatial dependence) can therefore have serious effects on statistical analyses, and hence can lead to misinterpretation. This is treated in more detail elsewhere in this book, including discussions in Chapters 5 and 10.

1.8 THE STRUCTURE OF THE BOOK

Chapter 2 covers methods of descriptive statistics – both visual and numerical approaches to describing data are covered. Chapters 3 and 4 provide the useful background on probability that facilitates understanding of inferential statistics. Inference about a population from a sample is carried out by first using the sample to make estimates of population characteristics. For example, a sample of individuals may be asked questions about their income; the sample mean provides an estimate of the unknown mean income of the entire population under study. Chapter 5 provides details on how these sample estimates can be used – to both construct confidence intervals that contain the true population value with a desired probability, and to formally test hypotheses about the population values. The chapter also contains details on the nature of sampling and the choice of an appropriate sample size.

Chapter 5 also contains descriptions of hypothesis tests designed to determine whether two populations could conceivably have the same population characteristic. For example, the two-sample difference of means test focuses upon the possibility that two samples could come from populations that have identical means (this objective was illustrated in the examples in sections 1.4.4.1 and 1.4.4.2). Chapter 6 covers the method of analysis of variance, which extends these two-sample tests to the case of more than two samples. For example, data on travel behavior (e.g., distance traveled to a public facility, such as parks or libraries) may be available for five different geographic regions, and it may be of interest to test the hypothesis that the true mean distance traveled was the same in all regions. In Chapter 7, we begin our exploration of methods that focus upon the relationship between two or more variables. Chapter 7 introduces the methods of correlation, and Chapter 8 extends this introduction to the topic of simple linear regression, where one variable is hypothesized to depend linearly on another. Regression is the most widely used method of inferential statistics, and it is given additional coverage in Chapter 9, where the linear dependence of one variable on more than one other variable (i.e., multiple linear regression) is treated.

One of the basic questions geographers face is whether geographic data exhibit spatial patterns. This is important both in its own right (where, e.g., we may wonder whether crime locations are more geographically clustered than they were in the past), and in addressing the fundamental problem of spatial dependence in geographic data when carrying out statistical tests. With respect to the latter, inferential statistical tests almost always assume that data observations are independent; this, however, is often not the case when data are collected at geographic locations. Instead, data are often spatially dependent – the value of a variable at one location is likely to be similar to the value of the variable at a nearby location. Chapter 10 is devoted to methods and statistical tests designed to determine whether data exhibit spatial patterns. Chapter 11 returns to the topic of regression, focusing upon how to carry out analyses of the dependence of one variable on others, when such spatial dependence in the data is present.

Finally, it is often desirable to summarize large datasets containing many observations and large numbers of variables. For example, it is often difficult to know where to begin when using census data for many different subregions (e.g., census tracts) to summarize the nature of a geographic region, in part because there are so many variables and many different subregions. Chapter 12 introduces factor analysis and cluster analysis as two approaches to summarizing large datasets. Factor analysis reduces the original number of variables to a smaller number of underlying dimensions or factors, and cluster analysis places the observations (i.e., the data for particular geographic subregions) into categories or clusters. The Epilogue contains some closing thoughts on new directions and applications.

1.9 DATASETS

1.9.1 Introduction

Some of the exercises at the end of each chapter make reference to various datasets. These datasets are available at the website associated with this book. Clearly the exercises constitute only a very small number of the many questions that could be asked, and the reader is encouraged to explore these datasets in more detail. The subsequent subsections describe these datasets.

1.9.2 Home Sales in Milwaukee, Wisconsin, USA in 2012

The City of Milwaukee (no date) maintains a database of home sales that is available to the public. It is accessible at http://assessments.milwaukee.gov/mainsales.html .

The dataset used here is an *Excel* file consisting of the 1,449 home sales reported in calendar year 2012. Each record, or row of the dataset, consists of the following column variables:

1. number of stories;

2. size of house, as measured by the number of finished square feet;

3. number of bedrooms;

4. number of bathrooms;

5. size of lot (in square feet);

6. date of sale;

7. sales price, in dollars;

8. political district (based on alderman district and represented by a number from 1 to 15;

9. age of house;

10. whether there is a full basement − 1 = yes; 0 = no;

11. whether there is an attic − 1 = yes; 0 = no;

12. whether there is a fireplace − 1 = yes; 0 = no;

13. whether the house is air-conditioned − 1 = yes; 0 = no;

14. whether the house has a garage − 1 = yes; 0 = no;

15. *x*-coordinate of house location;

16. *y*-coordinate of house location.

The complete online dataset covers the period from 2002 to present (2013), and includes information on a small number of other variables as well (taxkey number, address, style, type of exterior, number of half and full baths, and neighborhood number).

1.9.3 Singapore Census Data for 2010

This dataset consists of various census variables for the 36 planning areas of Singapore. The variables, which constitute the columns of the dataset, are as follows:

Name of planning area, total population, male population, female population, population over 65; Chinese population, Malay population, Indian population; population 5 and over, percent 5 and over who speak English, students 5 and over, students 5 and over with a long commute; population 15 and over, population over 15 who are in the labor force, unemployment rate, illiteracy rate among those 15 and over, population in various religious categories: not religious, Buddhist, Tao, Islam, Hindu, Sikh, Catholic, Other Christian, percent of those over 15 with no schooling, percent with university degree; population of working persons with monthly income under 1,000 Singapore dollars, population of working persons with monthly income over 8,000 Singapore dollars, fraction of those in the labor force with a long commute; total households, and percentage of householders who rent.

Many more variables are accessible at http://www.singstat.gov.sg/statistics/browse_by_theme/geospatial.html .

1.9.4 Hypothetical UK Housing Prices

This file is an *SPSS* formatted file consisting of 500 cases (rows) and 16 variables (columns). Each row represents a hypothetical home that has been sold, and variables consist of a mixture of regional location, housing attributes, and census attributes for the subregion that the house is located in.

1.9.4.1 Definition of variables

1.	**region**	a number between one and six, representing the region in which the house is located.
2.	**price**	price the house sold for in £.

3. **garage**   a dummy variable which takes the value:

   1 if a garage is present

   0 if a garage is not present.

4. **bedrooms**   number of bedrooms.

5. **bathrooms**   number of bathrooms.

6. **datebuilt**   year in which house was built.

7. **floor area**   floor area of the house in square meters.

8. **detached**   a dummy variable which takes the value:

   1 if the house is detached

   0 othewise.

9. **fireplace**   a dummy variable which takes the value:

   1 if a fireplace is present

   0 if a fireplace is not present.

10. **age<15** percentage of subregional population age less than 15.

11. **age65+** percentage of subregional population age 65 and over.

12. **nonwhite** percentage of subregional population non-white.

13. **unemploy** percentage of subregional population unemployed.

14. **ownocc** percentage of subregional housing that is owner-occupied.

15. **carsperhh** average number of cars per household in subregion.

16. **manuf** percentage of subregional population employed in manufacturing.

1.9.5 1990 Census Data for Erie County, New York

A 235 × 5 data table was constructed by collecting (from the 1990 US Census) and deriving the following information for the 235 census tracts in Erie County, New York (variable labels are in parentheses):

(a) Median household income (medhsinc).

(b) Percentage of households headed by females (femaleh).

(c) Percentage of high-school graduates who have a professional degree (educ).

(d) Percentage of housing occupied by owner (tenure).

(e) Percentage of residents who moved into their present dwelling before 1959 (lres).

2

DESCRIPTIVE STATISTICS

In Chapter 1, a fundamental distinction was drawn between descriptive and inferential statistics. We saw that describing data constitutes an important early phase of the scientific method. In this chapter, we will focus upon visual and numerical descriptive summaries of data.

We will begin by describing different types of data and by covering some of the visual approaches that are commonly used to explore and describe data. Following this, numerical measures of description are reviewed. Finally, description is discussed for the special context of spatial data.

2.1 TYPES OF DATA

Data may be classified as *nominal*, *ordinal*, *interval*, or *ratio*. Nominal data are observations that have been placed into a set of mutually exclusive and collectively exhaustive categories. Examples of nominal data include soil type and vegetation type. Ordinal data consist of observations that are ranked. Thus, it is possible to say that one observation is greater than (or less than) another, but with this much information it is not possible to say by how much an observation is greater or less than another. It is not uncommon to find ordinal data in almanacs and statistical abstracts; an example is data on the size of cities, by rank.

When it is possible to say by how much one observation is greater or less than another, data are either interval or ratio. With interval data, differences in values are identifiable. For example, on the Fahrenheit temperature scale, 44 degrees is 12 degrees warmer than 32 degrees. However, the 'zero' is not meaningful on the interval scale, and consequently ratio interpretations are not possible. Thus, 44 degrees is not 'twice as warm' as 22 degrees. Ratio data, on the other hand, *does* have a meaningful zero. Thus, 100 degrees Kelvin *is* twice as warm as 50 degrees Kelvin. Most numerical data are ratio data – indeed, it is difficult to think of examples for interval data other than the Fahrenheit and Celsius scales.

Data may consist of values that are either *discrete* or *continuous*. Discrete variables take on only a finite set of values – examples include the number of sunny days in a year, the annual number of visits by a family to a local public facility, and the monthly number of collisions between automobiles and deer in a region. Continuous variables take on an infinite number of values; examples include temperature and elevation.

2.2 VISUAL DESCRIPTIVE METHODS

Suppose that we wish to learn something about the commuting behavior of residents in a community. Perhaps we are on a committee that is investigating the potential implementation of a public transit alternative, and we need to know how many minutes, on average, it takes people to get to work by car. We do not have the resources to ask everyone, and so we decide to take a sample of automobile commuters. Let's say we survey $n = 30$ residents, asking them to record their average time it takes to get to work. We receive the responses shown in panel (a) of Table 2.1.

We may summarize our data visually by constructing a *histogram*, which is a vertical bar graph. To construct a histogram, the data are first grouped into categories. The histogram contains one vertical bar for each category. The height of the bar represents the number of observations in the category (i.e., the absolute frequency), and it is common to note the midpoint of the category on the horizontal axis. Figure 2.1 is a histogram for the hypothetical commuting data in Table 2.1, produced by *SPSS for Windows*. An alternative to the histogram is the *frequency polygon*; it may be drawn by connecting the points formed at the middle of the top of each vertical bar.

Data may also be summarized via *box plots*. Figure 2.2 depicts a box plot for the commuting data. The horizontal line running through the rectangle denotes the median (21), and the lower and upper ends of the rectangle (sometimes called the 'hinges') represent the 25th and 75th percentiles, respectively. Velleman and Hoaglin (1981) note that there are two common ways to draw the 'whiskers', which extend upward and downward from the hinges. One way is to send the whiskers out to the minimum and maximum values. In this case, the box plot represents a graphical summary of what is sometimes called a 'five-number summary' of the distribution (the minimum, maximum, 25th and 75th percentiles, and the median).

Table 2.1 Commuting data

	(a) Data on individuals		
Individual no.	Commuting time (min.)	Individual no.	Commuting time (min.)
1	5	16	42
2	12	17	31
3	14	18	31
4	21	19	26
5	22	20	24
6	36	21	11
7	21	22	19
8	6	23	9
9	77	24	44
10	12	25	21
11	21	26	17
12	16	27	26
13	10	28	21
14	5	29	24
15	11	30	23

(b) ranked commuting times
5, 5, 6, 9, 10, 11, 11, 12, 12, 14, 16, 17, 19, 21, 21, 21, 21, 21, 22, 23, 24, 24, 26, 26, 31, 31, 36, 42, 44, 77

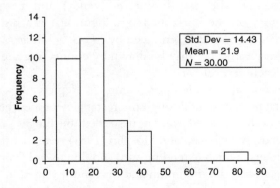

Figure 2.1 Histogram for commuting data

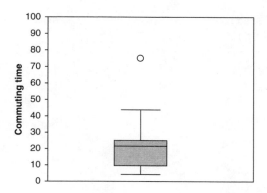

Figure 2.2 Boxplot for commuting data

There are often extreme outliers in the data that are far from the mean, and in this case it is not preferable to send whiskers out to these extreme values. Instead, whiskers are sent out to the outermost observations that are still within a distance from the hinge that is equal to 1.5 times the interquartile range (where the interquartile range is determined as the difference between the 75th and 25th percentiles, equivalent to the vertical length of the rectangle). All other observations beyond this are considered outliers, and are shown individually. In the commuting data, 1.5 times the interquartile range is equal to 1.5(14.25) = 21.375. The whisker extending downward from the lower hinge extends to the minimum value of 5, since this is greater than the lower hinge (11.75) minus 21.375. The whisker extending upward from the upper hinge stops at 44, which is the highest observation less than 47.375 (which in turn is equal to the upper hinge (26) plus 21.375). Note that there is a single outlier – observation 9 – and it has a value of 77 minutes.

A *stem-and-leaf* plot is an alternative way to display the frequencies of observations. It is similar to a histogram tilted onto its side, with the actual digits of each observation's value used in place of bars. The leading digits constitute the 'stem', and the trailing digits make up the 'leaf'. Each stem has one or more leaves, with each leaf corresponding to an observation. The visual depiction of the frequency of leaves conveys to the reader an impression of the frequency of observations that fall within given ranges. John Tukey, the designer of the stem-and-leaf plot, has said, 'If we are going to make a mark, it may as well be a meaningful one. The simplest – and most useful – meaningful mark is a digit' (Tukey 1972, p. 269).

For the commuting data, which have at most two-digit values, the first digit is the 'stem', and the second is the 'leaf' (see Figure 2.3). Note that for each item in the stem, the final digits are arranged in numerical order, from lowest to highest.

```
Frequency                        Stem & Leaf
  .00                            0 .
 4.00                            0 . 5569
 6.00                            1 . 011224
 3.00                            1 . 679
 9.00                            2 . 111112344
 2.00                            2 . 66
 2.00                            3 . 11
 1.00                            3 . 6
 2.00                            4 . 24
 1.00                     Extremes > =77)
Stem width:                      10. 00
Each leaf:                       1 case(s)
```

Figure 2.3   Stem-and-leaf plot for commuting data

To give another example that makes use of visual description, consider school district administrators, who often take censuses of the number of school-age children in their district, so that they may form hopefully accurate estimates of future enrollment. Table 2.2 gives the hypothetical responses of 750 households when asked how many school-age children are co-residents.

Table 2.2   Frequency of children in households

Number of children	Absolute frequency
0	100
1	200
2	300
3	100
4+	50
Total	750

The absolute frequencies may be translated into *relative frequencies* by dividing by the total number of observations (in this case, 750). Table 2.3 reveals, for example, that 26.7% of all households surveyed had one child. Note that the sum of the relative frequencies is equal to one. Note also that we can easily construct a histogram using the relative frequencies instead of the absolute frequencies (see Figure 2.4); the histogram in panel (b) has precisely

Table 2.3 Absolute and relative frequencies

Number of children	Absolute frequency	Relative frequency
0	100	100/750 = .133
1	200	200/750 = .266
2	300	300/750 = .400
3	100	100/750 = .133
4+	50	50/750 = .067
Total	750	

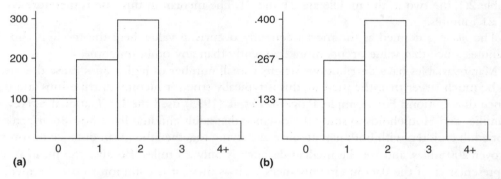

Figure 2.4 Number of children in households: (a) absolute frequency; (b) relative frequency

the same shape as that in panel (a); the vertical scale has just been changed by a factor equal to the sample size of 750.

2.3 MEASURES OF CENTRAL TENDENCY

We may continue our descriptive analysis of the data in Table 2.1 by summarizing the information numerically. The *sample mean* commuting time is simply the average of our observations; it is found by adding all of the individual responses and dividing by the number of observations. The sample mean is traditionally denoted by $\bar{x}$; in our example, we have $\bar{x} = 658/30 = 21.93$ minutes. In practice, this could sensibly be rounded to 22 minutes. We can use notation to state more formally that the mean is the sum of the observations, divided by the number of observations:

$$\bar{x} = \frac{\sum_{i=1}^{n} x_i}{n}$$

(2.1)

where x_i denotes the value of observation i, and where there are n observations. (A review of mathematical conventions and mathematical notation may likely be in order for many readers; see Appendix B.)

The *median* is defined as the observation that splits the ranked list of observations (arranged from lowest to highest, or highest to lowest) in half. When the number of observations is odd, the median is simply equal to the middle value on a ranked list of the observations. When the number of observations is even, we take the median to be the average of the two values in the middle of the ranked list.

Half of all respondents in our example have commutes that are longer than the median, and half have commutes that are shorter. When the responses are ranked as in panel (b) of Table 2.1, the two in the middle are 21 and 21. The median in this case is therefore equal to 21 minutes.

The *mode* is defined as the most frequently occurring value; here the mode is also 21 minutes, since that value occurs more frequently than any other outcome.

Many variables have distributions where a small number of high values cause the mean to be much larger than the median; this is typically true for income distributions and distance distributions. For example, Rogerson et al. (1993) used the US National Survey of Families and Households to study the distance that adult children lived from their parents. For adult children with both parents alive and living together, the mean distance to parents is over 200 miles, and yet the median distance is only 25 miles! Because the mean is not representative of the data in circumstances such as these, it is common to use the median as a measure of central tendency.

When data are available only for categories, grouped means may be calculated. This is achieved by assuming that all of the data within a particular category take on the midpoint value of the category. For example, Table 2.4 portrays some hypothetical data on income (units have been deliberately omitted to keep the example location-free!).

Table 2.4 Absolute frequencies associated with observations on income

Income	Frequency (Number of individuals)
< 15,000	10
15,000–34,999	20
35,000–54,999	30
55,000–99,999	15

The grouped mean is found by assuming that the ten individuals in the first category have an income of 7500 (the midpoint of the category), the 20 individuals in the second category each have an income of 25,000, the 30 individuals in the next category each have an income of 45,000, and those in the final category each have an income of 77,500. All of these individual values are added, and the result is divided by the number of individuals. Thus, the grouped mean for this example is:

$$\frac{10(7,500) + 20(25,000) + 30(45,000) + 15(77,500)}{10 + 20 + 30 + 15} = 41,167 \qquad (2.2)$$

More formally,

$$\overline{x}_g = \frac{\sum\limits_{i=1}^{G} f_i x_{i,mid}}{\sum\limits_{i=1}^{G} f_i} \qquad (2.3)$$

where $\overline{x}_g$ denotes the grouped mean, G is the number of groups, f_i is the number of observations in group i, and $x_{i,mid}$ denotes the value of the midpoint of the group.

The grouped mean may be seen as a weighted average, where the midpoint of each group is weighted by the frequency of observations in that group. After all of these weighted quantities have been calculated and summed, the last step in the calculation of a weighted average is to divide by the sum of the weights.

A similar way to look at the concept of the grouped mean is to imagine that we need to create a specific value for each observation, even though we don't have this information. In the example above, we only know that there are ten observations that are less than 15,000 – we don't know their individual values. Imagine starting a list – we will 'guess' that the first observation is 7,500, the second observation is 7,500, and so on. The beginning of our list thus has ten values of 7,500. Then we move on to the second category – there are 20 observations in the next category and so we add 20 instances of 25,000 to the list, since this would be our best guess for each of these observations. When we get done with the construction of this list, we simply add the numbers (which gives us the numerator in Equations 2.2 and 2.3), and we divide by the number of items on the list (which is the denominator in Equations 2.2 and 2.3).

It is not uncommon to find that the last category is open-ended; instead of the 55,000–99,999 category, it might be more common for data to be reported in a category labeled '55,000 and above'. In this case, an educated estimate of the average salary for those in this group should be made. It would also be useful to make a number of such estimates and repeat the calculation of the grouped mean, to see how sensitive the result is to different choices for the estimate.

2.4 MEASURES OF VARIABILITY

We may also summarize histograms and datasets by characterizing their variability. The commuting data in Table 2.1 range from a low of 5 minutes to a high of 77 minutes. The *range* is the difference between the two values: here it is equal to $77 - 5 = 72$ minutes.

The *interquartile range* is the difference between the 25th and 75th percentiles. With n observations, the 25th percentile is represented by observation $(n + 1)/4$, when the data have been ranked from lowest to highest. The 75th percentile is represented by observation $3(n + 1)/4$. These will often not be integers, and in that case interpolation is used, just as it is for the median when there is an even number of observations. For the commuting data, the 25th percentile is represented by observation $(30 + 1)/4 = 7.75$. Interpolation between the 7th and 8th lowest observations requires that we go 3/4 of the way from the 7th lowest observation (which is 11) to the 8th lowest observation (which is 12). This implies that the 25th percentile is 11.75 (since 11.75 is 3/4 of the way from 11 to 12; more formally, this can be found by (a) multiplying 3/4 by the difference in the two observations $(3/4 \times (12 - 11) = 3/4$, and then (b) adding the result to the smaller of the observations $(11 + 3/4 = 11.75)$. Similarly, the 75th percentile is represented by observation $3(30 + 1)/4 = 23.25$. Since both the 23rd and 24th observations are equal to 26, the 75th percentile is equal to 26. The interquartile range is the difference between these two values, or $26 - 11.75 = 14.25$.

The *sample variance* of the data (denoted s^2) may be thought of as the average squared deviation of the observations from the mean. To ensure that the sample variance gives an unbiased estimate of the true, unknown variance of the population from which the sample was drawn (denoted σ^2), s^2 is computed by taking the sum of the squared deviations, and then dividing the result by $n - 1$, instead of by n. Here the term *unbiased* implies that if we were to repeat this sampling many times, we would find that the average or mean of our many sample variances would be equal to the true variance. Thus, the sample variance is found by taking the sum of squared deviations from the mean, and then dividing by $n - 1$:

$$s^2 = \frac{\sum\limits_{i=1}^{n}(x_i - \bar{x})^2}{n - 1} \tag{2.4}$$

An approximate interpretation of the variance is that it represents the average squared deviation of an observation from the mean (it is an approximate interpretation because $n - 1$, instead of n, is used in the denominator).

In our example, $s^2 = 208.13$ minutes^2. The *sample standard deviation* is equal to the square root of the sample variance; here we have $s = \sqrt{208.13} = 14.43$ minutes. Note that the units for the standard deviation are the same as those for the variable itself — here, for example, the standard deviation is given in minutes. Since the sample variance characterizes the average *squared* deviation from the mean, by taking the square root to calculate the standard deviation, we are putting the measure of variability back on a scale closer to that used for the mean and the original data. Although the standard deviation is not equal to the average absolute deviation of an observation from the mean, it is usually close.

Variances for grouped data are found by assuming that all observations are at the midpoint of their category, and the calculation is based on the sum of squared deviations of these midpoint values from the grouped mean:

$$s_g^2 = \frac{\sum\limits_{i=1}^{G} f_i(x_{i,mid} - \overline{x}_g)^2}{(\sum\limits_{i=1}^{G} f_i) - 1} \tag{2.5}$$

For the data in Table 2.4, the grouped variance is

$$\left\{ \frac{\begin{aligned}&(10(7,500 - 41,167)^2 + 20(25,000 - 41,167)^2 + \\ &30(45,000 - 41,167)^2 + 15(77,500 - 41,167)^2)\end{aligned}}{(10 + 20 + 30 + 15) - 1} \right\} = 4.974 \times 10^8 \tag{2.6}$$

The square root of this, 22,301, is the grouped standard deviation.

2.5 OTHER NUMERICAL MEASURES FOR DESCRIBING DATA

2.5.1 Coefficient of Variation

Consider the selling price of homes in two communities. In community A, the mean price is 150,000 (units are again deliberately omitted, so that the illustration may apply to more than one unit of currency!). The standard deviation is 75,000. In community B, the mean selling price is 80,000, and the standard deviation is 60,000.

The standard deviation is an *absolute* measure of variability; in this example, such variability is clearly lower in community B. However, it is also useful to think in terms of *relative variability*. Relative to its mean, the variability in community B is greater than that in community A. More specifically, the *coefficient of variation* is defined as the ratio of the standard deviation to the mean. Here, the coefficient of variation in community A is 75,000/150,000 = 0.5; in community B, it is 60,000/80,000 = 0.75.

2.5.2 Skewness

Skewness measures the degree of asymmetry exhibited by the data and histogram. Figure 2.5 is clearly asymmetric, and it reveals that there are more observations below the mean than above it – this is known as positive skewness. Positive skewness can also be detected by comparing the mean and median. When the mean is greater than the median as it is here, the distribution is positively skewed. In contrast, when there are a small number of low observations and a large number of high ones, the mean is less than the median, and the data exhibit negative skewness (see Figure 2.6). The sample skewness is computed by first adding together the cubed deviations from the mean and then dividing by the product of the cubed standard deviation and the number of observations:

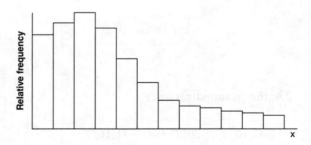

Figure 2.5   A positively skewed distribution

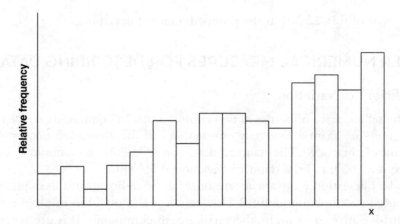

Figure 2.6   A negatively skewed distribution

$$\text{skewness} = \frac{\sum_{i=1}^{n}(x_i - \overline{x})^3}{ns^3} \qquad (2.7)$$

The 30 commuting times in Table 2.1 have a positive skewness of 1.86. If skewness equals zero, the histogram is symmetric about the mean.

2.5.3 Kurtosis

Kurtosis measures how peaked the histogram is. Its definition is similar to that for skewness, with the exception that the fourth power is used instead of the third:

$$\text{kurtosis} = \frac{\sum_{i=1}^{n}(x_i - \overline{x})^4}{ns^4}. \qquad (2.8)$$

(a)

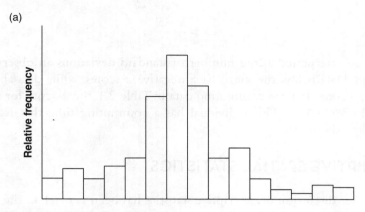

Figure 2.7(a) Leptokurtic distribution

(b)

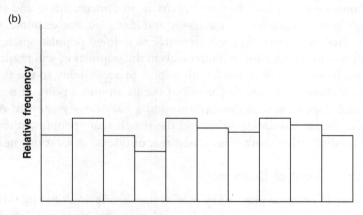

Figure 2.7(b) Platykurtic distribution

Data with a high degree of peakedness are said to be *leptokurtic*, and have values of kurtosis over 3.0 (see Figure 2.7a). Flat histograms are *platykurtic*, and have kurtosis values less than 3.0 (Figure 2.7b). The kurtosis of the commuting times is equal to 7.68, and hence the distribution of commuting times is relatively peaked.

2.5.4 Standard Scores

Since data come from distributions with different means and different degrees of variability, it is common to standardize observations. One way to do this is to transform each observation into a *z*-score by first subtracting the mean of all observations, and then dividing the result by the standard deviation:

$$z = \frac{x - \overline{x}}{s}.$$
(2.9)

z-scores may be interpreted as the number of standard deviations an observation is away from the mean. Data below the mean have negative z-scores, while data above the mean have positive z-scores. For the commuting data in Table 2.1, the z-score for individual 1 is $(5 - 21.93)/14.3 = -1.18$. This individual has a commuting time that is 1.18 standard deviations below the mean.

2.6 DESCRIPTIVE SPATIAL STATISTICS

To this point, our discussion of descriptive statistics has been general, in the sense that the concepts and methods covered apply to a wide range of data types. In this section, we review a number of descriptive statistics that are useful in providing numerical summaries of *spatial* data.

Descriptive measures of spatial data are important in understanding and evaluating such fundamental geographic concepts as *accessibility* and *dispersion*. For example, it is important to locate public facilities so that they are accessible to defined populations. Spatial measures of centrality applied to the location of individuals in the population will result in geographic locations that are in some sense optimal with respect to accessibility to the facility. Similarly, it is important to characterize the dispersion of events around a point. It is useful to summarize the spatial dispersion of individuals around a hazardous waste site. Are individuals with a particular disease less dispersed around the site than are people without the disease? If so, this could indicate that there is increased risk of disease at locations near the site.

2.6.1 The Measurement of Distance

The measurement of distance plays a key role in the description of spatial data. The shortest distance between two points on a global scale is called the *great circle distance*. This is represented by an arc that, if extended, would form a circle whose plane would pass through the center of the earth. Air travel over long distances is planned using these arcs.

The great circle distance between any two points on a sphere may be calculated as follows. Let (a_1, b_1) and (a_2, b_2) be the (latitude, longitude) pairs for points 1 and 2, respectively. The great circle distance between points 1 and 2 is

$$d = r\{\arccos[\sin a_1 \sin a_2 + \cos a_1 \cos a_2 \cos(b_2 - b_1)]\}$$
(2.10)

where the sphere has radius r and where the notation arccos[] signifies the angle whose cosine is equal to the term within the square brackets. (An equivalent alternative notation for $\arccos[x]$ is $\cos^{-1}[x]$.)

> ## Example
>
> Let's find the great circle distance between London (latitude a_1 = 51.51N, b_1 = longitude 0.13W) and New York City (latitude a_2 = 40.67N, longitude b_2 = 73.94W). The earth has a radius of approximately 3,963 miles. We must assume that it is a sphere (although all grade school children know that it is actually an oblate spheroid). Using Equation 2.10, this results in
>
> d = 3,963 {arccos[(0.7827)(0.6517)+(0.6224)(0.7585)(0.2788)]} = 3,963{arccos[0.6417]} = 3,963 (.8741) = 3,464 miles

The angle whose cosine is 0.6417 is 50.08 degrees. One final point about Equation 2.10 – it requires the use of angles in 'radians', and not degrees. To convert an angle from degrees to radians, multiply by 3.14/180. Here 50.08(3.14)/180 = 0.8741 radians. The final step is to multiply this result by r: (0.8741)(3,963) = 3,464 miles.

For the calculation of distances on a smaller scale (say, within a city) we can make the much more drastic assumption that the world is flat. Then we can superimpose a grid on a map of the area, and use the grid coordinates for our calculations. The two most common measures of distance on such a flat plane are Euclidean and Manhattan distances. Let (x_1, y_1) and (x_2, y_2) represent the x- and y-coordinates of points 1 and 2, respectively. The Euclidean distance is the shortest distance between two points and is calculated as

$$d = \sqrt{(x_2 - x_1)^2 + (y_2 - y_1)^2} \tag{2.11}$$

You may recognize this is as the Pythagorean formula for the length of the hypotenuse (the solid line in Figure 2.8) of a right triangle. The Manhattan distance is given by

$$d = |(x_2 - x_1)| + |(y_2 - y_1)| \tag{2.12}$$

where the vertical lines in Equation 2.12 indicate that the absolute value is to be taken. This is the distance between two points when travel is restricted to occur only in horizontal and vertical directions. It is equal to the sum of the length of the two dotted lines in Figure 2.8.

For example, suppose (x_1, y_1) = (8,4) and (x_2, y_2) = (7,10). Then the Euclidean distance is $d = \sqrt{(7 - 8)^2 + (10 - 4)^2} = \sqrt{37} = 6.08$ and the Manhattan distance is $d = |(7 - 8)| + |(10 - 4)| = 1+6 = 7$.

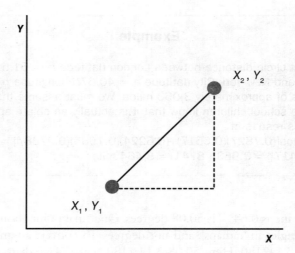

Figure 2.8   Manhattan and Euclidean distances

2.6.2 Mean Center

The most commonly used spatial measure of central tendency is the *mean center*. For point data, the *x*- and *y*-coordinates of the mean center are found by simply finding the mean of the *x*-coordinates and the mean of the *y*-coordinates, respectively.

For areal data, the mean center can be found by assuming that we know the *x*- and *y*-coordinates associated with the *centroids* of each subregion. *Geographic centroids* are point locations representing the balance point of the region. Thus if you were holding up a physical map of the region (say made out of metal, plastic, wood, or some other material), the geographic centroid is that location where you could balance the map on one finger. A *population centroid* is the same as the population-weighted center of population – the very quantity we are discussing in this section!

Given the *x*- and *y*-coordinates of the centroids for each suregion, it is often useful to attach weights to them. To find the mean center of population for instance, the weights are taken as the number of people living in each subregion. The weighted mean of the *x*-coordinates and *y*-coordinates then provides the location of the mean center of population. More specifically, when there are *n* subregions,

$$\bar{x} = \frac{\sum_{i=1}^{n} w_i x_i}{\sum_{i}^{n} w_i}; \qquad \bar{y} = \frac{\sum_{i=1}^{n} w_i y_i}{\sum_{i}^{n} w_i} \tag{2.13}$$

where the w_i are the weights (e.g., population in region *i*) and x_i and y_i are the coordinates of the centroid in region *i*. Conceptually, this is identical to assuming that all individuals living in a particular subregion live at a prespecified point (in this case, the centroid) of that subregion. The mean center is identical to the location that would be found if the x- and y-coordinates for each individual were first written out, and then the mean of all the x's and all the y's were found. Equation 2.13, through its use of a weighted mean, merely provides a quicker way to arrive at the solution. The mean center of population in the USA has migrated west and south over time (see Figure 2.9).

The weighted mean center has the property that it minimizes the sum of squared distances that individuals must travel (assuming that each person travels to the centralized facility located at the mean center). Although it is easy to calculate, this interpretation is a little unsatisfying – it would be nicer to be able to find a central location that minimizes the sum of distances, rather than the sum of squared distances.

2.6.3 Median Center

The location that minimizes the sum of distances traveled is known as the *median center*. Although its interpretation is more straightforward than that of the mean center, its calculation is more complex. Calculation of the median center is iterative, and one begins by guessing an initial location (a convenient initial location is the weighted mean center). Then the new x- and y-coordinates (denoted by x' and y', respectively) are updated using the following:

$$x' = \frac{\sum\limits_{i=1}^{n} \frac{w_i x_i}{d_i}}{\sum\limits_{i=1}^{n} \frac{w_i}{d_i}}; \qquad y' = \frac{\sum\limits_{i=1}^{n} \frac{w_i y_i}{d_i}}{\sum\limits_{i=1}^{n} \frac{w_i}{d_i}} \qquad (2.14)$$

where d_i is the distance from point i to the specified initial location of the median center. This process is then carried out again – new x- and y-coordinates are again found using these same equations, with the only difference being that d_i is redefined as the distance from point i to the most recently calculated location for the median center. This iterative process is terminated when the newly computed location of the median center does not differ significantly from the previously computed location.

In the application of social physics to spatial interaction, population divided by distance is considered a measure of population 'potential' or accessibility. If the w's are defined as populations, then each iteration finds an updated location based upon weighting each point or areal centroid by its accessibility to the current median center. The median center is the fixed point that is 'mapped' into itself when weighted by accessibility. Alternatively stated, the median center is an accessibility-weighted mean center, where accessibility is defined in terms of the distances from each point or areal centroid to the median center.

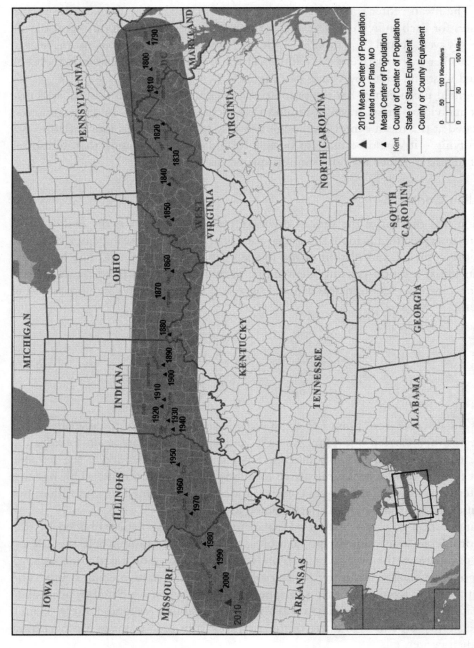

Figure 2.9 Mean Center of Population for the United States: 1790 to 2010

Source: US Bureau of the Census (2011)

2.6.4 Standard Distance

Aspatial measures of variability, such as the variance and standard deviation, characterize the amount of dispersion of data points around the mean. Similarly, the spatial variability of locations around a fixed central location may be summarized. The *standard distance* (Bachi 1963) is defined as the square root of the average squared distance of points to the mean center:

$$s_d = \sqrt{\frac{\sum\limits_{i=1}^{n} d_{ic}^2}{n}} \tag{2.15}$$

where d_{ic} is the distance from point i to the mean center and n is the number of locations around the central location.

Although Bachi's measure of standard distance is conceptually appealing as a spatial version of the standard deviation, it is not really necessary to maintain the strict analogy with the standard deviation by taking the square root of the average squared distance. With the aspatial version (i.e., the standard deviation), loosely speaking, the square root 'undoes' the squaring and thus the standard deviation may be roughly interpreted as a quantity that is on the same approximate scale as the average absolute deviation of observations from the mean. Squaring and taking square roots is carried out in part because deviations from the mean may be either positive or negative. But, in the spatial version, distances are always positive, and so a more interpretable and natural definition of standard distance would be to simply use the average distance of observations from the mean center:

$$s_d' = \frac{\sum\limits_{i=1}^{n} d_{ic}}{n} \tag{2.16}$$

In practice, the results found using Equations 2.15 and 2.16 will be quite similar.

Equations 2.15 and 2.16 represent definitions of standard distance when each point or location is given equal weight in the calculation. More commonly, each location will have an associated weight, w_i. In this case, the weighted standard distances are calculated using

$$s_d = \sqrt{\frac{\sum\limits_{i=1}^{n} w_i d_{ic}^2}{\sum\limits_{i=1}^{n} w_i}} \tag{2.17}$$

$$s_d' = \frac{\sum\limits_{i=1}^{n} w_i d_{ic}}{\sum\limits_{i=1}^{n} w_i}$$

The units for all of these definitions of standard distance are in the units of the distance variable (typically miles, kilometers, etc.).

2.6.5 Relative Distance

One drawback to the standard distance measure described above is that it is a measure of *absolute dispersion*; it retains the units in which distance is measured. Furthermore, it is affected by the size of the study area. The two panels of Figure 2.10 show situations where the standard distance is identical, but clearly the amount of dispersion about the central location, relative to the study area, is lower in panel (b).

   Relative distance is a measure of relative dispersion and it may be derived by dividing the standard distance by the radius of a circle with area equal to the size of the study area (McGrew and Monroe 2000). This makes the measure of dispersion unitless and standard-izes for the size of the study area, thereby facilitating comparison of dispersion in study areas of different sizes.

   For a circular study area, the relative distance is $s_{d,rel} = s_d/r$ and for a square study area, $s_{d,rel} = s_d \sqrt{\pi / s^2}$ (where r and s represent the radius of the circle and the side of the square, respectively). Note that the maximum relative distance for a circle is 1; this occurs when all points are located on the circumference of the circle. For a square, the maximum relative distance is $\sqrt{\pi / 2} = 1.253$, and this occurs when all points are located at corners of the square.

2.6.6 Illustration of Spatial Measures of Central Tendency and Dispersion

The descriptive spatial statistics outlined above are now illustrated using the data in Table 2.5. This is a simple set of ten locations; a simple and small dataset has been chosen

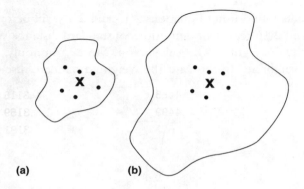

(a)   (b)

Figure 2.10   Illustration of standard distance. Note that dispersion relative to the study area is lower in panel (b)

deliberately, to facilitate the derivation of the quantities by hand, if desired. The study area is assumed to be a square with the length of each side equal to one. In this example, we assume implicitly that there are equal weights at each location (or, equivalently, that one individual is at each location).

The mean center is at (0.4513, 0.2642), and it is found simply by taking the mean of each column. The location of the median center is at (0.4611, 0.3312). Accuracy to three digits is achieved after 33 iterations. The first few iterations of Equation 2.14 are shown in Table 2.6.

Table 2.5 *x–y* coordinate pairs

x	y
.8616	.1781
.1277	.4499
.3093	.5080
.4623	.3419
.4657	.3346
.2603	.0378
.6680	.3698
.2705	.1659
.1981	.1372
.8891	.1192

Table 2.6 Convergence of iterations toward the median center

Iteration	x-coordinate	y-coordinate
1	.4512	.2642
2	.4397	.2934
3	.4424	.3053
4	.4465	.3116
5	.4499	.3159
6	.4623	.3191
.	.	.
.	.	.
.	.	.
.	.	.
.	.4611	.3312

It is interesting to note that the approach to the y-coordinate of the median center is monotonic, while the approach to the x-coordinate is a damped harmonic.

The sum of squared distances to the mean center is 0.8870; note that this is lower than the sum of squared distances to the median center (0.9328). Similarly, the sum of distances to the median center is 2.655, and this is lower than the sum of distances to the mean center (2.712).

The standard distance is 0.2978 (which is the square root of 0.8870/10); note that this is similar to the average distance of a point from the mean center (2.712/10 = 0.2712).

2.6.7 Angular Data

Angular data arise in a number of geographical applications; the analysis of wind direction, and the study of the alignment of crystals in bedrock provide two examples. The latter example has been particularly important in the study of continental drift, and in establishing the timing of reversals in earth's magnetic field.

Special considerations arise in the visual and numerical description of angular data. Consider the 146 observations on wind direction given in Table 2.7. A histogram

Table 2.7   Hypothetical data on wind direction

Direction	Angular direction	Frequency
North	0°	10
Northeast	45°	8
East	90°	5
Southeast	135°	6
South	180°	18
Southwest	225°	29
West	270°	42
Northwest	315°	28
Total		146

could be constructed, but it is not clear how the horizontal axis should be labeled. A histogram could arbitrarily start with North on the left, as in Figure 2.11a; another possibility is to arrange for the mode to be near the middle of the histogram, as in Figure 2.11b.

Neither of the options depicted in Figure 2.11 for constructing a histogram using angular data is ideal, since the observations at the far left of the horizontal axis are similar in

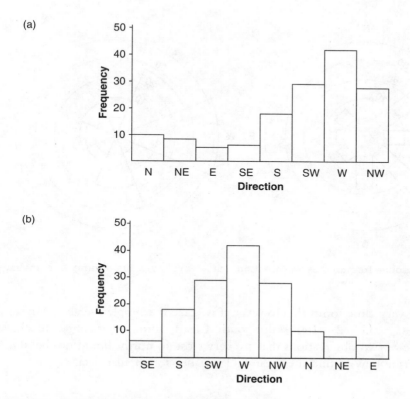

Figure 2.11 Absolute frequencies for directional data

direction to the observations on the far right of the horizontal axis. In particular, there is no provision for wrapping the histogram around on itself.

An alternative is the circular histogram (Figure 2.12a). Here bars extend outward in all directions, reflecting the nature of the data. As is the case with more typical histograms, the lengths of the bars are proportional to the frequency. A slight variation of this is the more common rose diagram (Figure 2.12b); here the rectangular bars have been replaced with pie- or wedge-shapes. The concentric rings shown in Figure 2.12b are often not displayed; they are shown here to emphasize how the relative frequencies are used to construct the diagram. The rose diagram is effective in portraying visually the nature of angular data.

There are also special considerations that are necessary when considering numerical summaries of angular data. Consider the very simple case where we have two observations – one observation is 1° and the other is 359°. If 0° is taken to be north, both of these

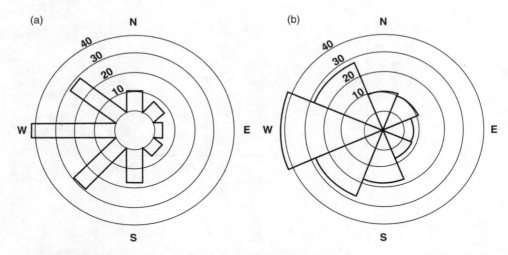

Figure 2.12   Absolute frequencies for directional data: (a) circular histogram; (b) rose diagram

observations are very close to north. However, if we take the simple average or mean of 1 and 359, we get $(1 + 359)/2 = 180°$ – due south! Clearly, some other approach is needed, since the average of two observations that are very close to north should not be 'south'.

Here we describe how to find the mean and variance for angular data.

Mean:

1. Find the sine and cosine of each angular observation.

2. Find the mean of the sines $(\bar{S})$ and the mean of the cosines $(\bar{C})$.

3. Find $\bar{R} = \sqrt{\bar{S}^2 + \bar{C}^2}$

4. The mean angle (say $\bar{\alpha}$) is the angle whose cosine is equal to $\bar{C} / \bar{R}$ and whose sine is equal to $\bar{S} / \bar{R}$. Thus, $\alpha = \arccos (\bar{C} / \bar{R})$ and $\alpha = \arcsin (\bar{S} / \bar{R})$.

Variance: A measure of variance for angular data (termed the *circular variance*) provides an indication of how much variability there is in the data. For example, if all observations consisted of the same angle, the variability, and hence the circular variance, should be zero.

The circular variance, designated by S_0, is simply equal to $1 - \bar{R}$. It varies from zero to one. A high value near one indicates that the angular data are dispersed, and come from many different directions. A value near zero implies that the observations are clustered around particular directions.

Readers interested in more detail regarding angular data may find more extensive coverage in Mardia and Jupp (1999).

Example 2.1

Three observations of wind direction yield measurements of 43°, 88°, and 279°. Find the angular mean and the circular variance.

Solution. We start by constructing the following table:

Observation	Cosine	Sine
43°	0.7314	0.6820
88°	0.0349	0.9994
279°	0.1564	−0.9877

The mean of the cosines is equal to $\bar{C} = (0.7314 + 0.0349 + 0.1564)/3 = 0.3076$. The mean of the sines is equal to $\bar{S} = (.6820 + .9994 - .9877)/3 = 0.2312$. Then $\bar{R} = \sqrt{0.3076^2 + 0.2312^2} = 0.3848$. The mean angle, α, is the angle whose cosine is equal to $\bar{C}/\bar{R} = 0.3076/0.3848 = 0.7994$, and whose sine is equal to $\bar{S}/\bar{R} = 0.2312/0.3848 = 0.6008$.

Using either a calculator or table, we find that the mean angle is arccos(0.7994) = arcsin(0.6008) = 37°. The circular variance is equal to $1 - \bar{R} = 1 - 0.3848 = 0.6152$. This value is closer to one than to zero, indicating a tendency for high variability – that is, the angles are relatively dispersed and are coming from different directions.

2.7 DESCRIPTIVE STATISTICS IN *SPSS 21 FOR WINDOWS*

2.7.1 Data Input

After starting *SPSS*, data are input for the variable or variables of interest. Each column represents a variable. For the commuting example set out in Table 2.1, the 30 observations were entered into the first column of the spreadsheet. Alternatively, respondent ID could have been entered into the first column (i.e., the sequence of integers, from 1 to 30), and the commuting times would then have been entered in the second column. The order that the data are entered into a column is unimportant.

2.7.2 Descriptive Analysis

2.7.2.1 Simple descriptive statistics

Once the data are entered, click on Analyze (or Statistics, in older versions of *SPSS for Windows*). Then click on Descriptive Statistics. Then click on Explore. A split box will appear on the screen; move the variable or variables of interest from the left box to the box on the right that is headed `Dependent List` by highlighting the variable(s) and clicking on the arrow. Then click on OK.

Table 2.8   Descriptive summary of commuting data using SPSS

Case Processing Summary

	Cases					
	Valid		Missing		Total	
	N	Percent	N	Percent	N	Percent
VAR00001	30	100.0%	0	.0%	30	100.0%

Descriptives

			Statistic	Std. Error
VAR00001	Mean		21.9333	2.63397
	95% Confidence	Lower Bound	16.5483	
	Interval for Mean	Upper Bound	27.3204	
	5% Trimmed Mean		20.4259	
	Median		21.0000	
	Variance		208.133	
	Std. Deviation		14.42683	
	Minimum		5.00	
	Maximum		77.00	
	Range		72.00	
	Interquartile Range		14.25	
	Skewness		2.057	.427
	Kurtosis		6.434	.833

2.7.2.2 Other options

Options for producing other related statistics and graphs are available. To produce a histogram, for instance, before clicking OK above, click on `Plots`, and you can then check a box to produce a histogram. Then click on `Continue` and `OK`.

2.7.2.3 Results

Table 2.8 displays results of the output. In addition to this table, box plots (Figure 2.2), stem and leaf displays (Figure 2.3) and, optionally, histograms (Figure 2.1) are also produced.

SOLVED EXERCISES

1. A new park is planned for a community, and planners wish to consider the mean center of four residential areas as a possible location. Using the coordinates and residential populations listed below, find the weighted mean center.

x-coordinate	y-coordinate	residential population
2	1	200
4	4	100
4	8	50
8	2	200

Solution. We will first find the x-coordinate of the weighted mean center. Conceptually, this is the average x-coordinate among all residents. There are 200 people with an x-coordinate of 2, 100 with an x-coordinate of 4, and so on. If we totaled the x-coordinates among all residents, the result would be 200(2) + 100(4) + 50(4) + 200 (8) = 400 + 400 + 200 + 1600 = 2600. There are 200 + 100 + 50 + 200 = 550 residents, and so the average x-coordinate among all residents is 2600/550 = 4.73. Similarly, the y-coordinate of the mean center is

$$\frac{200(1) + 100(4) + 50(8) + 200(2)}{550} = \frac{1400}{550} = 2.55$$

Thus, the weighted mean center is at (4.73, 2.55).

2. Using the following data, find the mean, median, and range:

Data values: 29, 35, 17, 30, 231, 6, 27, 35, 23, 29, 13

Solution. The mean (or average) is equal to the sum of the observations (29 + 35 + 17 + ... + 29 + 13 = 475), divided by the number of observations ($n = 11$); thus the mean is 475/11 = 43.18. To find the median, first order the observations: 6, 13, 17, 23, 27, 29, 29, 30, 35, 35, 231. The median is observation number $(n + 1)/2$ on this ordered list. Since $n = 11$, $(n + 1)/2 = 6$; the 6th observation on the list (29) is the median. Finally, the range is equal to the largest observation (231), minus the smallest (6), and is therefore equal to 231 − 6 = 225.

3. Find the skewness and kurtosis for the following data:

4, 7, 8, 13.

(Continued)

(Continued)

Solution. For the skewness, we use Equation 2.7. The equation implies that we will need the mean and the standard deviation. The mean is equal to $(4 + 7 + 8 + 13)/4 = 32/4 = 8$. The standard deviation is the square root of the variance. The variance is found by finding the sum of the squared deviations from the mean, and then dividing the result by $n - 1$, where n is the number of observations. The squared deviations from the mean are $(4 - 8)^2$, $(7 - 8)^2$, $(8 - 8)^2$, and $(13 - 8)^2$; summing these yields $(-4)^2 + (-1)^2 + 0^2 + 5^2 = 16 + 1 + 0 + 25 = 42$. The variance is thus equal to $42/(4-1) = 14$, and the standard deviation is $s = \sqrt{14} = 3.74$. The numerator of the skewness is equal to the sum of the cubed deviations from the mean. The cubed deviations from the mean are $(4 - 8)^3$, $(7 - 8)^3$, $(8 - 8)^3$, and $(13 - 8)^3$, and the sum of these quantities is $(-4)^3 + (-1)^3 + 0^3 + 5^3 = (-64) + (-1) + 0 + 125 = 60$. The denominator is equal to the number of observations, multiplied by the cube of the standard deviation: $4(3.74^3) = 209.25$. The skewness is equal to $60/209.25 = 0.287$. The data exhibit a small amount of positive skewness.

We use Equation 2.8 to find the kurtosis. The deviations from the mean are first raised to the fourth power: $(4 - 8)^4$, $(7 - 8)^4$, $(8 - 8)^4$, and $(13 - 8)^4$; these are summed to find the numerator of the kurtosis, and this result is $(-4)^4 + (-1)^4 + 0^4 + 5^4 = 256 + 1 + 0 + 625 = 882$. The denominator of the kurtosis is equal to the number of observations, multiplied by the standard deviation raised to the fourth power (and note that this latter quantity is equal to the square of the variance). The denominator is therefore $4(3.74^4) = 4(14^2) = 784$. The kurtosis is equal to $882/784 = 1.125$. Note that since this is less than 3, the distribution can be described as flat, or platykurtic.

4. Find the grouped mean for the following data:

Category	Frequency
0–19.99	5
20–39.99	15
40–59.99	10
60–79.99	12

Solution. To find the grouped mean, we assume that all observations are at the midpoint of the category they are in. There are five observations that are less than 20. All we know is that they are between 0 and 20; we do not know their exact values. We assume that they are all midway between 0 and 20 – and therefore assign them values of 10. Similarly, the 15 observations in the next category are all assigned a value of 30, which is the midpoint of the category they are in (20–39.99). The mean of all of these observations, once they are assigned the midpoint of their categories, is found by adding all of their assigned values, and then dividing the result by the total number of observations.

Thus, we have $5 \times 10 = 50$ (which is the total of the five observations in the first category), plus $15 \times 30 = 450$ (which is the total of the 15 observations in the second category), plus 10×50, plus 12×70. The solution is therefore equal to this total of $50 + 450 + 500 + 840 = 1840$, divided by the total number of observations ($5 + 15 + 10 + 12 = 42$); this is equal to $1840/42 = 43.81$.

EXERCISES

1. The 236 values that appear below are the 1990 median household incomes (in dollars) for the 236 census tracts of Buffalo, New York.

   (a) For the first 19 tracts, find the mean, median, range, interquartile range, standard deviation, variance, skewness, and kurtosis, using only a calculator (though you may want to check your results using a statistical software package). In addition, construct a stem-and-leaf plot, a box plot, and a histogram for these 19 observations.

   (b) Use a statistical software package to repeat part (a), this time using all 236 observations.

   (c) Comment on your results. In particular, what does it mean to find the mean of a set of medians? How do the observations that have a value of 0 affect the results? Should they be included? How might the results differ if a different geographic scale was chosen?

   22342, 19919, 8187, 15875, 17994, 30765, 31347, 27282, 29310, 23720, 22033, 11706, 15625, 6173, 15694, 7924, 10433, 13274, 17803, 20583, 21897, 14531, 19048, 19850, 19734, 18205, 13984, 8738, 10299, 10678, 8685, 13455, 14821, 23722, 8740, 12325, 10717, 21447, 11250, 16016, 11509, 11395, 19721, 23231, 21293, 24375, 19510, 14926, 22490, 21383, 25060, 22664, 8671, 31566, 26931, 0, 24965, 34656, 24493, 21764, 25843, 32708, 22188, 19909, 33675, 15608, 15857, 18649, 21880, 17250, 16569, 14991, 0, 8643, 22801, 39708, 17096, 20647, 30712, 19304, 24116, 17500, 19106, 17517, 12525, 13936, 7495, 10232, 6891, 16888, 42274, 43033, 43500, 22257, 22931, 31918, 29072, 31948, 36229, 33860, 32586, 32606, 31453, 32939, 30072, 32185, 35664, 27578, 23861, 18374, 26563, 30726, 33614, 30373, 28347, 37786, 48987, 56318, 49641, 85742, 43229, 53116, 44335, 30184, 36744, 39698, 0, 21987, 66358, 46587, 26934, 27292, 31558, 36944, 43750, 49408, 37354, 31010, 35709, 32913, 25594, 25612, 28980, 28800, 28634, 18958, 26515, 24779, 21667, 24660, 29375, 29063, 30996, 45645, 39312, 34287, 35533, 27647, 24342, 22402, 28967, 39083, 28649, 23881, 31071,

(Continued)

(Continued)

> 27412, 27943, 34500, 19792, 41447, 35833, 41957, 14333, 12778, 20000,
> 19656, 22302, 33475, 26580, 0, 24588, 31496, 30179, 33694, 36193, 41921,
> 35819, 39304, 38844, 37443, 47873, 41410, 34186, 36798, 38508, 38382,
> 37029, 48472, 38837, 40548, 35165, 39404, 34281, 24615, 34904, 21964,
> 42617, 58682, 41875, 40370, 24511, 31008, 16250, 29600, 38205, 35536,
> 35386, 36250, 31341, 33790, 31987, 42113, 37500, 33841, 37877, 35650,
> 28556, 27048, 27736, 30269, 32699, 28988, 22083, 27446, 76306, 19333

2. Ten migration distances corresponding to the distances moved by recent migrants are observed (in miles): 43, 6, 7, 11, 122, 41, 21, 17, 1, 3. Find the mean and standard deviation, and then convert all observations into *z*-scores.

3. By hand, find the mean, median, and standard deviation of the number of bedrooms, for the first ten observations of the Milwaukee dataset.

4. Using *SPSS* or *Excel*, answer the following questions using the full Milwaukee house sale dataset:

   (a) Make histograms of (i) house sale values for those observations that have more than two bedrooms, and (ii) those observations that have fewer than three bedrooms. Comment on the differences.

   (b) Find the mean, median, and standard deviation for sales price, number of bedrooms, age of house, and lot size.

5. Using *SPSS* or *Excel*, answer the following questions using the full Hypothetical UK Housing Prices dataset:

   (a) Provide descriptive information on house prices, number of bedrooms, number of bathrooms, floor area, and date built. For each, provide the mean, median, standard deviation, and skewness. Also for each, provide a box plot.

   (b) What percentage of homes have garages?

   (c) What percentage of homes were built during each of the following time periods: pre-war, inter-war, and post-war?

6. For the first ten observations of the Milwaukee dataset:

   (a) find the mean center;

   (b) find Bachi's standard distance.

Exercises 7–11 are questions pertaining to notation; see Appendix B for a review.

7. Given $a = 3$, $b = 4$, and

Observation	x	y
1	3	2
2	5	4
3	7	6
4	2	8
5	1	10

Find the following:

(a) Σy_i

(b) Σy_i^2

(c) $\Sigma ax_i + by_i$

(d) Πx_i

(e) $\Sigma_{i=2}^{3} y_i$

(f) $3x_2 + y_4$

(g) $32! / (30!)$

(h) $\Sigma_k x_k y_k$

8. Let $a = 5$, $x_1 = 6$, $x_2 = 7$, $x_3 = 8$, $x_4 = 10$, $x_5 = 11$, $y_1 = 3$, $y_2 = 5$, $y_3 = 6$, $y_4 = 14$, and $y_5 = 12$. Find the following:

(a) Σx_i

(b) $\Sigma x_i y_i$

(c) $\Sigma (x_i + ay_i)$

(d) $\Sigma_{i=1}^{3} y_i^2$

(e) $\Sigma_{i=1}^{i=n} a$

(f) $\Sigma_k 2(y_k - 3)$

(g) $\Sigma_{i=1}^{i=5} (x_i - \bar{x})$

9. Find $8!/3!$

(Continued)

(Continued)

10. Find $\binom{10}{5}$

11. Use the following table of commuting flows to determine the total number of commuters leaving each zone and the total number entering each zone. Also find the total number of commuters. For each answer, also give the correct notation, assuming y_{ij} denotes the number of commuters who leave origin i to go to destination zone j.

Origin zone	Destination zone			
	1	2	3	4
1	32	25	14	10
2	14	33	19	9
3	15	27	39	20
4	10	12	20	40

12. The following data represent stream link lengths in a river network:

    100, 426, 322, 466, 112, 155, 388, 1155, 234, 324, 556, 221, 18, 133, 177, 441.

    Find the mean and standard deviation of the link lengths.

13. For the following annual rainfall data, find the grouped mean and the grouped variance.

Rainfall	Number of years observed
< 20″	5
20–29.9″	10
30–39.9″	12
40–49.9″	11
50–59.9″	3
≥ 60″	2

Notice that assumptions must be made about the 'midpoints' of the open-ended age groups. In this example, use 15″ and 65″ as the midpoints of the first and last rainfall groups, respectively.

14. A square grid is placed over a city map. What is the Euclidean distance between two places located at (1,3) and (3,6)?

15. In the example above, what is the Manhattan distance?

16. Use a rose diagram to portray the following angular data:

Direction	Frequency	Direction	Frequency
N	43	S	60
NE	12	SW	70
E	23	W	75
SE	45	NW	65

17. Draw frequency distributions that have (a) positive skewness, (b) negative skewness, (c) low kurtosis and no skewness, and (d) high kurtosis and no skewness.

18. What is the coefficient of variation among the following commuting times: 23, 43, 42, 7, 23, 11, 23, 55?

19. A square grid is placed over a city map. There are residential areas at the coordinates (0,1), (2,3) and (5,6). The respective populations of the three areas are 2500, 2000, and 3000. A centralized facility is being considered for either the point (4,4) or the point (4,5). Which of the two points is the better location for a centralized facility, given that we wish to minimize the total Manhattan distance traveled by the population to the facility? Justify your answer by giving the total Manhattan distance traveled by the population to each of the two possible locations.

20. Is the following data on incomes positively or negatively skewed? You do not need to calculate skewness, but you should justify your answer.

    Data in thousands: 45, 43, 32, 23, 45, 43, 47, 39, 21, 90, 230.

21. (a) Find the weighted mean center of population, where cities' populations and coordinates are given as follows:

City	x	y	Population
A	3.3	4.3	34,000
B	1.1	3.4	6,500
C	5.5	1.2	8,000
D	3.7	2.4	5,000
E	1.1	1.1	1,500

   (b) Find the unweighted mean center, and comment on the differences between your two answers.

   (c) Find the Euclidean distances of each city to the weighted mean center of population.

(Continued)

(Continued)

(d) Find Bachi's (weighted) standard distance.

(e) Find the relative standard distance by using the standard distance in part (d) and by assuming that the study area is a rectangle with coordinates of (0,0) in the southwest and (6,6) in the northeast.

(f) Repeat part (e), this time assuming that the coordinates of the rectangle range from (0,0) in the southwest to (8,8) in the northeast.

22. Find the angular mean and the circular variance for the following sample of nine angular measurements: 43°, 45°, 52°, 61°, 75°, 88°, 88°, 279°, and 357°.

23. A public facility is to be located as closely as possible to the weighted mean center of five residential areas. Given the following data:

x-coordinate	y-coordinate	Population
3	7	40
2	2	10
1	1	50
6	2	30
2	5	20

(a) Find the weighted mean center. What is the aggregate travel distance?

(b) Find Bachi's standard distance, using the population weights. Compare this with the alternative measure of standard distance, which is simply a weighted average of the distances.

(c) Assuming that the study area is bounded by (0,0) in the southwest and by (6,7) in the northeast and using the answer from part (b), find the relative distance.

24. For the Milwaukee housing sales dataset,

(a) find the mean and standard deviation associated with the age of housing. Do this by hand for the first 15 houses in the dataset, and repeat using statistical software for the entire dataset. Also find the mean center by using the x- and y-coordinates (again by hand for the first 15 observations, and using statistical software for the entire dataset).

(b) Use software such as *SPSS* or *Excel* to produce a histogram of housing prices.

On the Companion Website

The website includes a number of resources utilised in Chapter 2. Both the **Home Sales in Milwaukee, Wisconsin** dataset and the **Hypothetical UK Housing Prices** dataset are available within the 'Datasets' section of the website, including information about their respective fields and formats. The website also provides links to other websites and videos that also explain the concepts described here, to supplement the text.

3

PROBABILITY AND DISCRETE PROBABILITY DISTRIBUTIONS

3.1 INTRODUCTION

In Chapter 1, we had our first glimpse into some of the concepts that are used to both describe sample data and make inferences from it. We saw that the field of probability provides a foundation for inferential statistics. In this chapter, we will learn more about this foundation. This will be accomplished by starting from some basic ideas and concepts, and then using them to learn more about *probability distributions*. Quite simply, probability distributions may be thought of as histograms depicting relative frequencies. Histograms of this type that have particular shapes arise repeatedly in applications, and we will review several examples of these in Sections 3.3 through 3.6.

In addition to serving as a foundation for inferential statistics and hence the material covered in later chapters of this book, probability concepts are also used directly in the development and use of models of spatial phenomena (recall both the examples in Section 1.4, and, from Section 1.2, the idea that models, as simplifications of reality, are an integral part of the scientific method). This function of probability in spatial statistical analysis is illustrated in the final section of the chapter.

3.2 SAMPLE SPACES, RANDOM VARIABLES, AND PROBABILITIES

Suppose we are interested in the likelihood that current residents of a suburban street are new to the neighborhood during the past year. To keep the example manageable, we shall assume that just four households are asked about their duration of residence. There are several possible questions that may be of interest. We may wish to use the sample to estimate the probability that residents of the street moved to the street during the past year. Or we may want to know whether the likelihood of moving onto that street during the past year is any different than it is for the entire city.

This problem is typical of statistical problems in the sense that it is characterized by the *uncertainty* associated with the possible outcomes of the household survey. We may think of the survey as an experiment of sorts. The experiment has associated with it a *sample space*, which is the set of all possible outcomes. Representing a recent move with a '1' and representing longer-term residents with a '0', the sample space is enumerated in Table 3.1 (the number of recent movers associated with each outcome is given in the table in parentheses). These 16 outcomes represent all of the possible results from our survey. The individual outcomes are sometimes referred to as *simple events* or *sample points*.

Random variables are functions defined on a sample space. This is a rather formal way of saying that associated with each possible outcome is a quantity of interest to us. In our example, we are unlikely to be interested in the individual responses, but rather the total number of households that are newcomers to the street. Portrayed in Table 3.1 is the sample space with the variable of interest, the number of new households (which is simply equal to the number of 'ones' associated with the corresponding outcome), given in parentheses.

In this instance, the random variable is said to be *discrete*, since it can take on only a finite number of values (namely, the non-negative integers 0–4). Other random variables are *continuous* – they can take on an infinite number of values. Elevation, for example, is a continuous variable.

Associated with each possible outcome in a sample space is a *probability*. Each of the probabilities is greater than or equal to zero, and less than or equal to one. Probabilities may be thought of as a measure of the likelihood or relative frequency of each possible outcome. The sum of the probabilities over the sample space is equal to one.

Table 3.1 Possible outcomes, with the number of new households in parentheses

0000 (0)	0100 (1)	1000 (1)	1100 (2)
0001 (1)	0101 (2)	1001 (2)	1101 (3)
0010 (1)	0110 (2)	1010 (2)	1110 (3)
0011 (2)	0111 (3)	1011 (3)	1111 (4)

There are numerous ways to assign probabilities to the elements of sample spaces. One way is to assign probabilities on the basis of relative frequencies. Given a description of the current weather pattern, a meteorologist may note that in 65 out of the last 100 times that this pattern prevailed, there was measurable precipitation the next day. The possible outcomes – rain or no rain tomorrow – are assigned probabilities of 0.65 and 0.35, respectively, on the basis of their relative frequencies.

Another way to assign probabilities is on the basis of subjective beliefs. The description of current weather patterns is a simplification of reality, and may be based upon only a small number of variables, such as temperature, wind speed and direction, barometric pressure, etc. The forecaster may, partly on the basis of other experience that is more difficult to quantify, assess the likelihoods of precipitation and no precipitation as 0.6 and 0.4, respectively.

Yet another possibility for the assignment of probabilities is to assign each of the n possible outcomes a probability of $1/n$. This approach assumes that each sample point is equally likely, and it is an appropriate way to assign probabilities to the outcomes in special kinds of experiments. If, for example, we flipped four coins, and let '1' represent 'heads' and '0' represent tails, there would be 16 possible outcomes (identical to the 16 outcomes associated with our survey of the four residents above). If the probability of heads on each toss is $1/2$, and if the outcomes of the four tosses are assumed independent from one another, the probability of any particular sequence of four tosses is given by the product $1/2 \times 1/2 \times 1/2 \times 1/2 = 1/16$. Similarly, if the probability that an individual resident is new to the neighborhood is $1/2$, we would assign a probability of $1/16$ to each of the 16 outcomes in Table 3.1.

It is important to note that if the probability of heads differs from $1/2$, the 16 outcomes would not be equally likely. If the probability of 'heads' or the probability that a resident is a newcomer is denoted by p, the probability of tails and the probability the resident is *not* a newcomer is equal to $(1 - p)$. In this case, the probability of a particular sequence is again given by the product of the likelihoods of the individual tosses. Thus, the likelihood of '1001' (or 'HTTH' using H for heads and T for tails) is equal to $p \times (1 - p) \times (1 - p) \times p = p^2 (1 - p)^2$.

Another classic example that is used to illustrate the concepts of sample spaces, random variables, and probability is the tossing of a die, or dies. Suppose that we have the usual 'fair' six-sided die, where the probability of obtaining each of the numbers one through six is equal to $1/6$. Also suppose that our experiment consists of tossing two of these dies. There are 36 possible outcomes, and these are listed in Table 3.2. Each of these outcomes is equally likely and can therefore be assigned a probability of $1/36$. (Another way to arrive at the probability of $1/36$ for each outcome is to recognize that the result observed on the second die is independent of the result observed on the first die; when events are independent, we calculate the probability of both occurring as the product of each individual event. Thus we have here, for any of the simple events in our sample space, $1/36 = (1/6)$

(1/6). More generally, for the case of independent events, $P(A \cap B) = P(A)P(B)$ where the term on the left represents the intersection of events A and B, and is read as 'the probability that both A and B occur'). Note that the probabilities assigned to the outcomes sum to one – it is certain that *one* of the outcomes will occur.

Table 3.2   Possible outcomes from tossing two dice

1	1	2	1	3	1	4	1	5	1	6	1
1	2	2	2	3	2	4	2	5	2	6	2
1	3	2	3	3	3	4	3	5	3	6	3
1	4	2	4	3	4	4	4	5	4	6	4
1	5	2	5	3	5	4	5	5	5	6	5
1	6	2	6	3	6	4	6	5	6	6	6

Now suppose that we are interested in the likelihood (i.e., probability) that the sum of these two dies is equal to nine or more. To determine this we can sum the probabilities associated with each of the relevant outcomes in the sample space. There are ten outcomes in the sample space that would give us a sum of nine or more: {36 63 45 54 46 64 56 65 55 66}. Therefore the answer is equal to $(1/36 + 1/36 + \ldots + 1/36) = 10/36 = 5/18$.

The probability of our event of interest, A = {sum greater than or equal to 9} is equal to the sum of the relevant simple events (A_i) in the sample space. Thus $pr(A) = pr(A_1) + pr(A_2) + \ldots$, where these simple events (A_i) are the outcomes within the sample space that have sums greater than or equal to 9.

What is the chance that the sum is equal to 9 or more, OR the sum is an even number? Let C be the set containing all of the outcomes that satisfy us and let B be the set containing the elements where the sum is an even number (the reader may recognize or wish to verify that $pr(B)$ is equal to 1/2). As before, A is the set of outcomes where the sum is greater than or equal to 9.

To arrive at the solution, we use the rule that $Pr(C) = Pr(A) + Pr(B) - Pr(A \cap B)$, where $Pr(A \cap B)$ denotes the probability that both A and B will occur on the same toss. The last term subtracts out those outcomes that are double-counted when summing the first two terms. In our example here, the intersection of sets A and B is $A \cap B = \{55\ 66\ 46\ 64\}$ and hence $Pr(A \cap B) = 4/36 = 1/9$. Thus $Pr(C) = 10/36 + 18/36 - 4/36 = 24/36 = 2/3$.

3.3 BINOMIAL PROCESSES AND THE BINOMIAL DISTRIBUTION

Returning to the example of whether the four surveyed households are newcomers, we are likely to be more interested in the random variable defined as the number of new households than in particular sample points. If we want to know the likelihood of receiving two 'successes', or two new households out of a survey of four, we must add up all of the probabilities associated with the relevant sample points. In Table 3.3, we use an asterisk to designate those outcomes where two households among the four surveyed are new ones.

If the probability that a surveyed household is a new one is equal to p, the likelihood of any particular event with an asterisk is $p^2 (1 - p)^2$. Since there are six such possibilities, the desired probability is $6p^2(1 - p)^2$.

Note that we have assumed (a) that the probability p is constant across households, and (b) that households behave independently. These assumptions may or may not be realistic. Different types of households might have different values of p (for example, those who live in bigger houses may be more (or less) likely to be newcomers). The responses received from nearby houses may also not be independent. If one respondent was a newcomer, it might make it more likely that a nearby respondent is also a newcomer (if, for example, a new row of houses has just been constructed).

Table 3.3   Asterisked outcomes indicating outcomes of interest

0000	0100	1000	1100*
0001	0101*	1001*	1101
0010	0110*	1010*	1110
0011*	0111	1011	1111

Under these assumptions, we have a *binomial process* – the number of households who are newcomers is a *binomial variable*, and the probability that it takes on a particular value is given by the *binomial distribution*. We can find the probability that the random variable, designated X, is equal to 2, using the binomial formula:

$$pr(X = 2) = \binom{4}{2} p^2(1-p)^2 = 6p^2(1-p)^2 \tag{3.1}$$

The binomial coefficient, $\binom{4}{2}$ provides a means of counting the number of relevant outcomes in the sample space:

$$\binom{4}{2} = \frac{4!}{2!2!} = \frac{24}{(2)(2)} = 6. \tag{3.2}$$

The binomial process may be summarized as resulting from situations where:

a) the process of interest consists of a number (n) of independent trials (in our example, the independent trials were the independent responses of the $n = 4$ residents);

b) each trial results in one of two possible outcomes (e.g., a newcomer, or not a newcomer); these outcomes can be given the general labels of 'success' and 'failure';

c) the probability of each outcome is known, and is the same for each trial; it is usual to designate the probability of 'success' as p and the probability of 'failure' as $1 - p$; and

d) the random variable of interest (x) is the number of successes.

For binomial processes, the probability of x successes is given by the binomial distribution:

$$pr(X = x) = \binom{n}{x} p^x (1 - p)^{n-x} . \tag{3.3}$$

The $p^x (1 - p)^{n-x}$ part of the equation is the probability of getting one particular ordering of x successes and $n - x$ failures. That is, it is the probability attached to a particular outcome of the sample space. The $\binom{n}{x}$ part of the equation represents the number of such outcomes in the sample space; it is the number of possible rearrangements of x successes and $n - x$ failures.

You should recognize that for given values of n and p, we can generate a histogram by using this formula to generate the expected frequencies associated with different values of x. This histogram, unlike those in the previous chapter, is not based upon observed data. This is instead a theoretical histogram. It is also known as the *binomial probability distribution*, and it reveals how likely particular outcomes are.

For example, suppose that the probability that a surveyed resident is a newcomer to the neighborhood is $p = 0.2$. Then the probability that our survey of four residents will result in a given number of newcomers is:

$$pr(X = 0) = \binom{4}{0} .2^0 .8^4 = 0.4096$$

$$pr(X = 1) = \binom{4}{1} .2^1 .8^3 = 0.4096$$

$$pr(X = 2) = \binom{4}{2} .2^2 .8^2 = 0.1536$$

$$pr(X = 3) = \binom{4}{3} .2^3 .8^1 = 0.0256 \tag{3.4}$$

$$pr(X = 4) = \binom{4}{4} .2^4 .8^0 = 0.0016$$

The probabilities may be thought of as relative frequencies. If we took repeated surveys of four residents, 40.96% of the surveys would yield no newcomers, 40.96% would reveal one newcomer, 15.36% would reveal two newcomers, 2.56% would yield three newcomers, and 0.16% would result in four newcomers. Note that the probabilities or relative frequencies sum to one. The binomial distribution depicted in Figure 3.1 portrays these results graphically. If we multiplied the vertical scale by n, the histogram would represent absolute frequencies expected in each category.

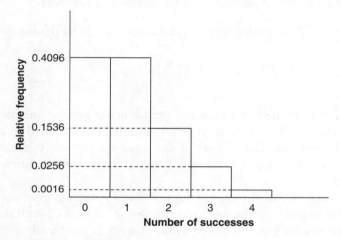

Figure 3.1   Binomial distribution with $n = 4, p = 0.2$

While the actual number of newcomers is determined from the survey, we can also define an *expected value*, or *theoretical mean*. In our present example, we would expect, on average, two-tenths of the four people to be newcomers. The expected value is therefore given simply as np, and in this example is equal to $(4)(0.2) = 0.8$. If this 'experiment' was repeated a large number of times, sometimes we would observe no newcomers, sometimes we would observe one newcomer, etc. The average result of a large number of such experiments would be 0.8 newcomers. This concept is a fairly intuitive one, and can be also illustrated with the following example. If we toss a fair coin (where the probability of getting heads on a single toss is equal to $p = 0.5$) a total of $n = 100$ times, the actual number of heads could be anything between 0 and 100, but the expected value or theoretical mean – that is, the average result we would expect from a large number of such experiments – would be $np = 100(0.5) = 50$ heads.

The *theoretical variance* associated with binomial processes is equal to $np(1 - p)$. If each person in a statistics class carried out the mini-survey described above by interviewing four residents, and if all of the responses were reported back to the class, we have already seen that the average result would be 0.8 new residents. There would of course be variability in these results, and there are two ways to use the concept of variance to summarize this variability. One could compute the sample variance from the actual results using Equation 2.4, or one could find the theoretical variance from $np(1 - p) = 4(0.8)(0.2) = 0.64$. If the statistics class was large (i.e., the survey was repeated a large number of times), the sample variance and theoretical variance would be very similar. We see that an attractive feature of specifying processes and distributions is that they have particular characteristics – we can say something about the mean and variance without actually seeing the data. Just as the shape of the histogram gives us visual information about the likelihood of particular outcomes, the theoretical mean and variance give us numerical information summarizing the expected outcomes.

Example 3.1

The annual probability of flooding in a community is 0.25. What is the probability of three floods in the next four years?

Solution: In this case, each year represents an independent trial. Each trial results in either a success (a year with flooding, though flooding is hardly considered a success), or a failure (a year without flooding). The probability of 'success' in each year is equal to $p = 0.25$. The probability of $X = 3$ successes in $n = 4$ years is

$$pr(X = 3) = \binom{4}{3} 0.25^3 0.75^1 = 0.0496 \tag{3.5}$$

3.4 THE GEOMETRIC DISTRIBUTION

The binomial distribution is used when there are n independent trials, when there are two possible outcomes on each trial, and where the probability of success is constant and equal to p on each trial. The discrete random variable of interest is the number of successes that occur. In some cases, however, we are interested in other characteristics of this same process. For example, we may be interested in how long it takes to observe the first 'success'. Suppose that an individual has a constant probability of moving in each year. The probability that the individual's next move is during the coming year is p; the likelihood that they move next in the second year of observation is $(1 - p)p$ (since we observe that they do not move in year 1, and they do move in year 2). We may generalize further: the probability that the next move occurs in year 3 is $(1 - p)(1 - p)p$, or $(1 - p)^2 p$. The probability that the next move occurs in year x is $(1 - p)^{x-1}p$, which corresponds to $x - 1$ years of no move, followed by a year with a move. The geometric distribution is applicable under the same conditions as the binomial, except that the discrete random variable of interest has shifted from the number of successes to the number of the trial on which the first 'success' occurs. In geographical applications, we may, for example, be interested in the time until the first flood, the time until an individual changes residence, the time until a house is sold, or the time until an accident occurs at an intersection.

The probability that a random variable X, representing the number of the trial on which the first success occurs, takes on a value of x, is

$$pr(X = x) = (1 - p)^{x-1} p; \; x \geq 1, \tag{3.6}$$

where p is again the probability of success on a given trial. This formula makes intuitive sense – observing the first success on trial number x can only happen by first observing $x - 1$ failures (each with probability $1 - p$), and then observing a success (which occurs with probability p). Unlike the binomial distribution, there is no need to worry about combinations and factorials, since the only event of interest in the sample space is the one consisting of $x - 1$ failures followed by a success. Note also that for the geometric distribution, x never takes on a value of zero.

Example 3.2

The probability of a residential move in a given year is $p = 0.18$. Make a table displaying the probability that an individual's next move will be in year $x = 1, 2, 3, 4, 5, 6, 7,$ or 8.

x	$p(X = x)$	$p(X = x)$
1	p	.18
2	$(1 - p)p$	$(.82)^1 .18 = .1476$
3	$(1 - p)^2 p$	$(.82)^2 .18 = .1210$
4	$(1 - p)^3 p$	$(.82)^3 .18 = .0992$
5	$(1 - p)^4 p$	$(.82)^4 .18 = .0814$
6	$(1 - p)^5 p$	$(.82)^5 .18 = .0667$
7	$(1 - p)^6 p$	$(.82)^6 .18 = .0547$
8	$(1 - p)^7 p$	$(.82)^7 .18 = .0449$

Figure 3.2 shows a histogram of the outcomes and their associated probabilities. Note that the probabilities decrease with increasing x. This decline is characteristic of the geometric distribution, which always has the general shape shown in Figure 3.2. The value of p governs the steepness of the decline; the higher the value of p, the steeper the decline.

The theoretical mean of a geometric random variable is $1/p$, and the theoretical variance is equal to $(1 - p)/p^2$. Thus, in our present example, if we observe a large number of individuals,

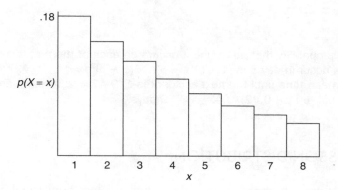

Figure 3.2   The geometric distribution

we would expect some to move in year 1, some to move in year 2, and so on. The average time until a move would be $1/p = 1/0.18 = 5.56$ years. The observations also display variability; not everyone moves for the first time in the same year. The variance of our observations for a large number of individuals would be $0.82/0.18^2 = 25.3$ years. The standard deviation is equal to the square root of 25.3, or 5.03 years.

Example 3.3

The probability of observing a flood during the year on a particular floodplain is 0.24. What is the probability that the next flood occurs three or more years from now? What is the mean time until the next flood?

Solution: We seek $pr(X \geq 3)$, which is equal to $1 - [pr(X = 1) + pr(X = 2)]$. Using the geometric distribution, the latter is equal to $1 - [p + (1 - p)p] = 1 - [0.24 + 0.76(0.24)] = 0.5776$. The mean time until a flood occurs is $1/0.24 = 4.17$ years. Note that this question is identical to the question, 'What is the probability of no floods during the next two years?' It could have therefore also been answered using the binomial distribution, with $n = 2$, $p = 0.24$, and $x = 0$.

Example 3.4

The daily probability of an accident at an intersection is $p = 0.42$. What is the probability that the next accident occurs three days from now? What are the mean and variance of the time until the next accident?

(Continued)

(Continued)

Solution: The probability that no accident occurs on each of the first two days, and then one *does* occur in day 3 is $(1 - p)(1 - p)p = (1 - p)^2 p = (1 - 0.42)^2(0.42) = 0.1413$. The mean time until the next accident is $1/0.42 = 2.38$ days, and the variance is equal to $(1 - 0.42)/0.42^2 = 3.29$ days.

3.5 THE POISSON DISTRIBUTION

Assume that you are a transportation planner, and you are concerned about safety at a particular intersection. During the last 60 days, there were three accidents, each occurring on separate days. You are asked to estimate the number of accidents that will occur during the next month (30 days), and you are also asked to estimate the probability that there will be more than three accidents during the next month.

One approach would be to use the binomial distribution. Each day may be thought of as a trial, and so $n = 30$. The daily probability of success (an accident at the intersection!) would be based on the observed frequency of accidents and would be set equal to $p = 0.05$ (which is equal to 3/60, the probability that there is an accident on any given day). The number of accidents that would be expected over the course of the next month would be equal to $np = 30(0.05) = 1.5$. This is precisely one-half of the number of accidents observed during the previous 60-day period. The probability of more than three accidents would be estimated, using the binomial distribution, as $1 - (pr(X = 0) + pr(X = 1) + pr(X = 2) + pr(X = 3))$, which is equal to:

$$1 - \left\{ \binom{30}{0} 0.05^0 0.095^{30} + \binom{30}{1} 0.05^1 0.095^{29} + \binom{30}{2} 0.05^2 0.095^{28} + \binom{30}{3} 0.05^3 0.095^{27} \right\} = 0.0608 \qquad (3.7)$$

However, if a trial is defined as a day, this approach really provides the answer to a slightly different question – namely, 'What is the probability that there are more than three days with accidents during the next month?' It ignores the possibility that one day could contain two or more accidents.

A related point is that the choice of day for the time unit is artificial. Suppose that, instead of observing the next 30 days, we observed the next 60 'half-days'. The probability of an accident during each half-day is 3/120 (or 0.025, since there were three half-days with accidents during the previous – 60 days or 120 half-day periods), and therefore the probability of more than three accidents during the next month is now

$$1- \left\{ \binom{60}{0} 0.025^0 0.975^{60} + \binom{60}{1} 0.25^1 0.975^{59} + \binom{60}{2} 0.25^2 0.975^{58} + \right.$$

$$\left. \binom{60}{3} 0.25^3 0.975^{57} \right\} = 0.0632 \qquad (3.8)$$

which is slightly different from our previous answer. To avoid this arbitrariness, we continue to divide the day into smaller and smaller time periods, each with a smaller probability of witnessing an accident (thus, there are 120 six-hour periods in a 30-day month, 240 three-hour periods, etc.). When the number of trials becomes very large, and when the probability of success (here, an accident) in any particular trial becomes very small, we make the transition from a discrete-time problem to a continuous time problem. We also make the transition from a binomial problem to a Poisson problem. For a continuous time scenario, the probability of x events during a period of time is equal to

$$pr(X = x) = \frac{e^{-\lambda} \lambda^x}{x!} \qquad (3.9)$$

where λ is the mean number of events one would expect during the time period of interest, and e is a constant, and always equal to 2.718. In our example, we expect the mean number of accidents during the 30-day period to be 1.5. The probability of observing no accidents during the month is $e^{-1.5} 1.5^0/0! = 0.2231$; the probability of observing one, two, or three accidents is:

$$pr(X = 1) = \frac{e^{-1.5} 1.5^1}{1!} = 0.3347$$

$$pr(X = 2) = \frac{e^{-1.5} 1.5^2}{2!} = 0.2510 \qquad (3.10)$$

$$pr(X = 3) = \frac{e^{-1.5} 1.5^3}{3!} = 0.1255$$

Thus, the probability of observing more than three accidents is $1 - (0.2231 + 0.3347 + 0.2510 + 0.1255) = 0.0657$. Note that this result is similar to that found in Equation 3.7 and even more similar to that found in 3.8.

One perspective of the Poisson distribution, then, is as a limiting case of the binomial distribution. The Poisson distribution is more appropriate than the binomial distribution when the number of trials is arbitrary, as it is here, and where a continuous time frame is more natural.

Example 3.5

Suppose that traffic accidents occur once every ten days. Find the probability that there are three accidents during the next 25-day period.

Solution: The first step is to find λ, which is the expected number of events during the time period of interest. If there is one accident every ten days, we would expect λ = 2.5 accidents every 25 days. (More formally, this is the solution to $\lambda/25 = 1/10$.) The probability of three accidents is, using Equation 3.9, $e^{-2.5}(2.5^3)/3! = 0.2138$. It is possible to approximate this solution with the binomial approach, using $n = 25$ and $p = 1/10$. This binomial probability is $\binom{25}{3} 0.1^3 0.9^{25-3} = 0.2265$.

It is also possible to use the Poisson distribution to approximate the binomial distribution. This is useful when n is large, when p is small, and when np is less than about 5. For example, suppose that we want to know the probability that fewer than two people walk to work out of a sample of 50. We are told that the probability that an individual walks to work is 0.04. Assuming a binomial process (where individuals make decisions independently, and all have the same probability of walking to work), the desired probability is

$$pr(X = 0) + pr(X = 1) = \binom{50}{0} 0.04^0 .96^{50} + \binom{50}{1} 0.04^1 0.96^{49} = 0.400.$$

A Poisson approximation to this probability is found by first realizing that the expected number of events, that is, people walking to work, is equal to $np = \lambda = 50(0.04) = 2$. This value of λ is also referred to as the expected value, or the theoretical mean. Next, using Equation 3.9, we find that the Poisson probability is $e^{-2}2^0/0! + e^{-2}2^1/1! = 0.4060$. This serves as an approximation to the binomial probability.

Example 3.6

Suppose that we are interested in the frequency of flooding along a creek that runs through a residential area. It would be useful to know how likely floods were, whether we were purchasing a house in the area, setting flood insurance premiums, or designing a flood control project.

Floods can of course be of different magnitudes. The magnitude of an n-year flood is such that it is exceeded with probability $1/n$ in any given year. Thus, the probability of a 50-year flood in any given year is 1/50. A 100-year flood is larger, and occurs less frequently; a 100-year flood occurs in any given year with probability 1/100. According to the US Army Corps of Engineers, hurricane Katrina, which hit the US Gulf Coast in

2005, was a 396-year storm, meaning that a storm of that intensity could be expected only once every 396 years.

What is the probability that there will be exactly one 50-year flood during the next 50-year period?

Solution: The Poisson probability is found by first recognizing that the expected number of floods during this period is equal to $\lambda = 1$ (if you have trouble deciding upon the correct value of λ, it may be useful to realize that, because it is a mean, you can think in terms of the binomial equivalent of np; in this case, we have $n = 50$ years, and the probability of a flood in a given year is $1/50$, so that $np = \lambda = 1$). Then the probability of observing exactly one such flood is $pr\,(X = 1) = e^{-1}1^1/1! = 0.368$. The binomial approximation $\binom{50}{1}(1/50)^1 (49/50)^{49}$ is equal to 0.3716; this is actually the probability that there is precisely one year in which (at least) one 50-year flood is observed.

To this point, our examples illustrating the Poisson distribution have involved questions where the variable of interest is the number of events that occur in (continuous) time. Because the Poisson distribution is a result of dividing time up into a very large number of independent trials, the Poisson distribution is essentially modeling the number of events that occur randomly in time. This distribution lies at the core of queueing theory – the theory of queues, or lines. If cars arrive randomly at a traffic light at the rate of three every ten seconds, we can use the Poisson distribution to determine, e.g., the probability that six cars arrive in the next ten seconds ($= e^{-3}\, 3^6/6! = 0.0504$).

We have already noted that the expected value or theoretical mean of the Poisson distribution is equal to λ. The theoretical variance of the Poisson distribution is also equal to λ; if we observed the actual counts that occurred during a large number of time periods, we would find that the sample variance associated with these counts would be close to this theoretical variance.

Importantly, the Poisson distribution may also be used to model events that occur randomly in space. Suppose that towns are randomly located throughout an area, with a mean density of two towns every 30 square kilometers. How many towns would we expect in a 50 square kilometer area? What is the probability that we find fewer than two towns in a 50 square kilometer area? The answers to these questions are found from proceeding exactly as before, except we now use spatial units instead of temporal ones. The mean number of towns in a 50 square kilometer area is found from $\lambda/50 = 2/30$, and this implies that $\lambda = 3.33$. To answer the second question, we use this value of λ in the Poisson distribution:

$$e^{-3.33}3.33^0 / 0! + e^{-3.33}3.33^1 / 1! = 0.1550 \qquad (3.11)$$

Just as the use of the Poisson distribution in a temporal context presumes that events occur randomly in time, the use of the Poisson distribution in a spatial context presumes that events occur randomly in space.

> ### Example 3.7
>
> A disease occurs randomly in space, with one case every 16 square kilometers. What is the probability of finding four cases in a 30 square kilometer area?
>
> **Solution:** First find $\lambda = 1.875$ from $\lambda/30 = 1/16$. Then the probability of four cases in our 30 square kilometer region is equal to $e^{-1.875}$ $1.875^4/4! = 0.0790$.

In Chapter 5, we will see how such questions are used to test hypotheses. For example, if a disease is presumed to occur randomly in space, and a large number of cases are observed in a very small geographic area, either (a) disease cases do occur randomly, and a very rare event has occurred, or (b) the presumption of spatial randomness is incorrect, and the disease may be exhibiting geographic clustering (due to, e.g., environmental causes).

3.6 THE HYPERGEOMETRIC DISTRIBUTION

The hypergeometric distribution has geographical applications to problems ranging from disease clustering to residential segregation. It is used to model processes that can be described as follows.

A population has N elements, and the elements may be classified into one of two categories (to follow this description more readily, imagine a box with N balls, where each ball is either red or blue). The number of elements falling into the first category (say the number of red balls) is designated by r; therefore, $N - r$ is the number of elements falling into the other category. The experiment consists of drawing a sample of n elements from the population at random, without replacement. The random variable of interest, X, is the number of elements in the sample that fall into the first category (e.g., X is the number of red balls that are chosen when a sample of n balls is taken from the box).

The probability that the sample of n contains x items of interest (e.g., x red balls), is

$$pr(X = x) = \frac{\binom{r}{x}\binom{N-r}{n-x}}{\binom{N}{n}}. \tag{3.12}$$

This is the hypergeometric distribution. The theoretical mean of the hypergeometric distribution is nr/N; the variance is equal to $\{nr(N - r)(N - n)\}/\{N^2(N - 1)\}$.

Example 3.8

A box contains $N = 5$ balls; $r = 3$ are red, and $N - r = 2$ are blue. A sample of $n = 2$ balls is drawn from the box without replacement. Find the probability that exactly one of the balls is red.

Solution: One approach is to write out the elements of the sample space, since the number of possible outcomes is small.

A simple way to write out the sample space is {RB RR BB BR}, where R designates a red ball, and B a blue ball. The probability of obtaining the first outcome, RB, is equal to $(3/5)(2/4) = 3/10$, since the probability of first drawing a red ball is 3/5 (there are 5 balls in the box, and 3 are red), and then drawing a blue is 2/4 (since at that point there would be 4 balls in the box, and 2 would be blue). Similarly, the probability of RR is $(3/5)(2/4) = 3/10$; the probability of BB is $(2/5)(1/4) = 1/10$, and the probability of BR is $(2/5)(3/4) = 3/10$. The probability of obtaining exactly one red ball is the sum of the probabilities associated with RB and BR, or $3/10 + 3/10 = 3/5 = 0.6$.

An alternative and more detailed specification of the sample space occurs if we label each of the balls so that they are identified as {R1, R2, R3, B1, and B2}; the sample space is then:

(R1, R2)	(R2, B1*)
(R1, R3)	(R2, B2*)
(R1, B1*)	(R3, B1*)
(R1, B2*)	(R3, B2*)
(R2, R3)	(B1, B2)

where we have ignored the order in which the balls are chosen (we could include information on ordering; this would double the number of elements in the sample space, and halve the probability associated with each outcome). There are ten possible outcomes, and each is equally likely. Six of these (noted here with asterisks in the sample space) have exactly one red ball; therefore, the probability of selecting one red ball in a sample of two is again found to be 0.6.

This second approach facilitates the understanding of the hypergeometric distribution. There are a total of $\binom{N}{n}$ ways to choose a sample of n items from a population of N items; in this example, $\binom{5}{2}$ is equal to 10, and this is the number of elements in the detailed sample space. In our example there are six of these elements that have exactly one red ball – there are $\binom{3}{1} = 3$ possibilities for choosing the one red ball (red ball number 1, 2, or 3), and $\binom{2}{1} = 2$ possibilities for choosing the one blue ball (blue ball number 1 or 2). More generally, there are $\binom{r}{x}$ ways of choosing the x 'successes' from the set of r, and there are $\binom{N-r}{n-x}$ ways to choose the other items (in this case, blue balls).

3.6.1 Application to Residential Segregation

Suppose that $N = 20$ people move into a community, and $r = 6$ of them are minorities. Furthermore, we wish to determine whether there is segregation in a particular residential block within the community. In particular, suppose that $n = 10$ people of the 20 people moving into the community move onto this specific residential block, and we observe that $x = 1$ of them is a minority. We may initially be somewhat surprised – after all, 30% (6/20) of the movers are minorities, but only 10% (1/10) of those moving onto this residential block are minorities. We wish to determine how likely it is that, out of the 10 people moving onto the block, at most one would be a minority. We can use the hypergeometric distribution as follows:

$$pr(X = 0) + pr(X = 1) = \frac{\binom{6}{0}\binom{14}{10}}{\binom{20}{10}} + \frac{\binom{6}{1}\binom{14}{9}}{\binom{20}{10}} \tag{3.13}$$

$$= 0.0054 + 0.065 = 0.0704.$$

Note that the use of the hypergeometric distribution yields the probability that zero or one minorities would end up on the block *if* the sample of ten individuals choosing to move onto the block was randomly chosen from the group of 20 people. The fact that this event has only a 7% chance of occurring implies that either (a) there is no segregation, and, just by chance, a small number of minorities has moved onto this block, or (b) the sample of ten individuals has not been taken randomly from the group of 20 individuals (which would be the case if segregation exists). How do we decide between (a) and (b)? Typically we set a threshold – often 5%. If the calculated probability is greater than 5%, as it is here, we go with option (a). If the calculated probability is lower than the threshold (say 5%), indicating that it would indeed be quite unusual, we decide on option (b). These issues are discussed further in Chapter 5.

3.6.2 Application to the Space-Time Clustering of Disease

A question that arises in the geographic analysis of disease is whether disease cases that are close together in time (i.e., the dates of onset or dates of diagnosis are similar) are also close together in space (i.e., their geographic locations are similar). If this was the case, the existence of such space–time clusters could potentially provide some insight into the nature of the disease. Our variable of interest, then, will be the number of pairs of cases that are, simultaneously, close together in time and space, thereby potentially constituting a space–time cluster.

To be more specific, assume that, for a number of people with a particular disease, we have data on their residential locations and their dates of diagnosis. A decision must first be made on what is meant by 'close in time' (e.g., it might be decided that cases that arise within three months of one another are close in time), and what is meant by 'close in space' (e.g., it might be decided that cases that are within 2 km of one another are close in space).

Then, all pairs of cases are examined. To take a simple hypothetical example, suppose that we have six disease cases – this implies that there are $N = \binom{6}{2} = 15$ pairs of cases. Of these pairs, let us assume that $r = 2$ pairs are close together in time, and therefore $N - r = 13$ pairs are not close together in time. Furthermore, of the $n = 3$ pairs of cases that are close together in space, $x = 2$ are also close together in time, and one is not. This may all be summarized as in Table 3.4.

Table 3.4 Classifications of case pairs according to closeness in time and space

	Close in time	Not close in time	Total
Close in space	2	1	3
Not close in space	0	12	12
Total	2	13	15

The hypergeometric distribution can be used to give us the probability of observing two pairs that are close in space and time, when we examine $n = 3$ of the $N = 15$ pairs (i.e., those that are close together in space), where whether the pairs are close together in time is analogous to the previous drawing of a red ball from the box. Thus

$$pr(X = 2) = \frac{\binom{2}{2}\binom{13}{1}}{\binom{15}{3}} = 0.03 \tag{3.14}$$

As in the previous example, this gives us the probability of observing two pairs that are close in space and time by chance alone – assuming that three pairs of the 15 pairs are examined. This could only happen if, among the three 'close in space pairs', we observe both of the two 'close in time' pairs, and one of the 13 'not close in time' pairs. The answer, 0.03, is relatively small – we can therefore choose to either believe that (a) the characteristics 'close in time' and 'close in space' are independent, and we happen to have an unusual sample, or (b) the characteristics are not independent – pairs that are close in space are actually relatively more likely to also be close in time. As in the previous example, we decide between these two options by comparing our result (0.03) with some threshold value (often 0.05); see Chapter 5 for more discussion of this.

In this particular example, it would also be correct to set the problem up as follows. Of the $N = 15$ pairs, there are $r = 3$ pairs that are close together in space, and $N - r = 12$ pairs that are not close together in space. Of the $n = 2$ pairs that are close together in time, both (i.e., $x = 2$) are also close together in space. The probability of observing $x = 2$ pairs that are close together in both time and space is found to be the same as before:

$$pr(X = 2) = \frac{\binom{3}{2}\binom{12}{0}}{\binom{15}{2}} = 0.03 \qquad (3.15)$$

When two pairs are observed from the 15 (i.e., those that are close together in time), we happen to obtain two of the three that are close together in space, and none of the 12 that are not close together in space.

To ensure that the hypergeometric probability is set up correctly, note that the numerator always has two terms, and the top portion of these two combinations sums to the top portion of the combination in the denominator (e.g., in Equation 3.15, 3 + 12 = 15). Also, the sum of the bottom portions of the combinations in the numerator is equal to the bottom portion of the combination in the denominator (in Equation 3.15, 2 + 0 = 2).

3.7 BINOMIAL TESTS IN *SPSS 21 FOR WINDOWS*

SPSS provides the probability of x or more successes in n trials, for a specified probability of success on each trial. Suppose we observe four successes in ten trials, and the probability of success on each trial is $p = 0.1$. What is the probability of obtaining this, or an even more extreme outcome? We need to find $pr(X = 4) + pr(X = 5) + \ldots + pr(X = 10)$. In *SPSS*, the outcomes from the ten trials are entered in a column using zeros and ones; the usual convention is to use ones to designate successes and zeros to designate failures. With a column containing four ones and six zeros (the order does not matter), we click on Analyze, then select Nonparametric Tests, Legacy Dialogs, and then Binomial. A window opens and we move the column variable over to the empty box on the right, with the heading Test Variable List. Then enter p (0.1 in this case) in the box labeled Test Proportion. Click on OK. The last column in the resulting output shows the exact significance as 0.013 – this is the probability of four or more successes in ten trials, when $p = 0.1$.

SOLVED EXERCISES

1. The probability of flooding on the Ellicott Creek in any given year is 0.19. What is the probability that the next flood will occur four years from now?

Solution. The variable defined as the time until the next flood follows a geometric distribution, since (although not explicitly stated in the problem) we will assume that, in addition to a constant annual probability of flooding (equal to 0.19), the probability of flooding in any year is independent of the probability of flooding in any other year. The only possible way that the next flood can occur four years from now is to observe three years without a flood, followed by a year with a flood. The desired probability is therefore equal to the probability of observing three 'failures', followed by a 'success'; from Equation 3.6, this is equal to $(1 - 0.19)^3(0.19) = 0.101$, where the probability of 'success' (i.e., a flood) is equal to 0.19, and x, the year (or trial number) in which the flood occurs is equal to 4.

2. The following absolute relative frequency distribution characterizes distances from individuals' homes to the nearest county park:

Distance (miles)	Absolute Relative Frequency
<5	0.15
5–9.99	0.1
10–14.99	0.15
15–19.99	0.25
20–24.99	0.15
25–29.99	0.20

Note that the relative frequencies sum to one.

(a) What is the probability that five individuals chosen at random will each travel five miles or more?

Solution. The probability that a single individual travels five miles or more is $1 - 0.15$, or 0.85. We can use this as p, the probability of success. Each individual can be considered as a single trial; hence there are $n = 5$ trials. We are interested in the probability of $x = 5$ successes. Using the binomial distribution (Equation 3.3), the solution is equal to $\binom{5}{5}0.85^5\,0.15^0 = 0.4437$.

Note that this is equivalent to the simpler expression, $0.85^5 = 0.4437$, which results from simply multiplying together the individual probabilities of success (here it is not necessary to calculate the combinatorial terms, since there is only one way in which five successes can occur).

(Continued)

(Continued)

(b) What is the likelihood that exactly one individual in a set of five surveyed individuals will travel less than five miles?

Solution. Here a 'success' is an individual traveling less than five miles; this is equal to $p = 0.15$. There are again $n = 5$ trials or individuals, and we are now interested in the probability of $x = 1$ success. The desired probability is therefore $\binom{5}{1}0.15^1\,0.85^4 = 0.3915$.

(c) What is the probability that at least two individuals in a group of six that have been interviewed travel less than ten miles?

Solution. The probability that an individual travels less than ten miles is equal to $0.15 + 0.1 = 0.25$; this is the probability of success, p, on an individual trial. We now have $n = 6$ individuals, and are interested in the probability of *at least* 2 successes, so $x = 2,3,4,5,$ or 6. Either there will be 2 or more successes, or there will be one or fewer successes. It will be quicker to first calculate the probability of either zero successes or one success, and then subtract this result from one. The probability that the number of successes is equal to either 0 or 1 is $\binom{6}{0}0.25^0 0.75^6 + \binom{6}{1}0.25^1\,0.75^5 = 0.534$. The probability of two or more successes is one minus this, or $1 - 0.534 = 0.466$.

3. A box has five green balls and six orange balls. Three balls are drawn, without replacement.

(a) What is the probability that all of the balls are green?

Solution. The probability that the first ball is green is 5/11, since there are five green balls in a box of eleven balls. If a green ball is drawn on the first try, then the box contains ten balls, and four of them are green. The probability of drawing a green ball on the second try is thus 4/10. Finally, if the first two balls are green, then, when preparing for the third draw from the box, there will be nine balls in the box, and three of them will be green. The probability of obtaining a green marble on this last draw is therefore 3/9. The answer to the question is the product of these three probabilities: $(5/11) \times (4/10) \times (3/9) = 10/165 = 0.0606$. The solution may also be found using the hypergeometric distribution. There are $N = 11$ balls in the box, and a sample of $n = 3$ will be taken. We are interested in the number of balls in the sample that are green (out of the total of five green balls in the box). We want all three to be green, and none to be orange (among the six orange balls in the box). The hypergeometric probability is

$$\frac{\binom{5}{3}\binom{6}{0}}{\binom{11}{3}} = \frac{(10)(1)}{165} = 10/165 = 2/33 = 0.0606$$

(b) What is the probability that the last ball is orange, *given* that the first two contain one green and one orange?

Solution. If the first two balls drawn contain one green and one orange, this implies that the box now contains nine balls, and five of them are orange. The probability of obtaining orange on the last draw, given the stated conditions (that among the first two drawn, one is green and the other is orange) is therefore simply five out of nine, or 5/9.

5. A left-turn lane has the capacity for two cars. Each second, a car wishing to turn left arrives with probability 0.15. If the light turns red, and stays red for 20 seconds, what is the probability that the lane capacity will be exceeded?

Solution. First, realize that the lane capacity will be exceeded if three or more cars arrive while the light is red; i.e., if three or more cars arrive during a 20-second period. There are two ways to think about this problem – one is from the perspective of the binomial distribution, and the other is from the perspective of the Poisson distribution.

We can use the binomial distribution if we think of each one second period as a trial. The probability of 'success' (i.e., the arrival of a car whose driver wishes to turn left) for a trial is $p = 0.15$. There are $n = 20$ trials. The probability of three or more successes can be found by first finding the probability of two or fewer successes, and then subtracting that result from one. Thus:

$$pr(X \geq 3) = 1 - pr(X \leq 2) = 1 - \left\{ \binom{20}{0}.15^0.85^{20} + \binom{20}{1}.15^1.85^{19} + \; = \binom{20}{2}.15^2.85^{18} + \right\}$$

$$1 - 0.4049 = .5951$$

Alternatively, we can use the Poisson distribution. During each 20-second period, we expect $np = 20(0.15) = \lambda = 3$ cars to arrive in the left-turn lane. The probability that two or fewer cars arrive is equal to:

$$(e^{-3}3^0/0!) + (e^{-3}3^1/1!) + (e^{-3}3^2/2!) = 0.4232$$

and the desired probability is found by subtracting this from one: $1 - 0.4232 = 0.5768$.

6. Category 4 hurricanes strike Florida at an average rate of one every four years. What is the probability that Florida will be struck by two or more such hurricanes over the next decade?

Solution. We can use the Poisson distribution to answer this question. To use that distribution, we need to know the number of events expected to occur on average during the time period of interest. If one hurricane occurs every four years, we can expect two every eight years. To determine the precise number expected over a ten-year period, we can set up the following equation:

$$\frac{1}{4} = \frac{\lambda}{10}$$

Solving this equation, we use $\lambda = 2.5$. To find the probability of two or more hurricanes, we first find the probability of the other possibilities – namely one or fewer hurricanes. This is equal to:

$$(e^{-2.5}2.5^0/0!) + (e^{-2.5}2.5^1/1!) = 0.2873$$

(Continued)

(Continued)

The solution is found by subtracting this from one; thus the probability of two or more hurricanes is equal to $1 - 0.2873 = 0.7127$.

7. Eighteen people move out of a neighborhood; six are minorities. Of the 18, eight move onto a block with new housing, and one of the eight is a minority. How likely is it that, if there were no discrimination, one or fewer people out of eight people on a new block would be minorities? If the resulting probability is less than 0.05, evidence for discrimination exists. Does such evidence exist in this case?

Solution. We are interested in the characteristics of the eight people moving onto a new block. We are essentially taking a sample of eight out of a total 'population' of 18, without replacement. This suggests use of the hypergeometric distribution. The characteristic or outcome of interest is whether individuals in the sample of eight are minorities, or not. The total number of ways to choose a sample of eight from a total of 18 is $\binom{18}{8}$. Of these possibilities, the number of ways to have no minorities (out of the six minorities in the population), and eight non-minorities (out of the 12 non-minorities in the population) is $\binom{6}{0}\binom{12}{8}$ and the number of ways to have one minority in the sample and seven non-minorities is $\binom{6}{1}\binom{12}{7}$. The desired probability is therefore:

$$\frac{\binom{6}{0}\binom{12}{8} + \binom{6}{1}\binom{12}{7}}{\binom{18}{8}} = \frac{5{,}247}{43{,}758} = 0.1199$$

Since this is greater than 0.05, evidence for discrimination does not exist. The outcome is likely due to chance (this outcome, or one more extreme, would occur 11.99% of the time, if there was no discrimination).

8. Individuals move each year with probability $p = 0.16$. What is the probability that an individual will make his or her first move in the seventh or eighth year of observation? What are the mean and variance associated with the time until the next move?

Solution. A question that often arises is, 'which distribution should I use? The answer to that can usually be found by thinking about the process, and the variable of interest. Here we are interested in the time until the first 'success', or move, and thus we can use the geometric distribution (Equation 3.6).

The probability that the first move occurs in the seventh year of observation is found by finding the probability of six 'failures' (i.e., non-moves), followed by a 'success' (move). This is equal to $(1 - 0.16)^6 (0.16) = 0.0562$. Similarly, the probability that the first move occurs in the eighth year of observation is equal to $(1 - 0.16)^7 (0.16) = 0.0472$. The answer is equal to the sum of these: the probability of a first move in either the seventh or eighth year is $0.0562 + 0.0472 = 0.1034$. Since the mean of a geometric variable is equal to $1/p$, the mean time until the next move is equal to $1/p = 6.25$ years. The variance of a geometric variable is equal to $(1 - p)/p^2$, and so the variance associated with the time to next move is equal to $(1 - 0.16)/0.16^2 = 0.84/(0.16)^2 = 32.81$ years.

9. A study of a disease reveals that there is an average of one case every 12 square miles. Residents of a town that has an area of 20 square miles are concerned because there are three cases in their area. The state's Department of Health has decided to investigate further if the probability of getting three or more cases in this town is less than 0.10. Does the Department of Health investigate further?

Solution. Since we are interested in the number of events (cases) that will occur, this suggests use of the binomial or Poisson distribution. We are interested in the number of events occurring in a continuous space, and this points us toward the Poisson distribution (alternatively stated, there is no natural definition of a 'trial', which would have pointed us toward the binomial distribution). To use the Poisson distribution, we need to know λ, which in this case is the mean or expected number of cases. We need to know the number of cases expected in our 20 square mile town, and we can find this by using our knowledge that there is an average of one case every 12 miles:

$$\frac{1}{12} = \frac{\lambda}{20}$$

This yields $\lambda = 1.67$. The probability of three or more cases is equal to one minus the probability of 0, 1, or 2 cases:

$$px(X \geq 3) = 1 - \left\{ \frac{e^{-1.67}1.67^0}{0!} + \frac{e^{-1.67}1.67^1}{1!} + \frac{e^{-1.67}1.67^2}{2!} \right\} = 1 - 0.7651 = 0.2349$$

Since this result is greater than 0.10, there is no need to investigate further.

EXERCISES

1. The probability of a dry summer is equal to 0.3, the probability of a wet summer is equal to 0.2, and the probability of a summer with normal precipitation is equal to 0.5. A climatologist observed the precipitation during three consecutive summers.

   (a) How many outcomes are in the sample space? Enumerate the sample space, and assign probabilities to each simple event.

   (b) What is the probability of observing two dry summers?

(Continued)

(Continued)

    (c) What is the probability of observing at least two dry summers?

    (d) What is the probability of not observing a wet summer?

2. If the probability that an individual moves outside of their county of residence in a given year is 0.15, what is the probability that:

    (a) less than three out of a sample of ten move outside the county?

    (b) at least one moves outside the county?

3. The annual probability that an individual makes an interstate move is 0.03. What is the probability that at least two out of a sample of ten people will make an interstate move next year?

4. Assume that the probability that an individual changes residence during the year is 0.21. A survey is taken of five individuals.

    (a) Write out the elements in the sample space.

    (b) What is the probability that exactly three out of the five individuals move during the year?

    (c) What is the probability that at least one of the five individuals moves during the year?

5. The probability that an individual commutes to work by car is 0.9. What is the probability that a sample of ten neighbors *all* commute by car? What is the probability that exactly eight of the ten commute by car?

6. What is the probability of getting three sixes in five tosses of a die?

7. What is the probability that the first six appears on the fifth toss of a die?

8. What is the probability of getting four heads when a fair coin is tossed ten times?

9. What is the probability that the first head will appear on the fourth toss of a fair coin?

10. Two evenly matched baseball teams play the first four games in a 'best-of-seven' series (where the first to win four games wins the series).

    (a) What is the probability that team A will win exactly one of the first four games?

    (b) What is the probability that the series will be over after four games?

    (c) What is the probability that the series will have one team leading three games to one, after four games?

    (d) What is the probability that the series will be tied at two games apiece, after four games?

11. The annual probability of moving is 0.17.

   (a) What is the probability that a person moves for the first time in the first year of observation?

   (b) What is the probability that the person moves for the first time in the fourth year of observation?

   (c) What is the mean time it takes for an individual to move?

   (d) What is the variance of the time it takes for a person to move?

   (e) Among ten people, what is the probability that three or fewer move in a given year?

12. 100 people each have a probability of 0.01 of contracting a disease. What is the probability that two or fewer out of 100 get the disease?

13. On average, there are two bicycle thefts every ten square miles during the course of a month.

   (a) What is the probability that there will be no bicycle thefts in a ten square mile area next month?

   (b) What is the probability that there will be no bicycle thefts in a seven square mile area next month?

   (c) What is the probability that there will be at least one bicycle theft in a 20 square mile area next month?

   (d) What is the probability that there will be exactly one bicycle theft in a ten square mile area over the next ten days?

14. The following cumulative distribution characterizes length of residence for a large number of individuals:

Years	Cumulative Probability
0–1	.1
1–2	.3
2–5	.5
5–10	.6
10–20	.9
20–40	1.0

   (a) What is the median length of residence?

   (b) What is the mean length of residence?

(Continued)

(Continued)

(c) What is the probability that an individual in this sample has length of residence greater than five years, and less than 20 years?

(d) What is the probability that an individual in this sample has length of residence greater than ten years, given that he or she has length of residence greater than two years?

(e) What is the probability that out of eight randomly selected individuals, all will have length of residence greater than one year?

(f) What is the probability that out of seven randomly selected individuals, at least two will have length of residence less than five years?

15. There is an average of four accidents per month at an intersection. Assume that a month has 30 days.

(a) What is the probability of three or more accidents next month?

(b) What is the probability of two or more accidents in a 23-day period?

(c) What is the probability of no accidents during the next seven days?

16. A left-turn lane has a capacity of six cars. Cars arrive into the lane at a rate of three every ten seconds. What is the longest amount of time the turn light can remain red, given that we want the probability of exceeding the capacity to be 0.05 or less?

17. A variable representing the number of 'successful' events has mean equal to 3. Suppose that there are 30 trials, with probability of success on each trial equal to 0.1. Compare binomial and Poisson probabilities of getting one success. Repeat using $n = 60$ and $p = 0.05$. Repeat using $n = 120$ and $p = 0.025$.

18. Volcanic eruptions on a particular mountain occur once every 32 years, on average.

(a) What is the probability that three eruptions will occur during the next 50 years?

(b) What is the probability that the next eruption will occur either 33, 34, or 35 years from now?

19. A group of 30 people moves out of a neighborhood. Twenty-five are white and the other five are minorities. Fifteen of the 30 people move to the adjacent town. What is the probability that either four or five of the 15 are minorities?

20. Two dice are rolled, and you find the product. What is the mean number of times you must roll the dice to get '18'?

21. The likelihood that a driver wishes to turn left at an intersection is 0.1.

    (a) Once cars begin arriving at the light, what is the probability that the first car wishing to turn left is the fifth car that we observe?

    (b) What is the mean of the random variable describing the number of cars we must observe in order to find one that turns left?

22. A mechanical part has a 0.7 chance of failing over the course of a year.

    (a) What is the probability that it will fail in year five?

    (b) What is the probability that the part lasts for more than two years?

23. A plant species has probability 0.18 of dying in any given year. What is the mean and variance of its lifetime?

24. The probability of commuting by train in a community is 0.1. A survey of residents in a particular neighborhood finds that four out of ten commute by train. We wish to conclude either that (a) the 'true' commuting rate in the neighborhood is 0.1, and we have just witnessed four out of ten as a result of sampling fluctuation, or (b) the 'true' commuting rate in the neighborhood is greater than 0.1, and it is very unlikely that we would have observed four out of ten train commuters if the true rate was 0.1. Decide which choice is best via the following steps, using the random number table in Appendix A.1:

    (i) Take a series of ten random digits, and then count and record the number of '0's; these will represent the number of train commuters in a sample of ten, where the 'true' commuting probability is 0.1.

    (ii) Repeat step (i) 20 times.

    (iii) Arrive at either conclusion (a) or (b). You should arrive at conclusion (b) if you had four or more commuters either once, or not at all, in the 20 repetitions (since one out of 20 is equal to 0.05, or 5%).

25. Thirty-two cases of a disease are found over an area of 100 square miles. If the cases are distributed at random within the area, what is the probability of finding four cases within an eight square mile area around a local power plant?

26. The probability of an accident at an intersection each month is 0.1. What is the probability that the next accident occurs in month 3 or 4?

27. What is the probability of an earthquake next year, if the average rate of earthquakes is once every three years?

28. The following data were derived by Rogerson et al. (1993) from the US National Survey of Families and Households.

(Continued)

(Continued)

Distance (in miles)	Cumulative relative frequency	Distance	Cumulative relative frequency
5	.236	250	.741
10	.338	300	.756
15	.418	350	.777
20	.460	400	.787
25	.492	450	.800
30	.522	500	.808
35	.546	1000	.876
40	.556	1500	.921
45	.565	2000	.945
50	.575	2500	.966
100	.643	3000	.974
150	.683	3500	.994

Cumulative Relative Frequency Distribution of Distance to Parents Among Adult Children (for those children with both parents alive and living together)

The questions below apply to adult children who have both parents alive and living together.

(a) What fraction of adult children have parents who live more than 30 miles away, but no more than 100 miles away?

(b) What fraction of adult children have parents who live more than 250 miles away?

(c) What is the likelihood that at least one of three randomly chosen adult children will have parents living more than 100 miles away? (i) Show the sample space. (ii) Assign probabilities to each element in the sample space. (iii) Give the desired probability.

(d) What is the median distance that characterizes the spatial separation of parents and their adult children?

(e) Determine the likelihood that two or more adult children out of a group of ten live more than 1000 miles from their parents.

29. The number of annual earthquakes along a given fault is found to be 1.2. Find the probability that there will be no earthquakes next year.

30. Drivers wishing to turn left at a particular intersection arrive at an average rate of five per minute. (i) If the left-turn arrow is red for 30 seconds, and there is room in the left-turn lane for five cars, what is the probability that the capacity of the lane will be exceeded for a given cycle of the signal? (ii) Given that transportation planners and traffic engineers wish to reduce the probability to less than 0.05 by shortening the length of the red signal, how would you determine the maximal time the signal could remain red?

31. Use the hypergeometric distribution to determine which lottery gives you the better chance to win:

    (a) Choose four balls from a box of 34. You win if you match three or four balls.

    (b) Choose five balls from a box of 25. You win if you match four or five balls.

32. A lottery consists of drawing five balls from a box of 50. What is the probability of matching either four or five numbers on a ticket containing five numbers?

On the Companion Website

The website contains links to a number of other websites and videos which deal with the concepts and methods discussed in this chapter. These resources, available under the 'Further Resources' section of the website, supplement and expand upon the material presented here.

4

CONTINUOUS PROBABILITY DISTRIBUTIONS AND PROBABILITY MODELS

4.1 INTRODUCTION

In the previous chapter, the random variable of interest was discrete, and the number of values the variable could assume was consequently finite. The number of successes in n trials is clearly a discrete variable, as is the number of accidents occurring at an intersection, since they can only take on non-negative integers. Many variables of interest are continuous in nature. The distance traveled by a consumer, the magnitude of a flood, the size of sand grains, and the distance moved by a migrant are all examples of continuous variables. We will have occasion to be interested in the frequency distributions or histograms of these variables, and hence the first part of this chapter is devoted to a discussion of widely used continuous distributions. The latter part of the chapter (Section 4.6) provides examples of how probability and probability distributions are used in modeling geographic phenomena.

4.2 THE UNIFORM OR RECTANGULAR DISTRIBUTION

The uniform distribution provides a useful starting point in the study of continuous probability distributions because it is perhaps the simplest to conceptualize. Although it does not occur as frequently as other continuous distributions in geographic applications, its simplicity facilitates the introduction and explanation of many basic ideas.

Quite simply, the uniform distribution is the outcome of processes that have equally likely outcomes. For example, suppose that the selling prices for homes in a residential area were equally likely to be anywhere within the interval between £60,000 and £90,000. If we collected data on the selling prices (x) of a number of houses, the histogram constructed from the data would be essentially flat. The probability distribution shown in Figure 4.1 reflects this – all outcomes in the range between the lowest and highest values are equally likely. If we define a as the lowest possible value, and b as the highest possible value, then the probability distribution is defined as

$$f(x) = \frac{1}{b-a}; \quad a \le x \le b \tag{4.1}$$

This is the equation of the line traced out in Figure 4.1; note that its height is constant, and is not a function of x. Note that the area of the rectangle in Figure 4.1 is one; the length of the rectangle is $(b - a)$, and the height of the rectangle is set equal to $1/(b - a)$, so that the product of the length and height is equal to one (corresponding to the fact that the probability of obtaining *some* value in this range is equal to one). This is a characteristic of all continuous probability distributions; the total area under the curve is equal to one.

One small point about notation – it is common to use $p(x)$ to describe discrete probability distributions, and $f(x)$ to describe continuous probability distributions.

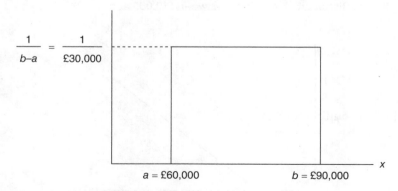

Figure 4.1 The uniform distribution

Desired probabilities are found as areas under the distribution. For example, to find the probability that a house sells for more than £70,000, the area of the shaded rectangle in Figure 4.2 is found as the product of its length (£90,000 − £70,000 = £20,000) and its height (1/{£90,000 − £60,000} = 1/£30,000). Hence, the desired probability is 20,000/30,000 = 2/3; this is the fraction of all houses that we expect will sell for more than £70,000.

As we saw in Chapter 2, it is also often useful to work with cumulative distributions. The cumulative distribution specifies the probability of obtaining a value less than x. For the uniform distribution, the cumulative distribution is a straight line (Figure 4.3); its equation is

$$F(x) = \frac{x - a}{b - a} \tag{4.2}$$

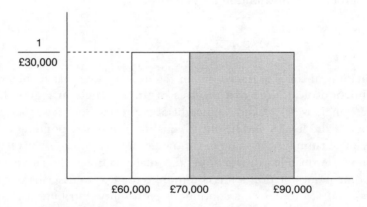

Figure 4.2   **Probability of an observation between £70,000 and £90,000**

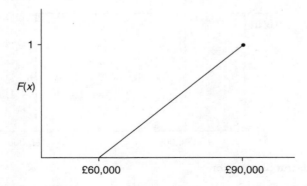

Figure 4.3   **Cumulative distribution for observations from a uniform distribution**

For example, the probability that a house sells for less than £67,000 is (£67,000 − £60,000) divided by (£90,000 − £60,000) = 7/30.

The probability that an observation from a uniform distribution falls between two values, say x_1 and x_2 is:

$$pr(x_1 \leq x \leq x_2) = \frac{x_2 - x_1}{b - a} \tag{4.3}$$

Example 4.1

Suppose that the mean annual temperature for a location was uniformly distributed, between 10°C and 18°C, implying that each year any temperature in this range is equally likely to be the mean for the year. (Note: This is an unrealistic example, in the sense that mean temperatures are not likely to have this kind of distribution.) What is the probability that the mean annual temperature is between 12° and 15°?

Solution: The range between lowest and highest temperatures is $b - a = 18 - 10 = 8$ degrees, and the range of interest is $X_2 - X_1 = 15 - 12 = 3$ degrees. From Equation 4.3, the probability is therefore 3/8. More formally, the height of the rectangle associated with the desired probability, using Equation 4.1, is $1/(18 - 10) = 1/8$; the length of the rectangle is $15 - 12 = 3$. The product of the height and length is 3/8.

Example 4.2

Using the same scenario, find the probability that the annual temperature is less than 14°.

Solution: Using Equation 4.2, the answer is $(14 - 10)/(18 - 10) = 4/8 = 0.5$.

4.3 THE NORMAL DISTRIBUTION

The most common probability distribution is the *normal distribution*.

Its familiar symmetric, bell-shaped appearance is shown in Figure 4.4. The normal distribution is a continuous one − instead of a histogram with a finite number of vertical bars, the relative frequency distribution is continuous. You can think of the distribution as a histogram with a very large number of very narrow vertical bars. The vertical axis is related to the likelihood of obtaining particular x values. As with all continuous frequency distributions, the area under the curve between any two x values corresponds to the probability of obtaining an x value in that range. For example, in Figure 4.4, the probability of obtaining

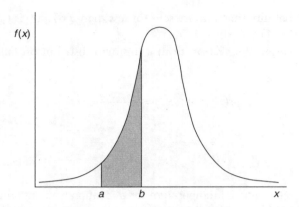

Figure 4.4   The normal distribution

a value of the variable between *a* and *b* is equal to the shaded area between *a* and *b*. The total area under the curve is equal to one.

The normal distribution is characterized by its mean (μ) and its variance (σ^2); the equation, which describes the curve shown in Figure 4.4, is

$$f(x) = \frac{1}{\sqrt{2\pi}\sigma} e^{-(x-\mu)^2/2\sigma^2}$$ (4.4)

Because this formula is complex, it is not used directly to find probabilities. Instead, to use the normal distribution to answer questions involving probability, a table that tabulates the probabilities is used. For example, suppose that annual precipitation is normally distributed, with a mean of 80 cm per year, and a standard deviation of 40 cm per year. What is the probability that a year has more than 120 cm of precipitation? The answer is equal to the shaded area in Figure 4.5. There are two steps involved in obtaining the answer – the first is to convert the normal distribution to a 'standard' normal distribution which has a mean of zero and a variance of one, and the second is to use Table A.2 in Appendix A to find the appropriate area under this standard normal distribution.

The first step is necessary because it is impossible to have separate tables for every possible combination of μ and σ. Instead, all problems can be converted to problems involving the standard normal distribution, by converting data to *z*-scores. For example, 120 cm may be converted to a *z*-score by first subtracting the mean, and then dividing the result by the standard deviation, as described in Chapter 2:

$$z = \frac{120 - 80}{40} = 1$$ (4.5)

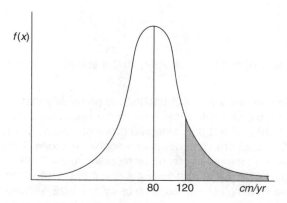

Figure 4.5 Probability of an observation greater than 120 cm/year

The z-score tells us how many standard deviations we are from the mean. In this example, 120 cm is one standard deviation above the mean. Asking how often we observe more than 120 cm of precipitation is the same as asking how often we observe an annual precipitation amount that is more than one standard deviation above the mean.

The second step is to use Table A.2. Our value of $z = 1.0$ is first located along the left-hand side of the table. The column headings give the second decimal place of the z-score, and hence we go over one column, to the column labeled '.00', since our z-value is 1.00. The corresponding entry in the table is 0.1587. The figure at the top of the table shows us that the entries in the table correspond to the area or probability associated with z-values that are higher than z. Thus, the probability of obtaining a z-score greater than one is 0.1587, or 15.87%. Alternatively stated, with a normal distribution, 15.87% of the time the observations will exceed values that are one standard deviation above the mean. In our example, precipitation greater than 120 cm will be observed 15.87% of the time.

Example 4.3

If commuting times are normally distributed with mean 30 minutes, and standard deviation 16 minutes, find the probability that a commute is shorter than 35 minutes.

Solution: A commute of 35 minutes is $z = (35 - 30)/16 = 0.3125$ standard deviations above the mean. Table A.2 reveals that the probability of a commute time *greater* than this would be 0.3783 (found by first rounding to $z = 0.31$, and then using the $z = 0.3$ row, and the 0.01 column of the table). Thus, the probability of a commute shorter than 35 minutes is equal to $1 - 0.3783 = 0.6217$.

Example 4.4

Using the same scenario, find the probability that a commute time is between 40 and 50 minutes.

Solution: We can find the answer by subtracting the probability that a commute is greater than 50 minutes from the probability that a commute is greater than 40 minutes. Forty minutes is $z = (40 - 30)/16 = 0.625$ standard deviations above the mean. Fifty minutes is $z = (50 - 30)/16 = 1.25$ standard deviations above the mean. The probability of a commute greater than 40 minutes is found by rounding to $z = 0.63$, using the $z = 0.6$ row and 0.03 column of the table, and finding a result of 0.2643. The probability of a commute greater than 50 minutes is found by using $z = 1.25$, yielding a result of 0.1056. Hence the probability of a commute that is between 40 and 50 minutes is equal to $0.2643 - 0.1056 = 0.1587$. A more precise answer could be found by interpolation between the elements in the table, instead of rounding the z-score. More specifically, the entry in the table associated with $z = 0.62$ is 0.2676, and the entry associated with $z = 0.63$ is 0.2643. We would like to find the value that is halfway from the entry for 0.62 to the entry for 0.63. Noting that the tabled values are declining, we subtract half of the difference between the two values from our starting value of 0.2676: $0.2676 - 0.5(0.2676 - 0.2643) = 0.26595$. It would be slightly more accurate to use this value; this would yield a final solution of $0.26595 - 0.1056 = 0.16035$.

Example 4.5

Using the same scenario, find the probability that a commute time is less than 20 minutes.

Solution: The standardized score, $z = (20 - 30)/16 = -0.625$, indicates that 20 minutes is 0.625 standard deviations below the mean. Because the normal distribution is symmetric, we still look up 0.625 in the z-table, and rounding to $z = 0.63$ yields a tabled value of 0.2643. This tells us that the probability of a commute time more than 0.63 standard deviations above the mean is 0.2643, but this is also equal to the probability that a commute time is more than 0.63 standard deviations below the mean. Hence, the probability of a commute time of less than 20 minutes is about 0.26.

Example 4.6

Using the same scenario, find the probability that a commute time is between 25 and 45 minutes.

Solution: Here we break the problem up into two parts, because some of the interval is below the mean, and some of the interval is above the mean. We will first find the probability of a commuting time between 25 and 30 minutes, and will add to that result the probability of a time between 30 and 45 minutes. Twenty-five minutes is 5/16, or 0.3125 standard deviations below the mean. Since the table gives tail probabilities, the probability of a commute time less than 25 minutes is 0.3783, which is the entry corresponding to $z = 0.31$. Because we know that the probability of observing a time less than the mean of 30 minutes is 0.5, this implies that the probability of observing a time between 25 and 30 minutes must be 0.5 – 0.3783, or 0.1217. A commute time of 45 minutes is (45 – 30)/16 = 0.9375 standard deviations above the mean; rounding to $z = 0.94$ yields a tabled value of 0.1736 (using the 0.9 row, and the 0.04 column of the table). This is the tail probability, or the probability of a commute time greater than 45 minutes. Since the probability of a time greater than the mean of 30 minutes is 0.5, the probability of a time between 30 and 45 minutes is 0.5 – 0.1736 = 0.3264. The solution to the problem is the sum of these two probabilities: 0.1217 + 0.3264 = 0.4481.

Example 4.7

Using the same scenario, find the 20th percentile of the distribution of commuting times.

Solution: The 20th percentile implies that 20% of all commuting times are less than that specific time. That is, the area under the normal distribution that is to the left of that commuting time is equal to 0.2000. We begin by using the standard normal table to find that z-value that is associated with a tail probability of 20%. The table gives only positive z-values, and these are associated with the right (upper) tail; when we are interested in the left, or lower tail (as we are here), we can simply use the right tail, and then put a negative sign in front of the appropriate z-value, recognizing that the distribution is symmetric. By searching within the body of the table for the entry 0.2000, we find that we are in the $z = 0.8$ row of the table. If we go over to the 0.04 column, we find an entry of 0.2005, which we will take as 'close enough'. Thus, 20% of all standard normal variables have a z value that is greater than 0.84 (and 20% of all z-values are less than $z = -0.84$.) The final step is to convert our z-value of –0.84 to a commuting time. We wish to find the commuting time that is 0.84 standard deviations less than the mean. This is equal to 30 – (0.84) (16), or 30 – 13.44 = 16.56 minutes. More formally, if $z = (x-\mu)/\sigma$, we can rearrange this to find $x = \mu + z\sigma$.

The normal distribution arises in a variety of contexts, and it is related to a variety of underlying processes. One way in which the normal distribution arises is through an approximation to binomial processes. Suppose that we use the residential mobility example from section 3.3, interviewing 40 residents instead of four. We could still use the binomial distribution to evaluate, for example, the probability that 11 or fewer households were new to the neighborhood, but that would entail a tedious calculation with large factorials:

$$pr(X \le 11) = pr(X = 0) + pr(X = 1) + ... + pr(X = 11)$$

$$= \binom{40}{0} 0.2^0 0.8^{40} + ... + \binom{40}{11} 0.2^{11} 0.8^{29}. \tag{4.6}$$

When the sample size is large, the binomial distribution is approximately the same as a normal distribution which has a mean of np and a variance of $np(1 - p)$. In our example, we would expect a mean of $np = (40)(0.2) = 8$ residents to indicate that they were newcomers. The variance, $np(1 - p) = 40(0.2)(0.8) = 6.4$, summarizes the variability we would expect in a summary of the results produced by many people who went out and surveyed 40 households.

The probability that 11 or fewer residents are newcomers, $p(X \le 11)$, may be determined by finding the shaded area under the normal curve shown in Figure 4.6.

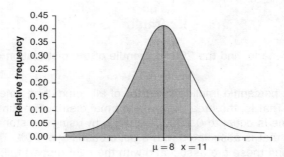

Figure 4.6   Probability of $X < 11$

We convert $x = 11$ into a z-score by first subtracting the mean, and then dividing the result by the standard deviation:

$$z = \frac{11 - 8}{\sqrt{6.4}} = 1.19 \tag{4.7}$$

We now find the probability that $z < 1.19$ from the normal table; it is equal to 0.8830. To be a bit more precise, we would also account for the fact that our variable of interest is a discrete one. The vertical bar associated with $x = 11$ on a histogram of the binomial distribution would stretch from $x = 10.5$ to $x = 11.5$. We can obtain a better approximation by finding the probability that $x < 11.5$ instead of the probability that $x < 11$. Converting

$x = 11.5$ to a z-score yields $z = 1.38$, and from the normal table the probability that $z < 1.38$ is 0.9162. For comparison, the binomial formula leads to a probability of 0.9125. Whether the binomial distribution may be approximated well by the normal distribution depends upon the values of n and p. The binomial distribution is only truly symmetric when $p = 0.5$, and so if p is near either 0 or 1, the normal approximation may not be accurate. The normal approximation also improves as n gets large. A common rule-of-thumb is that np should be greater than 5. In our example np was equal to 8, and the approximation was fairly accurate.

We will see in the next chapter that the normal distribution also characterizes the distribution of sample means. In particular, *sample means have a normal distribution*, with mean equal to the true population mean (μ) and variance equal to σ^2/n, where σ^2 is the population variance, and n is the sample size. If we repeatedly took samples of size n from a population, and then made a histogram of all of our sample means, the histogram would have the appearance of a normal distribution with mean (μ) and variance equal to σ^2/n.

4.4 THE EXPONENTIAL DISTRIBUTION

The normal distribution is an example of a symmetric distribution; a variable that is normally distributed will tend to have a symmetric, bell-shaped histogram when a large number of observations are collected and graphed. As noted in Chapter 2, many variables are not symmetric, and, in particular, many variables display positive skewness. Recall that variables with positive skewness have means that are greater than their medians. The histogram of such variables reveals that there is an extended right tail, with a large number of small values and a small number of large values. Examples of variables with positive skewness include personal income, commuting distances, and length of residential moves.

The exponential distribution is a common model for positively skewed distributions. The shape of its histogram or frequency distribution is depicted in Figure 4.7; it is characterized by a downward slope, where the steepness of the slope decreases with increasing values of the variable. The equation describing the distribution – i.e., the height of the curve, is

$$f(x) = \lambda e^{-\lambda x}, \; x \geq 0 \tag{4.8}$$

As was the case with the Poisson distribution (Section 3.5), the quantity e is equal to the constant, 2.718. The variable of interest, x, can only take on values that are non-negative. Equation 4.8 describes a curve that captures the exponential decay shown in the figure. Note that when $x = 0$, the height of the curve is equal to λ; for higher values of x, the height of the curve will be lower than this. Like all probability distributions, the area under the curve in Figure 4.7 is equal to one. There is one parameter (λ); high values of this parameter indicate a steep decline, and low values indicate a gentle decline in the distribution as values of x increase.

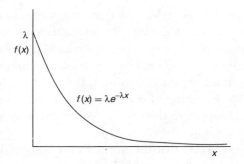

Figure 4.7   The exponential distribution

Unlike the normal distribution, the use of tables to find probabilities associated with the exponential distribution is not necessary. Because the equation describing the curve is relatively simple, it is possible to find probabilities of interest directly. For the exponential distribution, to find the probability that x falls in a certain range, it is useful to work instead with the cumulative distribution:

$$F(x) = pr(X < x) = 1 - e^{-\lambda x} \tag{4.9}$$

where the notation $F(x)$ is used for the cumulative distribution function of a continuous variable. Recall from Chapter 2 that a cumulative distribution gives the probability of observing a value of x or lower. Here this takes on a value of zero when x is equal to zero (since the probability of obtaining a value less than zero is equal to zero), and a value of close to one when x is very large (since the probability of obtaining a value of x that is less than some very large number is close to one). See Figure 4.8. It follows directly from this that the probability of obtaining an observation that is greater than x is simply:

$$1 - F(x) = pr(X > x) = e^{-\lambda x} \tag{4.10}$$

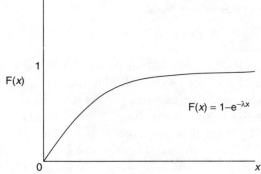

Figure 4.8   The cumulative distribution for exponential variables

Example 4.8

The distance of residential moves is found to be exponential, with $\lambda = 0.1$ km. Find the probability that a residential move is less than 5 km.

Solution: This is the shaded area shown in Figure 4.9a. Simply substitute $x = 5$ into Equation 4.9: $pr(X < 5) = 1 - e^{-0.1(5)} = 1 - 2.718^{-0.5} = 0.393$.

Example 4.9

For the scenario above, find the probability that a residential move is longer than 3 km.

Solution: This is the shaded area shown in Figure 4.9b. The probability that an exponential variable is greater than x is one minus the probability that it is less than x, or $\{1 - (1-e^{-\lambda x})\} = e^{-\lambda x}$. The probability of a move greater than 3 km is $e^{-0.1(3)} = 0.741$.

Example 4.10

For the scenario above, find the probability that a residential move is between 3 km and 8 km.

Solution: This is the shaded area in Figure 4.9c. It can be determined by subtracting the probability that a move is less than 3 km from the probability that a move is less than 8 km: $pr(3 < X < 8) = pr(X < 8) - pr(X < 3)$. This is equal to $(1 - e^{-0.1(8)}) - (1 - e^{-0.1(3)}) = 0.551 - 0.259 = 0.292$. Alternatively, the desired probability may be thought of as the probability that a move is greater than 3 km, minus the probability that a move is greater than 8 km: $pr(3 < X < 8) = pr(X \geq 3) - pr(X \geq 8)$, and this is equal to $e^{-0.1(3)} - e^{-0.1(8)} = 0.292$.

The expected value or theoretical mean of the exponential distribution is $1/\lambda$. The reciprocal relationship between the parameter and the theoretical mean makes intuitive sense, since one would expect that high values of λ (and hence steeper slopes, which in turn imply a larger majority of small values) would be associated with low means. The theoretical variance is $1/\lambda^2$.

You may at this point be wondering where the value of λ comes from — it was simply given in the examples above. Since we know that its value describes the steepness of the histogram, it makes sense, for a given set of data, to choose it in a way so Equation 4.8 will trace out a curve that will match the observed histogram as closely as possible. If λ is chosen

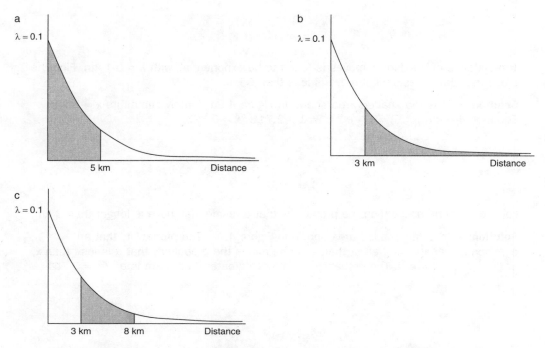

Figure 4.9 Probabilities of observations (a) less than 5 km; (b) greater than 3 km; (c) between 3 km and 8 km

to be too high, the curve traced out by the equation will be steeper than the actual histogram, and if it is chosen to be too low, the curve will be less steep than the actual histogram. The reciprocal relationship between the parameter and theoretical mean suggests a simple way to estimate the parameter. If you have a set of observations that is positively skewed, and would like to fit an exponential distribution to it, all that you need to do is find the sample mean. The estimated value of λ is the inverse or reciprocal of the sample mean.

Example 4.11

Information is collected on the commuting distances for 40 residents; the sample mean is found to be 7 km. A histogram is made, and it appears to be approximately exponential in shape. Find the appropriate value of λ.

Solution: The estimate of λ is $1/\bar{x} = 1/7 = 0.143$. This value could then be used to estimate probabilities, such as the probability that a resident has a commuting distance that is longer than 15 km.

The exponential distribution has a number of other interesting characteristics and properties. It is very similar to the geometric distribution, and in fact is the continuous version of that discrete distribution. Recall that the geometric distribution was used to model the time until the first success, where there were a discrete number of trials. In addition, like exponential variables, the mean of the geometric distribution is also equal to the reciprocal of its parameter.

If events occur randomly in (continuous) time, the time until the next event has an exponential distribution. When the times between random events are collected and visualized using a histogram, it will be evident that these times display an exponential distribution. Figure 4.10 depicts random events, and the time intervals between these events. The times t_1, t_2, ... have an exponential distribution.

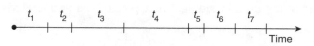

Figure 4.10   Events occurring randomly in time

Furthermore, the exponential distribution is characterized by what is known as the *memoryless property*. Specifically, the time until the next event is independent of the time since the last event. If the times between accidents at an intersection are modeled with an exponential distribution, the time until the next accident is totally independent of the time since the last accident (and this *should* be the case, since the events are occurring randomly in time; this independence between events is a defining characteristic of randomness).

Example 4.12

Accidents occur randomly at an intersection, at a rate of two per day. What is the probability that the time until the next accident is more than three days?

Solution: The time between accidents follows an exponential distribution with a mean of 0.5 days. The inverse of this is $\lambda = 1/0.5 = 2$. The probability that the variable of interest is greater than three is $1 - [1 - e^{-2(3)}] = e^{-2(3)} = 0.0025$. Note that the Poisson distribution could also have been used here. The mean number of accidents in a three-day period is $\lambda = 6$ (the 'λ' notation is commonly used for both exponential and Poisson distributions. Note that its meaning for each distribution is distinct.) The probability that no accident occurs in three days, using the Poisson distribution, is $e^{-6}6^0/0! = 0.00248$.

Example 4.13

Assume that the time between residential moves for individuals follows an exponential distribution, and that the mean time between moves is five years. (a) What is the probability that the person moves within the next year? (b) What is the probability that the person moves within the next year, given that they have been in their current residence for four years? (c) What is the probability that the person moves within the next year, given that they have been in their current residence for ten years?

Solution: First, $\lambda = 1/5 = 0.2$. For part (a), $pr(X < 1) = 1 - e^{-0.2(1)} = 0.1813$. For part (b), we desire the conditional probability that $4 < X < 5$, given that we know $X > 4$. In Figure 4.11, we know that X is going to be greater than 4. Given this, we want to know the probability that X is between 4 and 5. Thus, we want to know the shaded area in the figure, divided by the area to the right of $X = 4$. This is evaluated as

$$\frac{e^{-0.2(4)} - e^{-0.2(5)}}{e^{-0.2(4)}} = 0.1813 \qquad (4.11)$$

A similar argument reveals that the answer to part (c) is also equal to 0.1813. This example illustrates the memoryless property of the exponential distribution – the likelihood of an event occurring in the next unit of time (in this case, a residential move within the next year) does not depend on how long it has been since the last event.

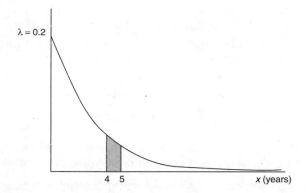

Figure 4.11 Probability of an outcome of between four and five years

4.5 SUMMARY OF DISCRETE AND CONTINUOUS DISTRIBUTIONS

Table 4.1 provides a summary of the discrete and continuous probability distributions covered in previous sections.

Table 4.1 Summary of discrete and continuous distributions

Distribution	Probability function	Cumulative probability function	Mean	Variance
Binomial	$pr(x) = \dbinom{n}{x} p^x (1-p)^{n-x}$	–	np	$np(1-p)$
Poisson	$pr(x) = \dfrac{e^{-\lambda} \lambda^x}{x!}; \quad x \ge 0$	–	λ	λ
Geometric	$pr(x) = (1-p)^{x-1} p; \, (x > 0)$	–	$\dfrac{1}{p}$	$\dfrac{1-p}{p^2}$
Hypergeometric	$pr(x) = \dfrac{\dbinom{r}{x}\dbinom{N-n}{n-x}}{\dbinom{N}{n}}$		$\dfrac{nr}{N}$	$\dfrac{nr(N-r)(N-n)}{N^2(N-1)}$
Normal	$f(x) = \dfrac{1}{\sqrt{2\pi}\sigma} e^{-\frac{(x-\mu)^2}{2\sigma^2}}$		μ	σ^2
Uniform/ Rectangular	$f(x) = \dfrac{1}{b-a}; \quad a \le x \le b$	$F(x) = \dfrac{x-a}{b-a}$	$\dfrac{a+b}{2}$	$\dfrac{(b-a)^2}{12}$
Exponential	$f(x) = \lambda e^{-\lambda x}; \quad x \ge 0$	$F(x) = 1 - e^{-\lambda x}$	$\dfrac{1}{\lambda}$	$\dfrac{1}{\lambda^2}$

Summary for each distribution:

Binomial:	random variable of interest is the number of successes in n trials, when the probability of success on a given trial is equal to p.
Poisson:	random variable of interest is the number of events in a specified time period (or a specified spatial area), where the mean number of events per time period (or the mean number of events per unit area) is equal to λ.
Geometric:	random variable of interest is the number of the trial on which the first success occurs, when trials are independent, and when the probability of success on a given trial is equal to p.

Hypergeometric: the random variable of interest is the number of trials that lead to a particular outcome of interest, where sampling occurs without replacement, and where there are two possible outcomes (for example, interest might be in the number of red balls (x) that occur in a sample of n balls taken from a box containing r red balls and $N - r$ balls of another color).

Normal: generated by sums of large numbers of independent variables; also used as an approximation to the binomial distribution.

Uniform/rectangular: generated by processes that have equally likely outcomes.

Exponential: generated by, among other things, the time between random events.

4.6 PROBABILITY MODELS

Probability is used as a basis for statistical inference. In college math course sequences, probability comes before statistics, since probability concepts form the foundation of statistical tests and statistical inference. In addition to forming the basis of standard statistical tests, probability is used to develop models of geographic processes. Though the primary emphasis of this book is on the use of probability in statistical inference, in this section we provide some examples of probability modeling.

The outline of the scientific method in Chapter 1 indicated that models are used as simplifications of reality. A simplified view of reality permits one to focus on the nature of the relationships between key variables. Uncertainty and probability are concepts that are central to the construction of many models in geography. A model that is particularly useful in illustrating both the nature of models and the manner in which probability is central is the *intervening opportunities model* (Stouffer 1940).

4.6.1 The Intervening Opportunities Model

The intervening opportunities model was originally used in the context of migration, but has since been used more widely in the field of transportation. The conceptual foundation rests on the idea that the movement behavior of individuals in space obeys the principle of least effort – individuals will consider opportunities that are closest to them first, and, if they find them unacceptable, they will go on to the next closest opportunity or opportunities.

This conceptual foundation is quite easy to develop into a probability model, which indicates how individual travel behavior might be organized. Suppose that an individual consumer is considering a purchase, and that there are several alternative stores in the vicinity where a purchase might be made. If there are n stores, we can arrange them in order of their distances from the individual. Let us call the store closest to the individual 'store 1', and the one that is furthest away 'store n'. The intervening opportunities model makes just two assumptions – that (a) individuals consider opportunities sequentially, in order of distance, and (b) individuals consider each opportunity, and find each opportunity acceptable with constant probability L.

These two assumptions imply that our individual starts by considering the closest opportunity. The probability that it is acceptable is L, so we may write the probability of stopping at the closest store as

$$pr(X = 1) = L \tag{4.12}$$

We use X to denote the random variable we are interested in – namely, the number of the store the individual purchases from. The probability that the individual finds the closest opportunity *unacceptable* is $1 - L$; if this occurs, the person goes on to the second closest opportunity, and accepts it with probability L. The probability the individual ends up purchasing at store 2 is therefore the product of these two independent terms, representing the likelihood of rejecting the first opportunity and accepting the second:

$$pr(X = 2) = (1 - L)L. \tag{4.13}$$

Similar reasoning implies that the probability of rejecting the first two opportunities and accepting the third is

$$pr(X = 3) = (1 - L)(1 - L)L = (1 - L)^2 L. \tag{4.14}$$

In general, the probability of accepting opportunity j is equal to the opportunity of rejecting the first $j - 1$ opportunities, multiplied by the likelihood of accepting the opportunity j:

$$pr(X = j) = (1 - L)^{j-i} L. \tag{4.15}$$

In this model, the variable X is a *geometric* variable that is characterized by a downward-sloping histogram. For example, if $L = 0.5$, the probabilities are $pr(X = 1) = 0.5$, $pr(X = 2) = 0.25$, $pr(X = 3) = 0.125$, $pr(X = 4) = 0.0625$, etc. Note how this simple model captures one of the most important of all geographical concepts – namely, that geographic interaction declines with increasing distance. Although the model is an oversimplification in many respects (e.g., the probability of accepting any given opportunity is probably *not* constant in most contexts), it does capture the most important feature of spatial interaction.

Once we have set out the main features of a probability model such as the intervening opportunities model, other questions naturally arise. Some of the questions that might arise here are:

1. Do the probabilities add to one, as they should?

2. Where does L come from?

3. What if we do not have data on the exact distances of the opportunities, and the data are arranged into zones around the origin?

We now consider each of these questions in turn.

4.6.1.1 Do the probabilities add to one?

A geometric random variable is used whenever the variable of interest is the number of the trial on which the first success occurs. Here the 'trials' are the opportunities, and 'success' refers to the selection of a particular opportunity. If the variable is allowed to take for its value any positive integer $(1, 2, \ldots)$, then the probabilities will add to one. In the intervening opportunities model, an individual will not have an infinite number of opportunities to consider, and therefore the probabilities will sum to less than one. That is,

$$\sum_{i=1}^{i=n<\infty} pr(X=i) < 1. \tag{4.16}$$

Consequently, to be more precise, we should adjust our probabilities accordingly. We may do so by dividing by their sum, to ensure that they add to one:

$$pr(X=j) = \frac{(1-L)^{j-1}L}{\sum_{i=1}^{n}(1-L)^{i-1}L}. \tag{4.17}$$

If either (a) n is large, (b) L is large, or (c) n is large *and* L is large, then such an adjustment is unnecessary, since the denominator in (4.17) will be close to one.

4.6.1.2 Where does L come from?

Suppose we have a set of data indicating the proportion of people leaving a particular residential origin who end up at each of the n destinations. We would like to choose (i.e., estimate) L in a way that is consistent with what we observe. That is, since we have the freedom to estimate L, clearly we should do so in a way that is consistent with our observed data.

There are many alternative approaches to choosing L, and here we will illustrate several of them. Suppose we observe the following proportions of people leaving an origin for one of the six potential stores in the area (arranged in terms of increasing distance away from the origin): $p_{obs}(X=1) = 0.55, p_{obs}(X=2) = 0.3, p_{obs}(X=3) = 0.1, p_{obs}(X=4) = 0.05,$ $p_{obs}(X=5) = 0.0,$ and $p_{obs}(X=6) = 0.0$ (where the notation p_{obs} emphasizes the fact that these are observed probabilities). Most individuals go to the closest store, and no one goes from our origin to the two stores that are farthest away. The nature of the observations therefore suggests that the intervening opportunities model might work well in replicating observed travel behavior.

One way to choose L would be to simply try several values, and see which one works best. We could define 'best' in different ways, but let's use the sum of the squared deviations between observed and predicted values. Thus, if we try $L = 0.5$, our criterion, the sum of squared deviations between observed and predicted proportions, is

$$\sum_{i=1}^{i=n} (p_{obs(i)} - pr(X=i)) = (0.55-0.5)^2 + (0.3-0.25)^2$$
$$+ (0.1-0.125)^2 + (0.05-0.0625)^2 \qquad (4.18)$$
$$+ (0-0.03125)^2 + (0-0.01563)^2$$
$$= 0.0070.$$

where we have used the simple form of the model given by Equation 4.15, which may be justified, not by the fact that n is large, but rather by the fact that L is likely to be large, since observed interaction falls off so sharply with distance). If we repeat this for many values of L, we obtain the graph in Figure 4.12, which shows that the deviations are minimized (and hence the fit of the model is best) at a value of $\hat{L} = 0.552$. The '^' notation is used to indicate that L is an estimate of the true, unknown value of L. With $\hat{L} = 0.552$, the predicted probabilities are

$$pr(X=1) = 0.552 \qquad pr(X=4) = .050$$
$$pr(X=2) = 0.247 \qquad pr(X=5) = .022$$
$$pr(X=3) = 0.111 \qquad pr(X=6) = .010 \qquad (4.19)$$

and these are reasonably close to the observed values.

An alternative way to estimate L is to make use of the fact that the mean of a geometric random variable is equal to the reciprocal of the probability of success (or, here, the reciprocal of L, the probability of accepting a given opportunity). The mean destination

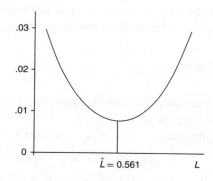

Figure 4.12   **Sum of squared errors as a function of L**

number of our sampled population is 1.65. To see this, suppose that 100 people left the origin. Using the observed data, 55 would stop at destination 1, 30 at destination 2, 10 at destination 3, and five at destination 4. If we make a list of the destination numbers for each of these 100 individuals, the total will be $(55 + (30)(2) + (10)(3) + (5)(4) = 165)$, and dividing by the total number of people in the sample yields $165/100 = 1.65$. Note that we would obtain the same estimate whether we chose 100 people, 1000 people, or any other number of people. Thus, we have $\hat{L} = 1/1.65 = 0.606$, which is similar to our previous estimate of L.

4.6.1.3 What if the opportunity data are organized into zones?

A common way of using the intervening opportunities model is to organize the potential destinations into zones around the origin. Transportation planners do not often need to know the number of individuals arriving at each destination; they are satisfied with more aggregate estimates of the number arriving within defined transportation zones.

We may specify the number of opportunities in destination zone j as d_j, and now arrange the zones in order of increasing distance around the origin. The probability of stopping somewhere in the zone closest to the origin is equal to the probability of accepting an opportunity somewhere (anywhere) within it. In turn, this may be thought of as one minus the probability of *not* finding any of zone 1's opportunities acceptable:

$$pr(X = 1) = 1 - (1 - L)^{d_1} \tag{4.20}$$

where $p(X = 1)$ now refers to the probability of stopping in zone 1. The probability of stopping in zone 2 is equal to the probability of going beyond zone 1, minus the probability of going beyond zone 2:

$$pr(X = 2) = (1 - L)^{d_1} - (1 - L)^{d_1 + d_2} \tag{4.21}$$

In general the probability of stopping in zone j is equal to the probability of going beyond zone $j - 1$, minus the probability of going beyond zone j:

$$pr(X = j) = (1 - L)^{\sum\limits_{i=1}^{j-1} d_i} - (1 - L)^{\sum\limits_{i-1}^{j} d_i} \tag{4.22}$$

L may be determined by minimizing the sum of squared errors, as described in the previous subsection. For example, suppose that $d_1 = 5$, $d_2 = 4$, $d_3 = 4$, $p_{obs}(1) = 0.5$, $p_{obs}(2) = 0.4$, and $p_{obs}(3) = 0.1$. By trying alternative L values, we find that $L = 0.132$ minimizes the sum of squared errors

$$\sum\limits_{i=1}^{3} [p_{obs}(i) - pr(X = i)]^2, \tag{4.23}$$

where the predicted values $pr(X = i)$ are found from Equations 4.20 to 4.22. The predicted probabilities are $pr(X = 1) = 0.508$, $pr(X = 2) = 0.213$, and $\hat{p}(X = 3) = 0.121$. Note that although these are somewhat near the observed probabilities, they do not sum to one. It is preferable to find the sum of squared errors for each value of L after first dividing all of the $pr(X = i)$'s by the sum of the $pr(X = i)$'s to derive new estimates of the predicted probabilities that will sum to one (in the previous example, the sum of the predicted probabilities was very close to one, and this step would not have made much of a difference − this occurs because there is very little chance of going to the outermost destinations). If this is done in the present example, the reader may wish to verify that minimizing the sum of squared errors leads to $L = 0.088$, with predicted probabilities that sum to one: $pr(X = 1) = 0.529$, $pr(X = 2) = 0.279$, and $pr(X = 3) = 0.193$.

4.6.2 A Model of Migration

It is possible to construct a simple model of the movement of people for a regional system that has been divided into n subareas. For illustrative purposes, we will look at the $n = 2$ case of movement between central city and suburbs. We focus only on the redistribution of people who are alive and living in the study region throughout the study period.

Suppose that each year 20% of central city residents move to the suburbs and 15% of all suburban residents move to the central city. If we start with 10,000 residents in each location, after the first year we have, for the suburban and central city populations:

$$P_{sub} = 0.85(10,000) + 0.2(10,000) = 10,500$$
$$P_{cc} = 0.8(10,000) + 0.15(10,000) = 9,500 \tag{4.24}$$

This is based on the fact that 85% of suburban residents and 80% of central city residents do not move in any given year.

Assuming that these probabilities of movement remain constant over time, the population at the end of year t can be written in terms of the populations at the end of the previous year:

$$P_{sub}(t) = 0.85P_{sub}(t-1) + 0.2P_{cc}(t-1)$$
$$P_{cc}(t) = 0.8P_{cc}(t-1) + 0.15P_{sub}(t-1) \tag{4.25}$$

For example, the populations at the end of the second year are:

$$P_{sub}(t) = 0.85(10,500) + 0.2(9,500) = 10,825$$
$$P_{cc}(t) = 0.8(9,500) + 0.15(10,500) = 9,175 \tag{4.26}$$

Note that the total population remains fixed at 20,000. This simple model is known as the *Markov model*, and it provides a useful short-run method for projecting the migration component of population change.

The model also has interesting long-run properties. If the model is allowed to run for a long time, the regional populations approach a constant, equilibrium value, i.e.,

$$P_{sub}(t) = 0.85 P_{sub}(t) + 0.2 P_{cc}(t)$$
$$P_{cc}(t) = 0.8 P_{cc}(t) + 0.15 P_{sub}(t)$$

(4.27)

The equilibrium populations may be determined by taking either of the two equations in 4.27 together with the fact that the total population is fixed at 20,000. For example:

$$P_{sub}(t) = 0.85 P_{sub}(t) + 0.2 P_{cc}(t)$$
$$P_{sub}(t) + P_{cc}(t) = 20,000$$

(4.28)

is a set of two equations and two unknowns. It may be solved to yield $P_{sub} = 11,429$ and $P_{cc} = 8,571$. Thus, if current probabilities of movement do not change, 4/7 of the regional population will reside in the suburbs, and 3/7 will reside in the central city. These equilibrium populations depend only upon the probabilities of movement between subareas; they do not depend upon the initial, subarea populations. Although, in reality, probabilities of movement do not remain constant for such long periods of time, the long-run equilibrium provides a useful (moving) target toward which the population distribution is heading.

The Markov model provides a good illustration of how elementary probability concepts may be used to model an important process. In this case, the model provides both useful short-range forecasts and understandable long-range consequences. For more details on this model, see Rogers (1975).

4.6.3 The Future of the Human Population

How long will the human species survive? This basic question has been the subject of much debate. Attention has been given to factors such as the rate at which we are deplenishing non-renewable resources, as well as how many people the earth can support (Cohen 1995).

An interesting approach that uses a simple probability argument to estimate the survival of the human species has been suggested by Gott (1993). Imagine a list of all humans who have ever lived, or will ever live. Since there is no reason to suppose that we occupy a special place on this list, there is only a 5% chance that we are listed among either the first 2.5% on the list or the last 2.5% on the list. As Gott notes:

Assume that you are located randomly on the chronological list of human beings. If the total number of intelligent individuals in the species is a positive integer $N_{tot} = N_{past} + 1 + N_{future}$, where N_{past} is the number of intelligent individuals born before a particular intelligent observer and N_{future} is the number born after, then we expect N_{past} to be the integer part of the number rN_{tot}, where r is a random number between 0 and 1.

We know that N_{past} is approximately 70 billion. If N_{future} turns out to be greater than about 2.8 trillion, that would mean we would be on the first 2.5% of the chronological list (since 70 billion/2.8 trillion = 0.025). Similarly if N_{future} turns out to be less than 1.75 billion, that would mean we would be on the last 2.5% of the chronological list (since 1.75 billion/70 billion = 0.025). If we do not occupy a special place on the chronological list, this means that the future population yet to be born is, with 95% confidence, in the range:

$$1.75 \text{ billion} \leq N_{future} \leq 2.8 \text{ trillion} \tag{4.29}$$

Using birth rates that are now slightly out-of-date, Gott translates this into a 95% confidence interval for the length of survival of the human species:

$$12 \text{ years} \leq t_{future} \leq 7.8 \text{ million years} \tag{4.30}$$

where t_{future} is the number of additional years that the human species will survive. Readers will probably be comfortable with the upper limit, but will be surprised by the lower limit! Even the upper limit is not large when considering, for example, the length of time that life has been present on the planet.

Different results are obtained when one uses different assumptions. Our species has been in existence for approximately t_{past} = 200,000 years. If we do not occupy a special place on the timeline of the past and future history of the human species, there is a 2.5% chance that we will live in years that are within the first 2.5% of the ultimate timeline, and a 2.5% chance that we will live in years that are within the last 2.5%. If t_{future} is less than 5000 years, this would mean that we would be in the last 2.5% of the timeline (since 5000/200,000 = 0.025), and if t_{future} is greater than 8 million years, this would mean that we would be in the first 2.5% of the timeline (since 200,000/8 million = 2.5%). Hence:

$$5000 \text{ years} \leq t_{furthur} \leq 8 \text{ million years} \tag{4.31}$$

So we get a little more time at the lower limit with this scenario.

SOLVED EXERCISES

1. A continuous random variable has a uniform distribution over the range $6 \leq x \leq 36$. What is the likelihood that a random draw from this distribution will be between 12.4 and 18.8? What is the mean and variance of this variable?

(Continued)

(Continued)

Solution. All values in the range of 6 to 36 are equally likely. The probability of obtaining a value in the range of 12.4 to 18.8 is equal to the width of that range, relative to the entire range. Thus, the likelihood that the random draw is in the range of 12.4 to 18.8 is equal to $(18.8 - 12.4)/(36 - 6) = 6.4/30 = 0.2133$. The mean of a uniform random variable is equal to the average of the lower and upper limits: $(6 + 36)/2 = 42/2 = 21$. The variance of a uniform variable is equal to one-twelfth of the square of the range: $(36 - 6)^2/12 = 900/12 = 75$.

2. The average time between floods in a small town in the Adirondacks is found to be 4.3 years.

   (a) Assuming that the distribution of times between floods is exponential, what is the likelihood that the time until the next flood will be longer than three years?

Solution. The parameter, λ, of an exponential distribution is equal to the reciprocal of the mean. In problems involving the exponential distribution, you will either be given the sample mean (in which case you can determine λ from $\lambda = 1/\bar{x}$, or you will be given the parameter of the distribution, λ, in which case you can find the sample mean from $\bar{x} = 1/\lambda$. Thus in this case $\lambda = 1/4.3$. The probability of obtaining a value less than x is equal to $1 - e^{-\lambda x}$, and therefore the probability of obtaining a value greater than x is equal to $e^{-\lambda x}$. Thus, the probability that the time until the next flood will be longer than $x = 3$ years is $e^{-(1/4.3)(3)} = e^{-3/4.3} = 0.4977$.

   (b) What is the likelihood that the time until the next flood is between four and seven years?

Solution. The likelihood that the time until the next flood is greater than four years, and less than seven years, is equal to the probability that the time is greater than four years, *minus* the probability that it is greater than seven years: $e^{-(1/4.3)4} - e^{-(1/4.3)7} = 0.39446 - 0.19634 = 0.19812$.

   (c) What is the likelihood that the time until the next flood is greater than seven years, given that it is greater than four years?

Solution. If we know that the time until the next flood is greater than four years, we are restricting our attention to the portion of the distribution that has area equal to $e^{-(1/4.3)4} = 0.3945$. Of this area, we would like to know what portion is associated with times greater than seven years. The portion of the original exponential curve that is associated with times greater than seven years is 0.1963. Thus, the desired probability is the ratio of these two values, or $0.1963/0.3945 = 0.4976$.

3. The average number of cars per day traveling on a country road is found to be normally distributed with mean 221.0 and standard deviation equal to 42.6.

   (a) What percentage of days have between 185 and 290 cars on the road?

Solution. The first step is to find the z-scores associated with 185 and 290. For the former, 185 is below the mean, and specifically, $z = (185 - 221)/42.6 = -36/42.6 = -0.845$. Thus, 185 is 0.845 standard deviations below the mean. Similarly, the z-score associated with 290 is equal to $z = (290 - 221)/42.6 = 69/42.6 = 1.62$. Therefore, 290 is 1.62 standard deviations above the mean. To find the desired probability, we break the problem into two parts. The probability of an observation between the mean ($z = 0$) and 290 ($z = 1.62$) is, from the standard normal table, $0.5 - 0.0526 = 0.4474$. Similarly, the probability of an observation between 185 ($z = -0.845$) and the mean ($z = 0$) is $0.5 - 0.1977 = 0.3023$ (where the z-score has been rounded to -0.85). The answer is the sum of these two probabilities: $0.4474 + 0.3023 = 0.7497$.

(b) What volume of traffic on the road is exceeded only 20% of the time?

Solution. We first seek the z-value associated with an area in the right tail of the distribution equal to 0.2. Searching the body of the z-table, we find that a z-value equal to 0.84 is very close to the desired area. 0.84 standard deviations above the mean is equal to $221 + 0.84 (42.6) = 221 + 35.78 = 256.78$.

4. Suppose we examine population movements in a city-suburb system, and assume that migration is limited to movement between these two regions. The probability of moving from the suburbs to the central city is 0.2 each year, and hence the probability of not moving to the central city is 0.8. Similarly, the probability of moving from the central city to the suburbs each year is 0.3. If there are 1000 residents in the central city, and there are 800 residents in the suburbs now, project the population for next year, and the year after, assuming that the migration probabilities remain the same over time.

Solution. Next year, the central city is expected to have seven-tenths of its initial population (0.7×1000) (since three-tenths move to the suburbs, the rest of the population does not move), plus two-tenths of the population that started in the suburbs (0.2×800); thus the forecast is for $700 + 160 = 860$. Similarly, the forecast for the suburbs consists of the proportion of suburban residents (800) who do not move (0.8), plus the proportion of central city residents (1000) who move to the suburbs (0.3); thus, the suburban population forecast is $640 + 300 = 940$. Note that the total population ($860 + 940$) is equal to 1800, which is the same as the initial population; we are ignoring births and deaths in this forecast. The same process can now be used to produce a forecast for the following year, now starting with 860 and 940 for the central city and suburban populations, respectively. So for the following year, the central city is predicted to have seven-tenths of its population of 860, plus two-tenths of the population now in the suburbs (940); $0.7(860) + 0.2(940) = 790$. Likewise, for the following year, the suburb is predicted to have eight-tenths of its population of 940, plus three-tenths of the population now in the central city (860); $0.8(940) + 0.3(860) = 1010$. Note that the total population is still equal to 1800 (= $790 + 1010$), as it should be.

(Continued)

(Continued)

5. A survey of residents reveals that 30% shop at the closest store, 20% at the second closest store, 10% at the third closest store, 0% at the fourth closest store, 10% at the fifth closest store, and 30% at the sixth closest store. For the intervening opportunities model, estimate L, the probability of stopping at an individual store.

Solution. The intervening opportunities model makes use of the geometric distribution; the probability of shopping at each store is constant. Here we are interested in estimating that probability, and we can do so by realizing that the mean of a geometric variable (i.e., the mean trial on which the first success is observed) is equal to $1/L$, where L is the probability of success (i.e., shopping at a store). Because the mean and L are reciprocals of one another, L can be found as the inverse of the (observed) mean, i.e., 1/(observed mean). The mean refers to the mean 'store number' that individuals are stopping at. If there were 100 individuals, 30% (or 30) would stop at store 1, 20 would stop at store 2, 10 at store 3, 10 at store 5, and 30 at store 6. If we added up all of the store numbers for these individuals, we would have $(30 \times 1) + (20 \times 2) + (10 \times 3) + (10 \times 5) + (30 \times 6) = 330$. Since there are 100 individuals, the mean store number is $330/100 = 3.3$. Assuming 100 individuals is clearly arbitrary, and the mean could also be found by weighting the store numbers directly by their probabilities: $(0.3 \times 1) + (0.2 \times 2) + (0.1 \times 3) + (0.1 \times 5) + (0.3 \times 6) = 3.3$. Our estimate of L is equal to 1/3.3 or 0.303.

EXERCISES

1. Snowfall for a location is found to be normally distributed with mean 96″ and standard deviation of 32″.

   (a) What is the probability that a given year will have more than 120″ of snow?

   (b) What is the probability that the snowfall will be between 90″ and 100″?

   (c) What level of snowfall will be exceeded only 10% of the time?

2. Assume that the prices paid for housing within a neighborhood have a normal distribution, with mean $100,000, and standard deviation of $35,000.

   (a) What percentage of houses in the neighborhood have prices between $90,000 and $130,000?

   (b) What price of housing is such that only 12% of all houses in the neighborhood have lower prices?

3. Residents in a community have a choice of six different grocery stores. The proportion of residents observed to patronize each are $pr(X = 1) = 0.4$, $pr(X = 2)$

= 0.25, $pr(X = 3) = 0.15$, $pr(X = 4) = 0.1$, $pr(X = 5) = 0.05$, and $pr(X = 6) = 0.05$, where the stores are arranged in terms of increasing distance from the residential community. Fit an intervening opportunities model to these data by estimating the parameter L.

4. The annual probability that suburban residents move to the central city is 0.08, while the annual probability that central city residents move to the suburbs is 0.11. Starting with respective populations of 30,000 and 20,000 in the central city and suburbs, forecast the population redistribution that will occur over the next three years. Use the Markov model assumption that the probabilities of movement will remain constant. Also, find the long-run, equilibrium populations.

5. The magnitude (Richter scale) of earthquakes along a Californian fault is exponentially distributed, with $\lambda = (1/2.35)$. What is the probability of an earthquake exceeding magnitude 6.3? If there is an earthquake with magnitude greater than 6.1, what is the probability that it exceeds magnitude 7.7?

6. A variable, X, is uniformly distributed between 10 and 24. (a) What is $p(16 \leq X \leq 20)$? (b) What is the mean and variance of X? (c) What is the 75th percentile for this variable?

7. The duration of residence for households is found to be exponentially distributed with $\lambda = 0.21$. What is the probability that a family is in their house for more than eight years? Between five and eight years? What is the probability of a move during the next year, given that they have been in the house for five years? What is the probability of a move during the next year, given that they have been in the house for eight years?

8. The mean value of annual imports for a country is normally distributed with mean $\mu = \$30$ million and standard deviation $\sigma = \$16$ million. What dollar value of imports is exceeded only 5% of the time? What fraction of years have import values between 29 and 45 million?

9. The number of customers at a bank each day is found to be normally distributed with mean $\mu = 250$ and standard deviation $\sigma = 110$. What fraction of days will have less than 100 customers? More than 320? What number of customers will be exceeded 10% of the time?

10. Incomes are exponentially distributed with $\lambda = 0.0001$. What fraction of the population has income (a) < \$8000? (b) > \$12,000? (c) between \$9000 and \$12,000?

11. The number of consumers purchasing a particular item on a given day is found to be uniform over the range (16,28). What is the mean daily number of consumers purchasing the item? What percentage of the time will there be more than 25 purchases? Less than 18?

(Continued)

(Continued)

12. Suppose that annual precipitation is normally distributed, with mean 50 inches, and standard deviation 12 inches. Find the 65th percentile of the precipitation distribution.

13. (a) Express the median of the exponential distribution in terms of its parameter, λ.

    (b) Show that the 64th percentile of the exponential distribution is also equal to its mean.

14. (a) Find the theoretical coefficient of variation for the uniform distribution.

    (b) Repeat part (a) for the special case where the lower limit of the uniform distribution, a, is equal to zero.

15. Find the 38th percentile of the standard normal distribution.

16. What is the theoretical coefficient of variation (see Chapter 2) for the exponential distribution? Note from your answer that no matter what the parameter of the exponential distribution is, observations from this distribution will always exhibit the same relative variability.

On the Companion Website

The 'Further Resources' section of the website contains links to other resources which supplement and expand upon the material presented in Chapter 4. These resources are designed to enrich the understanding of the methods and concepts presented in this chapter, and to encourage the student to engage with the material more meaningfully.

5

INFERENTIAL STATISTICS: CONFIDENCE INTERVALS, HYPOTHESIS TESTING, AND SAMPLING

5.1 INTRODUCTION TO INFERENTIAL STATISTICS

As noted in Chapter 1, the methods of inferential statistics are used to make inferences about a population from a sample. For example, we may interview 50 people in a city, and ask them how far they commute to work. The sample mean provides us with both a simple summary measure and our best estimate of what the 'true' average commuting distance is for the entire city. Because we only have a sample, we are aware that if we interviewed another 50 people, we would likely come up with a different estimate. In this chapter, we will see how we can make more detailed inferences from information we have collected through sampling.

With the tools of descriptive statistics from Chapter 2, and the foundation of probability from Chapters 3 and 4, we are now prepared to learn more about inferential statistics.

5.2 CONFIDENCE INTERVALS

5.2.1 Confidence Intervals for the Mean

Continuing with the example from the introduction, suppose that the mean commuting time for 50 randomly chosen residents is 10 km. Aside from noting that this is our best estimate of commuting distance for residents of the city, what else can we infer about commuting distances? In particular, how confident are we of this estimate? Ten kilometres is our best estimate of the true commuting distance, but how likely are other possibilities?

If we repeated this experiment many times, we could make a histogram of the results, based upon all of the means (each of which would be based upon a sample size of $n = 50$ residents). Because we know something about the nature of this histogram (or distribution) of sample means, we are able to make statements about our confidence in our estimate of the mean. In particular, the *central limit theorem* tells us about the nature of sample means. Any time we sum a large number of independent, identically distributed variables, the central limit theorem implies that the sum will have a normal, bell-shaped curve for its frequency distribution.

For example, if we sum the commuting times (and then simply divide by a constant, equal to the number of observations, to obtain the sample mean), the result may be thought of as an observation from a normal distribution. More specifically, if the process of interviewing 50 people was repeated many times, and if we made a histogram of the resulting means, the histogram would have the shape of a normal distribution.

In the central limit theorem, 'independent' implies that one individual's commuting time is unrelated to the commuting time of other individuals. 'Identically distributed' implies that each individual commuting time comes from the same frequency distribution (which may or may not be normal). In other words, there are not separate frequency distributions that govern separate subcategories of the population. Under these conditions, we would find that the frequency distribution of sample means (which could be constructed if we had the results of many surveys, each with its own sample mean) would follow a normal distribution.

Furthermore, the normal, bell-shaped curve representing the frequency distribution of the sample means will have a mean equal to the true mean, μ. Although it is unlikely that our own sample mean will be exactly equal to the true mean, it is certainly reassuring to know that, with a very large number of people repeating our survey of $n = 50$ individuals, the average of all of the sample means would approach the true mean. We can say that the sample mean is an *unbiased* estimate of the true mean.

Finally, we know something about the variability that we will observe among the sample means collected. In particular, the sample means will have a variance equal to σ^2/n, where σ^2 is the variance associated with the distribution governing the variable of interest. This is consistent with one's intuition that sample means will display more variability when the original data are inherently variable; high values of σ^2 will lead to high values of σ^2/n. If everyone in a class was told to increase the size of their survey (i.e., make n larger), the resulting distribution of sample means would display less variability.

Summarizing, we know that if others repeated our survey, and if we made a histogram using all of the many sample means that were collected, the histogram would have a roughly normal, bell-shaped appearance, the mean of the sample means would provide an unbiased estimate of the true mean, and the variance of the sample means would be equal to σ^2/n.

Since we know something about the distribution of sample means, we can now make statements about how confident we are that the true mean is within a given interval about our sample mean. For a normal distribution, 95% of the observations lie within about two standard deviations (actually 1.96 standard deviations) of the mean. This can be verified using the standard normal table (Table A.2 in Appendix A), which reveals that the probability of a z-score with absolute value less than 1.96 is 0.95. The standard deviation of the distribution of sample means is equal to the square root of the variance, or $\sigma / \sqrt{n}$. Since we usually don't know σ, we estimate this quantity with $s / \sqrt{n}$. This implies that our individual sample mean should, 95% of the time, lie within $\pm 1.96 s / \sqrt{n}$ of the true mean μ:

$$pr\left[(\mu - 1.96\frac{s}{\sqrt{n}}) \leq \bar{x} \leq (\mu + 1.96\frac{s}{\sqrt{n}})\right] = 0.95 \qquad (5.1)$$

The probability that $\bar{x}$ lies in the range described in parentheses is 0.95. Rearranging Equation 5.1 allows us to construct a confidence interval around our sample mean:

$$pr\left[(\bar{x} - 1.96\frac{s}{\sqrt{n}}) \leq \mu \leq (\bar{x} + 1.96\frac{s}{\sqrt{n}})\right] = 0.95 \qquad (5.2)$$

This tells us that 95% of the time, the true mean should lie within $\pm 1.96 s / \sqrt{n}$ of the sample mean. This is termed a 95% confidence interval; we don't know what the true mean is, but we are 95% sure that the true mean is within the range given in Equation 5.2.

A 90% confidence interval could be constructed by recognizing that the true mean would lie within 1.645 standard deviations of the mean 90% of the time. The value of 1.645 comes from the standard normal z-table found in Table A.2; it is the value of z associated with 5% of the area under each of the two tails of the distribution. It is also fairly common to use 99% confidence intervals, constructed by adding and subtracting $2.58s / \sqrt{n}$ to the sample mean.

More generally, a $(1 - \alpha)$% confidence interval around the sample mean is:

$$pr\left[(\bar{x} - z_{\alpha/2}\frac{s}{\sqrt{n}}) \leq \mu \leq (\bar{x} + z_{\alpha/2}\frac{s}{\sqrt{n}})\right] = 1 - \alpha \qquad (5.3)$$

where $z_{\alpha/2}$ is the value taken from the z-table that is associated with an area equal to $\alpha/2$ in the tail of the standard normal distribution.

Consider the case where, in addition to finding a mean commuting distance of 10 km in our sample of 50 individuals, we find a standard deviation of 9 km. Using Equation 5.2, a 95% confidence interval for the mean is $10 \,\text{km} \pm 1.96(9 / \sqrt{50})$ or 10 ± 2.49. This implies that we are 95% confident that the true mean lies between 7.51 km and 12.49 km.

Example 5.1

For the scenario just described above, find 90% and 85% confidence intervals.

Solution: For the 90% confidence interval, we use $z_{\alpha/2} = 1.645$ in place of the 1.96 we used for the 95% confidence interval. This leads to a 90% confidence interval of $10 \,\text{km} \pm 1.645(9 / \sqrt{50})$, or 10 ± 2.093. We are 90% sure that the true mean is between 7.907 and 12.093. Note that this interval is more narrow than the 95% confidence interval. For the 85% confidence interval, we first use the standard normal table (A.2) to find $z = 1.44$. This is found by searching within the body of the table for $\alpha/2 = (1 - 0.85)/2 = 0.075$. This tail area of 7.5% is found in the $z = 1.4$ row, and the 0.04 column, corresponding to a z-value of 1.44. Then the 85% confidence interval is $10 \,\text{km} \pm 1.44(9 / \sqrt{50})$, or $10 \,\text{km} \pm 1.833$. We are therefore 85% confident that the true mean lies in the interval 8.167 to 11.833. More specifically, 85% of the time, confidence intervals constructed in this way will contain the true mean.

5.2.2 Confidence Intervals for the Mean When the Sample Size is Small

The central limit theorem applies when the sample size is 'large'; only then will the distribution of means possess a normal distribution. When the sample size is not 'large', the frequency distribution of the sample means has what is known as the *t-distribution*; it is symmetric like the normal distribution, but it has a slightly different shape.

The areas under the t-distribution are given in the t-table (Table A.3 in Appendix A). The table is used with $n - 1$ *degrees of freedom*. For example, with a sample size of $n = 30$, 95% confidence intervals are constructed using $t = 2.045$ (found using the 0.025 column of Table A.3, with 29 degrees of freedom), instead of the value of $z = 1.96$ used above for the normal distribution.

For the commuting data in Table 2.1, the 95% confidence interval around the mean is therefore:

$$pr[\{21.93 - \frac{2.045(14.43)}{\sqrt{30}}\} \le \mu$$

$$\le \{21.93 + \frac{2.045(14.43)}{\sqrt{30}}\}] = 0.95$$

(5.4)

and thus we are 95% sure that the true mean is within plus or minus $2.045(14.43)/\sqrt{30} = 5.39$ of our sample mean of 21.93. The 95% confidence interval for the mean may be stated as (16.54, 27.32). More precisely, 95% of confidence intervals constructed from samples in this way will contain the true mean.

The t-distribution is used when the sample size is not large, but what constitutes 'large'? A common guideline is that if the sample size is over 30, the normal distribution can be used. By examining the columns of Table A.3, it is clear that the t-distribution begins to approach the normal distribution for degrees of freedom greater than about 30. For example, note that the last entry in the 0.025 column is the familiar 1.96 from the standard normal table; similarly, the last entry in the 0.05 column is the familiar value of 1.645 from the standard normal table.

Use of the t-distribution to construct confidence intervals requires one additional assumption. In particular, it is assumed that the data come from a normal distribution. In the previous example, we need to assume that the histogram associated with commuting time has the bell-shaped, symmetric curve of the normal distribution. However, it also turns out that the construction of accurate confidence intervals is relatively robust with respect to deviations from this assumption. Even if the distribution of commuting times is not precisely normal, we can still trust our confidence intervals to be reasonably accurate.

5.2.3 Confidence Intervals for the Difference between Two Means

There are many situations where we are interested in comparing two samples. Suppose, for example, we wish to compare the commuting times in two suburbs. For suburb 1, we find a mean commuting time of 15 minutes, with a standard deviation of 20 minutes, based on a sample of 30 individuals. For suburb 2, we find a mean commuting time of 22 minutes, with a standard deviation of 23 minutes, based on a sample of 40 individuals.

Our best estimate of the difference between commuting times in the two suburbs is that residents of community 2 have a longer commute, by 7 (= 22 − 15) minutes. However, this estimate is based upon a sample − the 'true' difference between the two communities may be more, or less, than this estimate.

It turns out that we know something about the distribution of the differences between sample means. If we repeatedly took samples, and each time looked at the differences in the sample means, and then made a histogram of these differences, that distribution would be centered upon the 'true' difference (whatever it was). In addition, if the individual sample sizes are large (say over 30), then the histogram of the differences in means would have the shape of a normal distribution, and it would have a standard deviation equal to $\sqrt{(s_1^2 / n_1) + (s_2^2 / n_2)}$. Therefore, a $100(1 - \alpha)\%$ confidence interval for the difference in sample means is:

$$(\bar{x}_2 - \bar{x}_1) \pm z_{\alpha/2}\sqrt{(s_1^2 / n_1) + (s_2^2 / n_2)}. \tag{5.5}$$

where $z_{\alpha/2}$ is the appropriate value taken from the standard normal table (where there is an area of $z_{\alpha/2}$ in each tail). For example, for a 95% confidence interval, we would use $z = 1.96$; for a 90% confidence interval, we would use $z = 1.645$.

Continuing with the example above, a 95% confidence interval for the difference in commuting times would be found from:

$$(22 - 15) \pm 1.96\sqrt{20^2 / 30 + 23^2 / 40} = 7 \pm 10.1 \tag{5.6}$$

Our confidence interval around the quantity $\bar{x}_2 - \bar{x}_1$ ranges from $7 + 10.1 = 17.1$ minutes (with community 2 having the longer commute), to $7 - 10.1 = -3.1$ minutes (with community 1 having the longer commute). 95% of all confidence intervals constructed in this way will contain the 'true' difference.

It does not matter whether we construct the confidence interval around $\bar{x}_2 - \bar{x}_1$ (as above, in Equation 5.5) or construct it around $\bar{x}_1 - \bar{x}_2$. In this latter case the confidence interval is constructed around $\bar{x}_1 - \bar{x}_2 = 15 - 22 = -7$. Thus we would have -7 ± 10.1; we are 95% sure that the true difference ($\mu_1 - \mu_2$) is between -17.1 (where community 2 has the longer commute by 17.1 minutes) and 3.1 (where community 1 has the longer commute by 3.1 minutes).

If the sample sizes are smaller, we use the t-table instead of the z-table, and the width of the confidence interval will depend upon what we assume regarding the variances of the two samples:

(a)   If we do not assume that the variances of the two samples are equal (this is the most conservative approach), then the width of the confidence interval is:

$$(\bar{x}_2 - \bar{x}_1) \pm t_{\min(n_1,n_2)-1df} \sqrt{(s_1^2 / n_2) + (s_2^2 / n_2)} \tag{5.7}$$

This is identical to the previous equation, with the exception that we use the t-table instead of the z-table, and we use it with degrees of freedom equal to the smaller of the two sample sizes, minus one. Thus, if we had $n_1 = 14$ and $n_2 = 17$, and we desired a 95% confidence interval, we would use a value of $t = 2.16$ in the expression above (based upon 13 df).

(b)   If we are willing to assume that the true variances of the samples are equal, then the width of the confidence interval is:

$$(\bar{x}_2 - \bar{x}_1) \pm t_{n_1+n_2-2df} \sqrt{s_p^2 / n_1 + (s_p^2 / n_2)} \tag{5.8}$$

where s_p^2 is a pooled estimate based upon the two sample variances; it can be found using Equation 5.26. Note that the degrees of freedom is now found from $n_1 + n_2 - 2$. Again, with $n_1 = 14$ and $n_2 = 17$, we would now use $t = 2.045$ for a 95% confidence interval, based on $14 + 17 - 2 = 29$ degrees of freedom.

It is reasonable to wonder why one would choose to make an assumption that the true variances were equal. After all, why not just go with option (a), where no such assumption needs to be made? The answer is that by making the assumption of equal variances in (b), the confidence intervals that are constructed will be more narrow than those that result when not making the assumption. Thus, when we make the assumption, we will be confident that the true difference in means is within a relatively more narrow range.

5.2.4 Confidence Intervals for Proportions

When we use samples to estimate proportions, we are also interested in confidence intervals. For example, if a survey of $n = 45$ people reveals that $x = 20$ moved last year, the estimated proportion of recent movers is $p = 20/45 = 0.444$. Although we don't know the 'true' proportion that characterizes the population, since we have not surveyed everyone, we can still place a confidence interval around our estimate.

The construction of confidence intervals for proportions relies on the fact that sample proportions also have normal distributions. Thus, if we repeated our survey of 45 people many times, we would get a range of proportions; if we made a histogram of these sample proportions, it would have the familiar bell-shaped normal distribution. Furthermore, the distribution would be centered on the 'true' population proportion (ρ), and the variability in the sample proportions could be measured by the variance, $\rho(1-\rho)/n$. Since the true proportion is unknown, we use the sample variance, $p(1-p)/n$ to construct confidence intervals. This can be done in a manner analogous to the construction of confidence intervals for the mean:

$$p \pm z_{\alpha/2}\sqrt{\frac{p(1-p)}{n}}$$

(5.9)

For example, we would expect that, 95% of the time, the true proportion would be within 1.96 standard deviations of the sample proportion:

$$p - 1.96\sqrt{\frac{p(1-p)}{n}} \leq \rho \leq p + 1.96\sqrt{\frac{p(1-p)}{n}}$$

(5.10)

To continue with the present example, a 95% confidence interval around our sample proportion is $0.444 \pm 1.96\sqrt{\frac{0.444(1-0.444)}{45}}$, or 0.444 ± 0.145. We are 95% certain that the true proportion is between 0.299 and 0.589. Alternatively stated, 95% of confidence intervals constructed in this way will contain the true proportion.

We may also construct a confidence interval around a difference in sample proportions. If two sample proportions, p_1 and p_2, are based upon respective sample sizes of n_1 and n_2, a 100 x (1 - α)% confidence interval is

$$(p_2 - p_1) \pm z_{\alpha/2} \sqrt{\frac{p_1(1-p_1)}{n_1} + \frac{p_2(1-p_2)}{n_2}} \tag{5.11}$$

We may state that 100 x (1 - α)% of the time, confidence intervals constructed in this way will contain the true difference, $\rho_2 - \rho_1$.

5.3 HYPOTHESIS TESTING

Hypothesis testing constitutes a fundamental way in which inferences about a population are made from a sample. In this section, we focus on the testing of hypotheses involving either one sample or two samples.

5.3.1 Hypothesis Testing and One-Sample z-Tests of the Mean

We will first describe some of the basic concepts of hypothesis testing and statistical inference through an example using a one-sample test involving a mean. Suppose we want to know whether the mean number of weekly shopping trips made by households in a particular neighborhood of an urban area differs from 3.1, which is the corresponding mean for the urban area as a whole. We do not wish to survey all households in the neighborhood to find the desired mean, since that would be too costly in terms of both time and money (and if we *had* the time and/or money to do this, it would be wasteful). Instead, we choose to take a random sample of households. In this example, it is assumed that the value of 3.1, which applies to the entire urban area, is known.

The first step is to set up a *null hypothesis*, where the mean number of shopping trips in the neighborhood is hypothesized to be equal to the mean for the entire urban area:

$$H_0: \mu = 3.1 \tag{5.12}$$

where μ is the hypothesized, true mean for the neighborhood. Null hypotheses are set up in this way; failing to reject it will be in keeping with the default option that the true neighborhood mean may be no different from the hypothesized mean. This would be a null result. Rejecting the null hypothesis occurs when we find evidence for a significant difference between our sample mean and the hypothesized mean.

The second step is to state an *alternative hypothesis*. Suppose that we are interested in this example because we suspect from other anecdotal evidence that the neighborhood of interest has a high number of shopping trips. In this case, our alternative hypothesis is that the true, unknown mean in the neighborhood is greater than 3.1:

$$H_A : \mu > 3.1 \tag{5.13}$$

This is known as a *one-sided* alternative hypothesis, since we suspect (before we carry out our survey) that if the true neighborhood mean *does* differ from 3.1, it will be greater than 3.1 and not less. If, on the other hand, we had no *a priori* idea about how the neighborhood mean might differ from that for the entire urban area, we would postulate the following:

$$H_A : \mu \neq 3.1 \tag{5.14}$$

Here we have a *two-sided hypothesis*; if the true neighborhood mean differs from 3.1, it could lie on either side of 3.1. In carrying out statistical tests, we need to recognize that we will be making decisions on the basis of a sample drawn from a larger population. We will never know for certain whether the null hypothesis is true or false. We base our decision on the evidence in favor of, or against, the null hypothesis. If we interview ten households, and the sample mean is 4.2 shopping trips/week, two conclusions are possible. One possibility is that the null hypothesis is true. In this case, the true mean among households in the neighborhood is 3.1, and we have obtained an unusual sample. The other possibility is that the null hypothesis is false. In this event, the true mean among households in the neighborhood is not equal to 3.1. We fail to reject the null hypothesis if, under H_0, the sample is not *too* unusual; otherwise we will reject H_0. The role of statistics in this case is to inform us regarding precisely how unusual it would be to obtain our sample if the null hypothesis were true.

In the course of this process, it is possible that one of two kinds of error might be made. A *Type I error* refers to rejecting a true null hypothesis, while a *Type II error* refers to failing to reject a false null hypothesis. The likelihood of making a Type I error is denoted by α and is referred to as the *significance level*. The analyst has control over α, and the third step in setting up a statistical test is to choose a significance level. Common values chosen for α are 0.01, 0.05, and 0.10. Though we of course wish to keep the likelihood of errors as small as possible, we cannot simply choose α to be exceedingly small. This is because there is an inverse relationship between α and β, the likelihood of making a Type II error. The lower our choice of α, the greater the chance that we will fail to reject a false null hypothesis. Figure 5.1 summarizes the four possible outcomes associated with statistical testing. If the null hypothesis is true (column 1), we either make a correct decision with probability $1 - \alpha$ or an incorrect decision with probability α. If the null hypothesis is false (column 2), we either make a correct decision (with probability $1 - \beta$), or, with probability β, we make a Type II error.

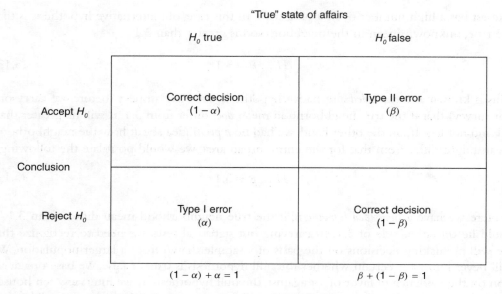

Figure 5.1 Four outcomes associated with statistical testing

   The fourth step in hypothesis testing is to choose a test statistic and to find the observed value of the test statistic.

   For one-sample tests involving the mean, we have sample data $x_1, x_2, \ldots x_n$. From sections 4.3 and 5.2.1, we have learned that sample means are normally distributed with mean μ and standard deviation $\sigma \sqrt{n}$. If we replace the unknown population standard deviation with its sample estimate (s), we can use the test statistic

$$z = \frac{\overline{x} - \mu}{s/\sqrt{n}} \tag{5.15}$$

   When the null hypothesis is true, this test statistic will have, for n greater than about 30, a standard normal distribution with mean 0 and standard deviation 1. The normal distribution is often denoted $N(\mu, \sigma^2)$, and so in this case of the standard normal distribution we can use the notation $N(0,1)$.

   Suppose that, for our example, we interview $n = 100$ individuals in the neighborhood and find that the sample mean is 4.2 shopping trips per week, with a sample standard deviation of $s = 5.0$. Then the observed z-statistic is

$$z = \frac{4.2 - 3.1}{5/\sqrt{100}} = \frac{1.1}{0.5} = 2.2 \tag{5.16}$$

Intuitively, the z-score is large in absolute value when the sample mean is far from the hypothesized mean, and it is in such cases that we will reject the null hypothesis for two-sided alternative hypotheses.

How large must the test statistic be before we reject H_0? The fifth step in hypothesis testing is to use α and our knowledge of the sampling distribution of the test statistic to determine the *critical value* of the test statistic. Critical values are those values of the test statistic where we are on the knife-edge between acceptance and rejection. If the observed test statistic is slightly to one side of the critical value, we reject H_0; if it is slightly to the other side, we fail to reject H_0. Returning to our example, where the sampling distribution is normal, if we have chosen $\alpha = 0.05$ and the two-sided alternative $H_A : \mu \neq 3.1$, the critical values of z are equal to -1.96 and 1.96 (Figure 5.2); under H_0, we would expect 5% of all experiments to result in $|z| > 1.96$. This critical value may be found from Table A.2; note that $z = 1.96$ is associated with a probability of 0.025 in each tail. Since our observed value of 2.2 is greater than 1.96, we reject H_0. How often would we observe a value as high (or higher) than our observed value of 2.2 under H_0? A check of the table of normal distribution probabilities reveals that such an event would occur with probability $(2)0.0139 = 0.0278$ (see Figure 5.3). The value of 0.0278 is referred to as the *p*-value; it tells us how likely a result equal to or more extreme than the one we observed would be, if the null hypothesis were true. Note that *p*-values less than α are consistent with rejection of the null hypothesis.

Also note that if we had used the one-tailed alternative $H_A : \mu > 3.1$, we would also have rejected H_0, since the critical value of z (denoted z_{crit}) would have been equal to 1.645 (Figure 5.4). In this case, the *p*-value would have been 0.0139 (we do not multiply by two because we are only interested in the one tail).

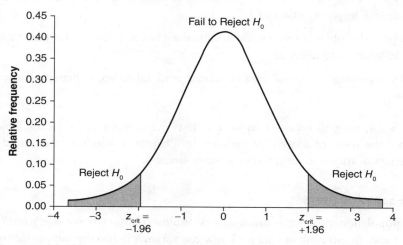

Figure 5.2 Critical regions of the sampling distribution of z-scores based on the sample mean

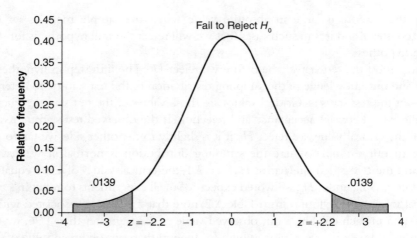

Figure 5.3   A p-value of 0.0278

The steps involved in hypothesis testing are summarized below:

1. State the null hypothesis, H_0.

2. State the alternative hypothesis, H_A.

3. Choose α, the probability of making a Type I error (rejecting a true H_0).

4. Choose a statistical test, and find the observed test statistic.

5. Find the critical value of the test statistic to determine which values of the observed statistic will imply rejection of H_0.

6. Compare the observed test statistic with the critical value of the test statistic, and decide whether to reject H_0.

7. Find the p-value as the tail area associated with values more extreme than the test statistic.

In Section 5.3.4, we will review two-sample tests for differences in means. This will lead naturally into the topic of analysis of variance in Chapter 6, which is concerned with possible differences in means among three or more samples.

5.3.2 One-Sample t-Tests

When the population variance is unknown (as it almost always is; we don't even know the population mean (μ), so how would we know the variance?) and the sample size is small, the sampling distribution of the mean is no longer normal, and hence we should not use the

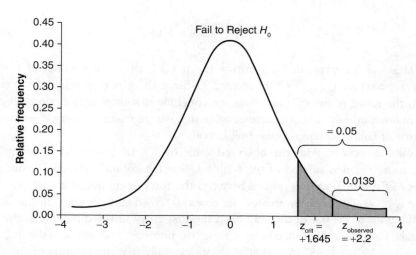

Figure 5.4 Critical and p-values in one-sided test

z-statistic. Instead, the sampling distribution of the mean follows a t-distribution, with $n - 1$ degrees of freedom. Degrees of freedom may be somewhat loosely thought of as the number of observations minus the number of quantities estimated. We have n observations, and we 'use up' one degree of freedom to estimate the mean (the sample mean is also used to estimate the sample variance). We have $n - 1$ degrees of freedom since if you were given the values of $n - 1$ observations, you could calculate the value of the nth observation without being told it, if you knew the sample mean.

In addition to the usual assumption that the observations are independent, use of the t-distribution requires the additional assumption that the observations come from a normal distribution.

The t-distribution is similar in shape to the normal distribution, though the tails of the distribution are slightly fatter in comparison with the normal distribution. The t-statistic is found in exactly the same way as the z-statistic:

$$t = \frac{\bar{x} - \mu}{s / \sqrt{n}} \tag{5.17}$$

To test a hypothesis about the mean, we compare our observed t-statistic with the critical value taken from a t-table (see, e.g., Table A.3 in Appendix A) with $n - 1$ degrees of freedom.

5.3.2.1 Illustration

Suppose that in our previous example we interviewed $n = 20$ people instead of $n = 100$, and found $\bar{x} = 4.5$ and $s = 5.5$. Our test statistic is:

$$t = \frac{4.5 - 3.1}{5.5 / \sqrt{20}} = 1.14 \tag{5.18}$$

For $\alpha = 0.05$ and the two-sided alternative $H_A : \mu \neq 3.1$, the critical values of t with $n - 1$ degrees of freedom are $t_{0.05,19} = -2.09$ and $+2.09$. Since the observed value of the test statistic falls within the range of the critical values, we conclude that there is not enough evidence to reject the null hypothesis. It is of course possible that we are making an error – specifically the Type II error of failing to reject a false null hypothesis.

The p-value associated with the observed value of $t = 1.14$ is found by using Table A.4, which is a more detailed version of the t-table. Using the column headed 19 (since we have 19 degrees of freedom), we interpolate between the rows beginning with 1.1 and 1.2 (since the t-value is 1.14). The entry in the former row is 0.85746 and the entry in the latter row is 0.87756. We wish to go 0.4 of the way from the first entry, in the direction of the latter. We therefore take 0.4 of the difference between the two numbers, and add it to the first entry: 0.4 (0.87756 − 0.85746) + 0.85746 = 0.8655. Note carefully that the heading of the table indicates that these entries represent the cumulative distribution (see Figure 5.5). This implies that the probability of observing a t-statistic greater than 1.14 is $1 - 0.8655 = 0.1345$. The p-value for this two-sided test is therefore $2(0.1345) = 0.269$. Since this is greater than 0.05, it is consistent with the fact that we have not rejected the null hypothesis. In situations where the null hypothesis is true, we would expect more extreme values of t (either higher than $t = 1.14$, or lower than $t = -1.14$) about 27% of the time. We would only reject the null hypothesis if our observed value of t was so extreme that it would be expected less than 5% of the time.

A 95% confidence interval for the mean in the case of small samples is found by replacing z with t in Equation 5.3:

$$(\bar{x} \pm t_{0.05,19} s / \sqrt{n} = 4.5 \pm 2.09(5.5) / \sqrt{20} = 4.5 \pm 2.57 = (1.93, 7.07) \tag{5.19}$$

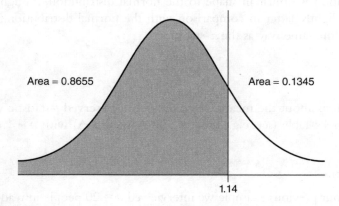

Area = 0.8655   Area = 0.1345

1.14

Figure 5.5   Cumulative distribution of t with 19 degrees of freedom

Note that this interval includes the hypothesized value of 3.1. In general, there is a correspondence between confidence intervals for the mean and the results of two-sided hypothesis tests. When the null hypothesis is rejected, the confidence interval around the sample mean will not include the hypothesized mean. The null hypothesis will not be rejected when the confidence interval includes the hypothesized mean.

5.3.3 One-sample Tests for Proportions

To test whether a proportion, rather than a mean, is different from some hypothesized value, we need to know the sampling distribution of sample proportions when the null hypothesis is true. Consider an example where we are interested in knowing whether the proportion of people who intend to move next year in our local community is consistent with the countywide proportion, which is known to be 0.15. Our null hypothesis is that the true proportion of people in our community who intend to move next year is 0.15. The two-sided alternative hypothesis is that this proportion differs from 0.15 – that is, it is either significantly lower, or significantly greater. We will use a Type I error probability of $\alpha = 0.05$.

Suppose now that we take a sample of $n = 10$ residents and that we ask each about their mobility intentions. Intuitively, we would want to reject the null hypothesis if we get results that are inconsistent with the null hypothesis. First note that in this example it would be difficult to reject the null hypothesis by claiming that the proportion in our community who intend to move is *less* than the countywide value of 0.15. This is because the probability of obtaining the extreme case of $X = 0$ residents indicating that they intend to move, when the null hypothesis is true, is found from the binomial distribution as:

$$pr(X = 0) = \binom{10}{0} 0.15^0 0.85^{10} = 0.1969 \tag{5.20}$$

Since this result is not that uncommon when the true probability is 0.15 (and, in particular, because it is greater than the probability of $\alpha/2 = 0.025$), we cannot reject the null hypothesis.

Consider now the case where a larger than expected number of people indicate that they intend to move. We will reject the null hypothesis if the probability of getting such a response, or an even more extreme response, is unlikely (and, more specifically, if the probability of getting such a response is less than $\alpha/2 = 0.025$). Accordingly, we find the binomial probabilities:

$$pr(X = 2^+) = 1 - pr(X = 0) - pr(X = 1) = 0.456$$
$$pr(X = 3^+) = 0.180$$
$$pr(X = 4^+) = 0.04997 \tag{5.21}$$
$$pr(X = 5^+) = 0.0099$$

We will therefore reject the null hypothesis if five or more respondents indicate that they plan to move – this probability is less than 0.025, and therefore quite unlikely if the null hypothesis is actually true.

Note that if we had set this exercise up with a one-sided alternative, where we were only interested in the possibility of a deviation from the null hypothesis in this direction, we could reject the null hypothesis with four or more individuals responding that they planned to move, since this would occur less than 5% of the time when the null hypothesis was true. Setting this particular problem up with a two-sided alternative does not make much sense, since, as we have seen, it is not possible to reject the null hypothesis in the direction of a true probability that is less than the one that has been hypothesized.

More commonly, we use the normal approximation to the binomial (introduced in Section 3.3) to approximate the distribution of sample proportions and, in turn, to test hypotheses about proportions. Suppose that the true proportion in a population is equal to ρ_0. Then the sampling distribution of proportions is approximately normal, with mean ρ_0 and standard deviation $\sqrt{\rho_0(1-\rho_0)/n}$. This implies that we may test hypotheses of the form

$$H_0 : \rho = \rho_0 \tag{5.22}$$

by using a z-statistic of the form

$$z = \frac{p - \rho_0}{\sqrt{\rho_0(1-\rho_0)/n}} \tag{5.23}$$

Even though we don't know the true value of ρ_0, when calculating z in Equation 5.23, we simply use the hypothesized value, ρ_0. When the null hypothesis is true, this statistic will have a standard normal distribution.

Note that the form of the z-statistic is always the same – the numerator is equal to the observed sample value minus the hypothesized value, and the denominator is equal to the standard deviation of the sampling distribution when H_0 is true. The z-statistic tells us how many standard deviations the observed value is away from the hypothesized value. For example, $z = 2$ would imply that our observed value was two standard deviations above the hypothesized value.

5.3.3.1 Illustration

Suppose we are interested in knowing whether the citywide proportion of households that own three cars differs from the nationwide figure of 0.2. We survey $n = 50$ households, and find $p = 16/50 = 0.32$. To test $H_0 : \rho = 0.2$ against the two-sided alternative $H_A : \rho \neq 0.2$ with $\alpha = 0.05$, we find:

$$z = \frac{0.32 - 0.2}{\sqrt{\dfrac{0.2(0.8)}{50}}} = 2.12 \tag{5.24}$$

Since the observed value of z falls outside the range of the two critical values, $z_{0.05/2} = \pm 1.96$, we reject the null hypothesis and conclude that the proportion of households that own three cars in this city is significantly higher than the nationwide proportion.

The p-value is found by using the z-table (Table A.2) to determine the likelihood of a more extreme z-value than the one observed. The table reveals that $pr(z>2.12) = 0.017$. Since the probability of a z-value less than -2.12 is also 0.017, the probability of getting a statistic more extreme than the one observed is $2(0.017) = 0.034$. Note that the p-value is less than 0.05, and this is consistent with rejecting H_0.

5.3.4 Two-sample Tests: Differences in Means

Often a sample mean is compared with another sample mean, rather than with some known population value. In this case, the t-test is appropriate, and the form of the two–sample t-test depends upon whether the variances of the two samples can be assumed equal. The assumption of equal variances is known as *homoscedasticity*. If the variances can be assumed equal, the t-statistic is:

$$t = \frac{\bar{x}_1 - \bar{x}_2}{\sqrt{\dfrac{s_P^2}{n_1} + \dfrac{s_P^2}{n_2}}} \tag{5.25}$$

where x_1 and x_2 are the observed means of the two samples, n_1 and n_2 are the observed sample sizes, and the pooled estimate of the standard deviation, s_p, is equal to:

$$s_p = \sqrt{\frac{(n_1 - 1)s_1^2 + (n_2 - 1)s_2^2}{n_1 + n_2 - 2}} \tag{5.26}$$

Here, s_1^2 and s_2^2 represent the observed variances of samples 1 and 2, respectively. The number of degrees of freedom associated with this test statistic is $n_1 + n_2 - 2$, since the total sample size is effectively reduced by two due to the estimation of two means. The quantity s_p is a pooled estimate of the (assumed) common variance and it is found in Equation 5.26 as a weighted average of the two individual variances.

If it cannot be assumed that the two samples have equal variances, then the appropriate t-statistic is:

$$t = \frac{\bar{x}_1 - \bar{x}_2}{\sqrt{\dfrac{s_1^2}{n_1} + \dfrac{s_2^2}{n_2}}} \tag{5.27}$$

In this case, the degrees of freedom are more difficult to calculate (see, e.g., Sachs 1984), and it is often suggested that the degrees of freedom be simply taken as the minimum of the two quantities $(n_1 - 1, n_2 - 1)$. The actual value for the degrees of freedom is higher than this, and so taking the degrees of freedom to be equal to min $(n_1 - 1, n_2 - 1)$ will be conservative, in the sense that the actual probability of committing the Type I error of rejecting a true hypothesis will be less than the stated value of α.

As is the case with the one-sample t-test, we must also assume that the populations from which the samples are taken are themselves normally distributed.

How do we know if we should assume that the variances are equal? In practice, one may use the F-test to determine whether the assumption of equal variances is justified. Under the null hypothesis of equal variances, the test statistic

$$F = \frac{s_1^2}{s_2^2} \tag{5.28}$$

has an F-distribution, with $n_1 - 1$, and $n_2 - 1$ degrees of freedom in the numerator and denominator, respectively. In carrying out this test, the sample with the larger variance is always designated as sample '1', and the sample with the lower variance is designated as sample '2'. Consequently, the numerator in Equation 5.28 will always be greater than the denominator, and hence the observed value of F will always be greater than one. The observed value of the F statistic, as calculated in Equation 5.28, is then compared with the critical value of F found in Table A.5 in Appendix A.

As was the case with confidence intervals, it does not matter whether we choose to use $\bar{x}_1 - \bar{x}_2$ or $\bar{x}_2 - \bar{x}_1$ in the numerator of Equation 5.25 or 5.27; in either case we will come to the same conclusion.

5.3.4.1 Illustration

Suppose we are interested in knowing whether differences in recreation behavior exist between the central city and suburban regions of a metropolitan area. In particular, suppose we are interested in swimming frequencies. Before collecting the data, we have no prior hypothesis regarding whether one region's individuals will have higher frequencies than the other, and so we will use a two-tailed test. The null and alternative hypotheses may be stated as:

$$H_0 : \mu_{cc} = \mu_{sub} \qquad H_A : \mu_{cc} \neq \mu_{sub} \tag{5.29}$$

where 'cc' and 'sub' refer to central city and suburbs, respectively.

We collect the hypothetical data shown in Table 5.1, based on a random sample of eight residents in each region. Because the sample size is small, we will use a t-test; if the sample size was larger (say 30 or so residents from each region), we would use a z-test. An examination of the sample means reveals that the annual frequency is higher in suburban locations. Is this difference 'significant', or might it have arisen by chance? With respect to the latter possibility, it could be the case that the observed difference is attributable to sampling fluctuation – if we took another sample of eight residents from each region, we might not find such a large difference. To proceed with the two-sample t-test, we should first decide whether the population variances are equal. Using the F-test with $\alpha = 0.05$, we have:

$$F = \frac{s_1^2}{s_2^2} = \frac{19.88^2}{12.66^2} = 2.47 < F_{crit} = F_{0.05,7,7} = 3.79 \tag{5.30}$$

Table 5.1 Annual swimming frequencies for eight central city and eight suburban residents

| | Annual swimming frequencies | |
	Central city	Suburbs
	38	58
	42	66
	50	80
	57	62
	80	73
	70	39
	32	73
	20	58
Mean	48.63	63.63
Standard Deviation	19.88	12.66

The critical value of $F = 3.79$ comes from the F-table for $\alpha = 0.05$ (Table A.5 in Appendix A). Note from the appendix that separate F tables are given for α values of 0.01, 0.05, and 0.10. Once the α value and a table are chosen, the critical value is determined by using the column of the table that is associated with the degrees of freedom for the numerator $(n_1 - 1)$, and the row of the table that is associated with the degrees of freedom for the denominator $(n_2 - 1)$. We therefore fail to reject the assumption of equal variances. The sampling distribution of the mean has the form of a t-distribution with $n_1 + n_2 - 2 = 14$ degrees of freedom. Using the t-table (Table A.3) with 14 degrees of freedom, and a two-tailed test with $\alpha = 0.05$ implies that the critical values of t are -2.14 and 2.14. Again, the sampling distribution may be thought of as the frequency distribution resulting from many replications of the experiment (where 'experiment' is defined here as surveying eight residents from each region) under the condition that the null hypothesis is true. If the null hypothesis of no difference between central cities and suburbs is true, then 5% of the time we can expect the observed t-value to either be less than -2.14 or greater than 2.14. In those 5% of the cases, we would be making a Type I error by rejecting a true hypothesis.

Using Equation 5.26 to find $s_p = 16.67$, we next use Equation 5.25 to find the observed t-statistic:

$$t = \frac{63.63 - 48.63}{\sqrt{\dfrac{16.67^2}{8} + \dfrac{16.67^2}{8}}} = 1.8 \tag{5.31}$$

Since our observed value of t is less than the critical value of 2.14, we fail to reject the null hypothesis.

We can also find the likelihood of obtaining a result that is more extreme than the one we observed, assuming that the null hypothesis is true (i.e., the p-value). Figure 5.6 shows that the p-value associated with the test is equal to $0.0467 + 0.0467 = 0.0934$. This is found by using a t-table with 14 degrees of freedom, and finding the area to the left of -1.8 and to the right of 1.8 (see Table A.4 in Appendix A). The p-value tells us the likelihood of getting a more extreme result than the one we observed, if H_0 is true. As a reminder, low p-values (i.e., lower than α) are coincident with rejection of H_0, since they imply that it would be quite unlikely to obtain a more extreme t-statistic than the one observed, if H_0 were true. In our case, we have failed to reject the null hypothesis. The p-value gives us added information about precisely how unlikely our results are under the null hypothesis – if H_0 is true, we would expect a t-statistic with absolute value 1.8 or greater about 9.34% of the time. So we have observed a t-value that would be a bit unusual if the null hypothesis were true, but it is not unusual enough to reject the null hypothesis.

If we had chosen to use $\bar{x}_{CC} - \bar{x}_{SUB}$ instead of $\bar{x}_{SUB} - \bar{x}_{CC}$ in the numerator of Equation 5.31, we would have found $t = -1.8$ and would have come to the same conclusion and found the same p-value.

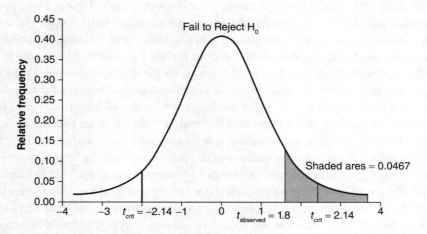

Figure 5.6 t-distribution with 14 degrees of freedom

It is interesting to see what would have happened had we not assumed the variances of the two columns of data were equal. In that case, we would have had:

$$t = \frac{63.63 - 48.63}{\sqrt{\dfrac{19.88^2}{8} + \dfrac{12.66^2}{9}}} = 1.8 \tag{5.32}$$

The observed t-value is the same, but the degrees of freedom associated with the t-distribution is now $\min(8 - 1, 8 - 1) = 7$. Consulting a t-table reveals that the critical values are now -2.36 and 2.36. Since $1.8 < 2.36$, we again fail to reject the null hypothesis. Although the conclusion is the same, note that the p-value of $0.0574 + 0.0574 = 0.1148$ (found from Table A.4) is larger than it was before. A larger p-value and an observed t-value farther from its critical value than it was before imply that we do not come as close to rejecting the null hypothesis as we did when we assumed equal variances. When the variances are not assumed equal, it is more difficult to reject the null hypothesis. This illustration emphasizes the desirability of using the homoscedasticity assumption. In fact, we could have concocted a more interesting example, where the observed value of t was equal to 2.2. Then we would have rejected H_0 under the first approach, assuming homoscedasticity, since $2.2 > 2.14$. We would not have rejected H_0 under the second approach (assuming unequal variances), since $2.2 < 2.36$.

To reiterate, the benefit that we obtain by making the assumption of homoscedasticity is an enhanced ability to reject null hypotheses when they are false (i.e., the benefit is greater statistical power).

5.3.5 Two-sample Tests: Differences in Proportions

When estimates of proportions are made from two samples taken from two identical populations, the distribution of differences in proportions is normal, with mean 0 and standard deviation equal to:

$$\hat{\sigma}_p = \sqrt{\frac{p(1 - p)}{n_1} + \frac{p(1 - p)}{n_2}} \tag{5.33}$$

where p is a pooled estimate of the true proportion:

$$p = \frac{n_1 p_1 + n_2 p_2}{n_1 + n_2} \tag{5.34}$$

Note that this pooled estimate is a weighted average of the individual proportions, where the weights are the individual sample sizes. This means that we can test null hypotheses of the form:

$$H_0 : \rho_1 - \rho_2 = 0 \tag{5.35}$$

or, equivalently,

$$H_0 : \rho_1 = \rho_2 \tag{5.36}$$

using a z-statistic

$$z = \frac{(p_1 - p_2) - (\rho_1 - \rho_2)}{\hat{\sigma}_{p_1 - p_2}} = \frac{p_1 - p_2}{\sqrt{\dfrac{p(1-p)}{n_1} + \dfrac{p(1-p)}{n_2}}} \tag{5.37}$$

Again, using $p_2 - p_1$ in the numerator of Equation 5.37 instead of $p_1 - p_2$ will yield identical conclusions.

5.3.5.1 Illustration

We are interested in knowing whether two communities have identical proportions of people who use mass transit. Before collecting the data, suppose that we expect that community A has a higher percentage of transit users than community B. The null and alternative hypotheses are:

$$H_0 : P_A - P_B = 0, \quad or \quad H_0 : P_A = P_B$$
$$H_A : P_A - P_B > 0, \quad or \quad H_A : P_A > P_B \tag{5.38}$$

We collect the following sample data:

$$p_A = 0.3 \quad n_A = 39$$
$$p_B = 0.2 \quad n_B = 50 \tag{5.39}$$

Using Equation 5.34, the pooled estimate of the proportion is:

$$p = \frac{0.3(39) + 0.2(50)}{39 + 50} = 0.244 \tag{5.40}$$

Note that the pooled estimate is approximately midway between the individual proportions of 0.2 and 0.3, but is slightly close to the estimate of 0.2 associated with community B's larger sample size. Using Equation 5.37, the z-statistic is:

$$z = \frac{0.3 - 0.2}{\sqrt{\dfrac{0.244(1-0.244)}{39} + \dfrac{0.244(1-0.244)}{50}}} = 1.09 \tag{5.41}$$

With $\alpha = 0.05$ the critical value of z, z_{crit}, for this one-sided test is 1.645. This is found from the z-table (Table A.2); 1.645 is the z-value that leaves an area of 0.05 in the right-hand tail of the distribution. Since our observed value of z (equal to 1.09) is less than this critical value, we fail to reject the null hypothesis and cannot conclude that there is a difference between the two communities. The p-value for this example is 0.138, since this corresponds to the area under the standard normal curve that is more extreme than the observed z-value. Note the fact that the p-value is greater than 0.05 is consistent with failing to reject the null hypothesis.

5.3.6 Type II Errors and Statistical Power

You may be aware that until this point, we have said little about Type II errors, where false null hypotheses are not rejected. The framework of hypothesis testing relies on the analyst making a choice for α, the probability of a Type I error. In our discussion of Figure 5.1, we learned that the smaller the value of β that we choose, the larger will be α, the probability of making a Type II error. Intuitively, when α is small, the critical values used in hypothesis testing are high, making it difficult to reject null hypotheses. We will have only a small chance of wrongly rejecting a true null hypothesis, but we will also be less likely to reject the null hypothesis when it is false. This latter quantity — the probability of rejecting a null hypothesis when it is false — is equal to $1-\beta$, and it is referred to as the *statistical power* of a test.

Many introductory texts on statistics say little else about Type II errors and statistical power; here we will explore these topics a bit further. To do so, we will focus upon one sample tests involving the mean. To study statistical power it is also helpful to focus on very specific alternative hypotheses. Suppose, for example, that our null hypothesis is H_0: $\mu = \mu_0$ and our alternative hypothesis is H_1: $\mu = \mu_1$. We are now specifying a particular value for the alternative hypothesis, unlike our previous specifications, where we simply postulate that the true mean is less than, greater than, or not equal to the mean associated with the null hypothesis.

Intuitively, statistical power will be greater, and β will be lower, when $\mu_1 - \mu_0 = \delta$ is larger. Thus if the means μ_1 and μ_0 are really very different, we will be able to correctly reject the null hypothesis when it is false. In Figure 5.7, we reject the one-sided null hypothesis when $(\bar{x} - \mu_0) / (\sigma / \sqrt{n}) > 1.645$ (using $\alpha = 0.05$). The probability that this occurs is equal to the area to the right of 1.645 in a standard normal distribution (denoted by $1-\Phi(1.645)$; the symbol Φ is used to mean the cumulative distribution of the standard normal) and this corresponds to the dark shaded area in the figure, representing 5% of the distribution on the left of the figure. This expression may be rewritten as $(\bar{x} - \mu_1) / (\sigma / \sqrt{n}) > 1.645 - \delta(\sigma / \sqrt{n})$. When the alternative hypothesis is true (i.e., when the true mean is equal to μ_1), the quantity on the left-hand side of this last expression has a

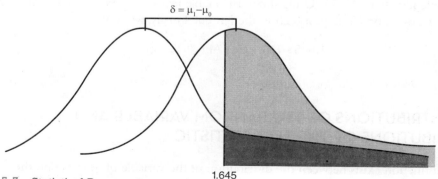

Figure 5.7 Statistical Power

standard normal distribution and the probability that it exceeds the quantity on the right is denoted by $1 - \Phi(1.645 - \delta / (\sigma / \sqrt{n}))$; it is equal to the statistical power, and it corresponds to the lightly shaded area (plus the darkly shaded area) of the distribution on the right of the figure.

For example, suppose that $\mu_0 = 15$, $\mu_1 = 16$, $\sigma = 4$, and $n = 49$. Then the power of the test — i.e., the probability that you correctly reject the null hypothesis that the mean is equal to 15 when it is really equal to 16, is equal to:

$$1 - \Phi(1.645 - (16 - 15) / (4 / \sqrt{49})) = 1 - \Phi(-0.105). \tag{5.42}$$

Rounding -0.105 to -0.11 and using the standard normal table gives the resulting power as 0.5438. The probability of making a Type II error in this situation is:

$$\beta = 1 - 0.5438 = 0.4562. \tag{5.43}$$

If this probability of a Type II error is unacceptably high, one solution is to increase the sample size. For instance if the sample size is increased to $n = 64$, the power is now equal to

$$1 - \Phi(1.645 - (16 - 15) / (4 / \sqrt{64}) = 1 - \Phi(-0.355). \tag{5.44}$$

Rounding the quantity to -0.355 to 0.36 and using the normal table reveals the power to now be 0.6406; therefore the probability of a Type II error has decreased to

$$\beta = 1 - 0.6406 = 0.3594. \tag{5.45}$$

If we wanted to reduce this probability to say $\beta = 0.25$, we could solve for the necessary sample size. We would first want to solve for x in the following:

$$1 - \Phi(x) = 0.75. \tag{5.46}$$

or equivalently, $\Phi(x) = 0.25$. Using the normal table reveals x to be equal to -0.675; it is this value that gives a probability of 0.25 in the left tail. Finally, we solve

$$1.645 - (16 - 15) / (4 / \sqrt{n}) = -0.675 \tag{5.47}$$

for n; this yields $n = 9.28^2$, or about 86.

5.4 DISTRIBUTIONS OF THE RANDOM VARIABLE AND DISTRIBUTIONS OF THE TEST STATISTIC

A key distinction exists between the distribution of the variable of interest and the sampling distribution of the test statistic. This distinction is often not fully appreciated.

Suppose that the distribution of distances traveled by park-goers from their residences to the park is governed by the 'friction of distance' effect. This effect is widely observed in many types of spatial interaction, where the distribution of trip lengths is characterized by many short trips, and relatively fewer longer ones (see Figure 5.8). If we want to test the null hypothesis that the mean trip distance is the same on weekdays as it is on weekends, we would take two samples. We might expect that both the weekday trip length distribution and the weekend trip length distribution would have shapes that are similar to the exponential distribution. To test the null hypothesis of equal mean trip lengths, we must compare the observed difference in mean distances with the sampling distribution of differences, derived by assuming H_0 to be true. The latter may be thought of as the histogram of differences in means when many samples are used to calculate many differences in means, when H_0 is true. We know from the Central Limit Theorem that the means of variables are normally distributed (given a large enough sample size), even when the underlying variables themselves are not normally distributed. We also know that the difference of two normally distributed variables is itself normally distributed. Hence, the two-sample difference of means test makes use of the fact that the sampling distribution of differences in means is normal (Figure 5.9). The important point is that there are two distributions to keep in mind – the distribution of the underlying variable (in this case exponential), and the distribution of the test statistic (in this case normal).

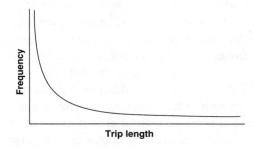

Figure 5.8   Distribution of trip lengths

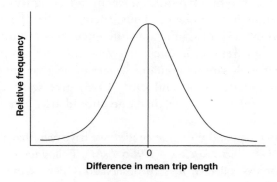

Figure 5.9   Distribution of differences in mean trip length

5.5 SPATIAL DATA AND THE IMPLICATIONS OF NONINDEPENDENCE

One of the assumptions of the statistical hypothesis tests described in this chapter is that observations are independent. This means that the observed value of one observation is not affected by the value of another observation. At first glance, this assumption sounds innocent enough, and it is tempting to simply ignore it, and hope that it is satisfied. However, spatial data are often *not* independent; the value of one observation is very likely to be influenced by the value of another observation. In the swimming example, two respondents who happen to live very near to one another might be more likely to have similar responses than two individuals who live far from one another. This could be because their accessibility to swimming pools is similar; the closer the two chosen individuals live together, the more similar is their distance to pools, and this would tend to make their swimming frequencies similar to one another. The closer two individuals live together, the more similar their incomes and lifestyles tend to be. This too would tend to yield similar swimming frequencies. This tendency for two nearby locations to yield similar values is often referred to as Tobler's First Law of Geography.

What are the consequences of a lack of independence among the observations? Because observations that are located near one another in space often exhibit similar values on variables, the effect is to reduce the effective sample size. Instead of n observations, the sample effectively contains information on less than n individuals. To take an extreme case, suppose that two individuals lived next door to one another, and 30 miles from the nearest pool. If we survey both of them, they are both likely to indicate that their swimming frequency was either zero or some very small number. The information contained in these two responses is essentially equivalent to the information contained in one response.

The implication of this is that when we carry out, for example, a two-sample t-test on observations that do not exhibit independence, our sample size is effectively smaller than we think it is, and we should really be using a critical value of t that is based on a smaller number of degrees of freedom. This in turn means that the critical value of t should be larger than the one that we use when we assume independence. Thus we should make it more difficult to reject the null hypothesis; we are rejecting too many null hypotheses if we incorrectly assume independence. Thus, there is a tendency to find significant results when in fact there are no significant differences in the underlying means of the two populations. The 'apparent' differences between the two samples can be attributed instead to the fact that each sample contains observations that are similar to each other because of spatial dependence. Cliff and Ord (1975) give some examples of this, and supply the correct critical values of t that one should use, given a specified level of dependence.

When data are independent and the variance is σ^2, we have seen that a 95% confidence interval for the mean, μ, is $(\bar{x} - 1.96\sigma / \sqrt{n}, \ \bar{x} + 1.96\sigma / \sqrt{n})$. Following Cressie's illustrative example (1993), suppose that we collected $n = 10$ observations. For example, we might collect air quality data systematically along a transect (Figure 5.10).

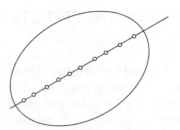

Figure 5.10 Systematic collection of data along a transect

Let us choose x_1 from a normal distribution with mean μ and variance σ^2. Then, instead of choosing x_2 from a normal distribution with mean μ and variance σ^2, choose x_2 as:

$$x_2 = \rho x_1 + \varepsilon \tag{5.48}$$

where ε comes from a normal distribution with mean 0 and variance $\sigma^2(1 - \rho^2)$, and σ is a constant between 0 and 1 indicating the amount of dependence (with $\rho = 0$ implying independence, and $\rho = 1$ implying a perfect dependence, so that $x_2 = x_1$). Cressie indicates that the variance of the mean of the x's is equal to σ^2/n only when data are independent ($\rho = 0$); more generally it is equal to:

$$\sigma_{\bar{x}}^2 = \frac{\sigma^2}{n}\left[1 + \frac{2\rho(n-1)}{n(1-\rho)} - \frac{2\rho^2(1-\rho^{n-1})}{n(1-\rho)^2}\right] \tag{5.49}$$

Cressie gives an example for $n = 10$ and $\rho = 0.26$; in this case Equation 5.49 yields $\sigma_x^2 = \frac{\sigma^2}{10}[1.608]$, implying that a correct 95% two-sided confidence interval for μ is $(\bar{x} - 2.458\sigma / \sqrt{n}, \ \bar{x} + 2.458\sigma / \sqrt{n})$. Note here that the value of 2.458 results from multiplying the usual value of 1.96 by the factor of 1.608 that emerges in this example from Equation 5.49. It is important to realize that this is *wider* than the usual confidence interval of $\bar{x} \pm 1.96\sigma$ that results from assuming independence.

If we write the variance of the mean as $\sigma_{\bar{x}}^2 = \frac{\sigma^2}{n}[f]$ where f is the inflation factor induced by the lack of independence, we can also write:

$$\sigma_{\bar{x}}^2 = \frac{\sigma^2 f}{n} = \frac{\sigma^2}{n'} \tag{5.50}$$

where $n' = n/f$ is the effective number of independent observations. With $n = 10$ and $\rho = 0.26$, $f = 1.608$ and $n' = 10/1.608 = 6.2$; this means that our 10 *dependent* observations are equivalent to a situation where we have $n' = 6.2$ independent observations.

5.6 FURTHER DISCUSSION OF THE EFFECTS OF DEVIATIONS FROM THE ASSUMPTIONS

The use of probability distributions and the implementation of hypothesis testing are predicated upon specific assumptions. When these assumptions are not met, decisions reached on the basis of assuming their validity may be called into question. This section provides some examples of how deviations from assumptions may affect conclusions reached on the basis of hypothesis testing.

5.6.1 One-sample Test of Proportions: Binomial Distribution – Assumption of Constant or Equal Success Probabilities

Use of the binomial distribution requires that each trial have the same probability of success (p). In many situations, this assumption may not be satisfied. What are the implications when this assumption is not met?

Suppose that we suspect that, for a particular neighborhood, the probability of commuting by train is greater than the region–wide proportion of 0.1. We take a small sample of ten individuals, and we find that four of the ten commute by train. The usual procedure would be to find the binomial probability of obtaining four or more such individuals under the null hypothesis that the probability of commuting by train is 0.1. If the resulting probability is very low, we would reject the null hypothesis in favor of the alternative that the actual probability of taking the train, for individuals in the neighborhood, was greater than 0.1. The reader may wish to verify that this probability is 0.0128.

Now suppose that the probability is not the same for each individual. Suppose that five of the individuals have a relatively high probability of taking the train (say 0.2; perhaps these are individuals in households with two workers and only one car), and the other five have a probability equal to zero (perhaps these are individuals with two or more cars). Note that the *average* probability among the individuals is still equal to 0.1 as before, but there is now heterogeneity among individuals. What effect does this heterogeneity have on the probability of observing four or more individuals taking the train? The probability of observing four people take the train is simply:

$$p(X = 4) = \binom{5}{4}0.2^4 0.8^1 = 0.0064 \tag{5.51}$$

since this can only occur when four of the five high-probability individuals take the train (the probability of the others taking the train is zero). Likewise, the probability that all five of the high-probability individuals taking the train is:

$$p(X = 5) = \binom{5}{5}0.2^5 0.8^0 = 0.0003 \tag{5.52}$$

Thus, the actual probability of observing four or more individuals taking the train is 0.0064 + 0.0003 = 0.0067, which is substantially lower than the previously calculated probability of 0.0128, under the assumption of homogeneous individual probabilities.

Note that the effect of heterogeneity is to lead to conservative decision-making, in the sense that the p-value is actually less than you think it is. Consequently, in situations where there is heterogeneity, and homogeneity is assumed, there will be significant relationships that are not uncovered with the usual statistical test.

5.6.2 One-sample Test of Proportions: Binomial Distribution – Assumption of Independence

Another assumption of the binomial distribution is that observations are independent; the outcome of one trial is not affected by, and does not affect, the outcome of other trials. This may be an unrealistic assumption in many actual applications, and it is especially important to consider its validity when data are collected for locations in space that may be near to one another. Whether one individual respondent takes the train may be affected by whether his or her neighbor takes the train. For instance, if one takes the train, because of the relatively high likelihood of communication among neighbors, the neighbor may be more likely than others to take the train.

Let us illustrate this using the same data used in the previous example. We interview ten individuals, and the probability of each taking the train is 0.1. We have already seen that under the usual assumptions of (a) an equal probability of success on each trial, and (b) independent trials, the binomial probability of four or more individuals taking the train is 0.0128.

Suppose, however, that the outcomes are not in fact independent. Specifically, assume that the first nine individuals make commuting decisions independently, and that the tenth person decides to commute by whatever mode the ninth person chooses (which might occur, e.g., if the ninth and tenth persons interviewed were spouses employed at the same location). Consider the possible outcomes by first enumerating the possible outcomes among the first eight individuals:

$$\binom{8}{0}0.1^0 0.9^8 = 0.4305$$

$$\binom{8}{1}0.1^1 0.9^7 = 0.3826$$

$$\binom{8}{2}0.1^2 0.9^6 = 0.1488 \tag{5.53}$$

$$\binom{8}{3}0.1^3 0.9^5 = 0.0331$$

For each of these possibilities, there are two further possibilities – the ninth and tenth people commute by train (which occurs with probability 0.1), or they do not (probability equal to 0.9). Outcomes with three or fewer people taking the train occur according to the following probabilities:

$$pr(X = 0) = \left\{ \binom{8}{0} 0.1^0 0.9^8 \right\} (0.9) = 0.3875$$

$$pr(X = 1) = \left\{ \binom{8}{1} 0.1^1 0.9^7 \right\} (0.9) = 0.3443$$

$$pr(X = 2) = \left\{ \binom{8}{2} 0.1^2 0.9^6 \right\} (0.9) + \left\{ \binom{8}{0} 0.1^0 0.9^8 \right\} (0.1) = 0.1770 \qquad (5.54)$$

$$pr(X = 3) = \left\{ \binom{8}{3} 0.1^3 0.9^5 \right\} (0.9) + \left\{ \binom{8}{1} 0.1^1 0.9^7 \right\} (0.1) = 0.0681$$

The first quantity in Equation 5.54, $pr(X = 0)$, refers to the probability that no one takes the train; none of the first eight take the train, and the ninth and tenth individuals do not take the train. The second quantity is the probability that one person takes the train – in this instance, one among the first eight individuals takes the train, and the ninth and tenth do not. There are two ways that two people could take the train, and these correspond to the two terms for $pr(X = 2)$. The first term captures the probability that two among the first eight take the train (and the ninth and tenth do not), and the second term captures the probability that none among the first eight take the train, and the ninth and tenth *do* take it. Finally, there are two ways that three of the ten could take the train: three among the first eight (and the ninth and tenth do not take it), and one among the first eight takes the train (and the ninth and tenth *do* take it). Thus, the actual probability that four or more take the train is one minus the sum of these probabilities, or $1 - (0.3875 + 0.3443 + 0.1770 + 0.0680) = 0.0232$.

The effect on decision making associated with a lack of independence is precisely the opposite of what it was in the previous example. If we proceed as if the assumption of independence is true, when it is not, we will think that the observed event is rarer under the null hypothesis ($p = 0.0128$) than it actually is ($p = 0.0232$). This may cause us to reject null hypotheses when we should not. In this illustration, if we had set $\alpha = 0.02$ as our acceptable level for a Type I error, and if we had proceeded as if the assumption of independence was true, we would have incorrectly rejected the null hypothesis (since we would have found $p = 0.0128$, when the *actual* probability of observing four or more take the train is 0.0232).

One way to gain an intuitive understanding of this is to realize that dependence among observations implies redundancy of information (as was noted in Section 5.5). In our example, we effectively have only nine independent observations, and not ten. We should not proceed as if we have ten observations; if we do, we may incorrectly reject a true null hypothesis, since it is easier to reject null hypotheses with larger sample sizes.

In this section, we have shown the consequences of incorrect assumptions. One should be aware that heterogeneity in binomial probabilities leads to conservative decision-making in the sense that the analysis will lead to p-values that are too high and the null hypothesis will not be rejected often enough; lack of independence leads to liberal over-rejection of the null hypothesis.

5.6.3 Two-sample Difference of Means Test: Assumption of Independent Observations

One assumption made in employing the two-sample t-test is that observations in each sample are independent. This assumption may often be called into question; if, for example, two soil samples are collected at nearby locations, we would not necessarily expect the measurement at one location to be unrelated to the measurement at the nearby location. To illustrate the effects of dependent observations on two-sample difference of means hypothesis tests, consider the following experiment. Fifty observations were generated for each sample; successive observations were generated from:

$$x_t = \rho x_{t-1} + \varepsilon \tag{5.55}$$

where ε is a normal variable with mean equal to 0 and variance equal to $\sigma^2(1 - \rho^2)$. The parameter ρ is a measure of dependence; it varies from zero (no dependence) to one (perfect dependence, where the next observation is equal to the previous one). By taking the first observation from a standard normal distribution, and then generating the next 49 observations using this equation, a collection of observations will be generated, and these will have a mean of zero, a variance of one, and dependence measured by ρ.

Suppose we proceed with our usual two-sample difference-of-means test in this situation. What are the consequences of assuming independence when dependence exists?

Equation 5.55 was used to generate two columns of 50 numbers; a one-sided, two-sample t-test was then carried out on the hypothetical data to test the null hypothesis of no difference in the column means (note that we know in this instance that the null hypothesis is true). For choices of $\alpha = 0.025, 0.05$, and 0.10, we have the usual critical values of 1.96, 1.645, and 1.28, respectively. The experiment was repeated 10,000 times for each value of ρ, and the percentage of times that the null hypothesis was rejected was noted.

Table 5.2 Fraction of null hypotheses rejected using usual critical values

α	.025	.05	.10
ρ			
0	.025	.054	.104
0.1	.038	.067	.121
0.2	.057	.091	.150
0.3	.078	.117	.177
0.4	.104	.142	.203
0.5	.132	.172	.226
0.6	.173	.215	.267
0.7	.209	.247	.295
0.8	.265	.299	.342
0.9	.355	.380	.408

Table 5.2 shows that if we were to employ the usual critical values, we would reject the (true) null hypothesis far more than the nominal percentage of the time. For instance, when $\rho = 0.2$, we would reject the null hypothesis 9.1% of the time, instead of the desired 5% of the time, when $\alpha = 0.05$.

Note that the over-rejection of the null hypothesis is more pronounced as the degree of dependence increases. Again, we see that the effect of ignoring dependence when it is present in the data is to reject null hypotheses more frequently than we should. Table 5.3 shows, for each ρ, the results of ranking the 10,000 observed t-statistics. In particular, the

Table 5.3 Appropriate critical values for one-sided difference of means test

α	.025	.05	.10
ρ			
0	1.97	1.68	1.30
0.1	2.16	1.81	1.41
0.2	2.44	2.04	1.58
0.3	2.69	2.26	1.77
0.4	3.09	2.55	1.99
0.5	3.52	2.95	2.24
0.6	4.15	3.41	2.65
0.7	4.97	4.11	3.17
0.8	6.30	5.27	4.07
0.9	10.64	8.84	6.72

9750th, the 9500th, and the 9000th highest *t*-values are shown; these are the critical values of *t* that *should* be used to achieve $\alpha = 0.025, 0.05$, and 0.10, respectively, for given amounts of dependence. Note that when $\rho = 0$, the entries in the table are near the critical values that come from the *t*-table. When dependence exists, higher critical values should be used.

5.6.4 Two-sample Difference of Means Test: Assumption of Homogeneity

Here we examine the consequences of assuming the variances in a two–sample *t*-test to be equal, when in fact they are not.

Sample one was created by taking 50 observations from a normal distribution with mean zero and standard deviation equal to one; sample two was created by taking 50 observations from a normal distribution with mean zero and standard deviation equal to *c*.

t-tests were carried out for the difference of means, assuming the variances to be equal.

Results are shown in Table 5.4. Heterogeneity causes slightly more null hypotheses to be rejected than desired, but the effect is not strong.

Table 5.4   Fraction of null hypotheses rejected for differing amounts of heterogeneity (*c*)

c	95th percentile	Fraction of hypotheses rejected using critical value of 1.645
2	1.670	.0527
5	1.674	.0533
10	1.664	.0526
20	1.695	.0544
200	1.661	.0523

5.7 SAMPLING

The statistical methods discussed throughout this book reply upon sampling from a larger population. The population may be thought of as the collection of all elements or individuals that are the object of our interest. The list of all elements in the population is referred to as the *sampling frame*. We may, for example, be interested in the commuting times of all individuals in a community, or in the migration distances of all people who have moved during the past year. Sampling frames may also consist of spatial elements – for example, all of the census tracts in a city. It is important to have a clear definition of the population, since this is the group about which we are making inferences. The inferences are made using information collected from a sample.

There are many ways to sample from a population. Perhaps the simplest sampling method is *random sampling*, where each of the elements has an equal probability of being selected from the population into the sample. For example, suppose we wish to take a random sample of size $n = 4$ from a population of size $N = 20$. (A common convention is to use upper case 'N' to denote population size and lower case 'n' to denote sample size.) Choose a random number from 1 to 20. Then select another random number from 1 to 20. If it is the same as the previous random number, discard it and choose another. Repeat this until four distinct random numbers, representing elements of the sampling frame, have been chosen. To illustrate, we will use the first two digits of the five-digit random numbers from Table A.1 in Appendix A. Beginning at the upper-left of the table, and proceeding down the column, the first two-digit number in the range 01–20 is 17. To complete our sample of $n = 4$, we proceed down the column and choose the next three numbers in this range – they are 04, 03, and 07. Our sample thus will consist of items 3, 4, 7, and 17 on the list of possible items to sample.

Choosing a *systematic sample* of size n begins by selecting an observation at random from among the first $[N/n]$ elements, where the square brackets indicate that the integer part of N/n is to be taken. Thus, if N/n is not an integer, one just uses the integer part of N/n. Randomly choose an element from among the first $[N/n]$ elements; call the label of this randomly chosen element k. The elements of the sampling frame that are in the sample are $k + i [N/n]$, $i = 0, 1,\ldots, n - 1$. With $N = 20$ and $n = 4$, the integer part of N/n is equal to 5. Suppose, from among the first five elements, we choose element $k = 2$ at random. The elements in the sample are items 2, $2 + 5 = 7$, $2 + 10 = 12$, and $2 + 15 = 17$. Note that it was necessary to choose only one random number.

When it is known beforehand that there is likely to be variation across certain subgroups of the population, the sampling frame may be *stratified* before sampling. For example, suppose that our $N = 20$ individuals can be divided into two groups: $N_m = 15$ men and $N_w = 5$ women. A *proportional, stratified* sampling of individuals is achieved by making the sample proportions in each strata equal. Thus, we could choose $n_m = 3$ men randomly from among the group of $N_m = 15$, and $n_w = 1$ woman randomly from the group of $N_w = 5$ women. For both men and women, the sampling proportion is 1/5.

When the sampled size of the stratum is small, it may be advantageous to obtain a *disproportional* random sample, where the small group is oversampled. In the case above, using $n_m = 2$ and $n_w = 2$ would result in unequal sample proportions, since $n_m/N_m = 2/15$ for men and $n_w/N_w = 2/5$ for women.

5.7.1 Spatial Sampling

When the sampling frame consists of all of the points located in a geographical region of interest, there are again several alternative sampling methods.

A *random spatial sample* consists of locations obtained by choosing x-coordinates and y-coordinates at random. If the region is a non-rectangular shape, x- and y-coordinates may be chosen by selecting them at random from the ranges (x_{min}, x_{max}) and (y_{min}, y_{max}), where 'min' and 'max' refer respectively to the smallest and largest coordinates in the region. If the pair of coordinates happens to correspond to a location outside of the study region, the point is simply discarded.

To ensure adequate coverage of the study area, the study region may be broken into a number of mutually exclusive and collectively exhaustive strata. Figure 5.11(a) divides a study region into a set of $s = mn$ strata. A *stratified spatial sample* of size mnp is obtained by taking a random sample of size p within each of the mn strata (Figure 5.11b). A *systematic spatial sample of size mnp* is obtained by (i) taking a random sample of size p within any individual stratum, and then (ii) using the sample spatial configuration of those p points within that stratum within the other strata (Figure 5.11c).

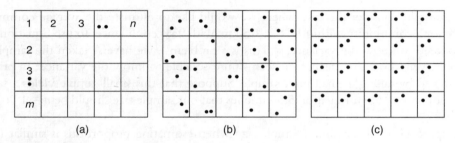

(a)   (b)   (c)

Figure 5.11 Examples of spatial sampling: (a) study region stratified into subregions; (b) stratified spatial sampling; (c) systematic spatial sampling

The question of which sampling scheme is 'best' depends upon the spatial characteristics of variability in the data. In particular, because values of variables at one location tend to be strongly associated with values at nearby locations, random spatial samples can provide redundant information when sample locations are close to one another. Consequently, stratified and systematic random sampling tend to provide better estimates of the variable's mean value. Thus, if one were to repeat the sampling many times, the variability associated with the means calculated using systematic or stratified sampling would be less than that found with random spatial sampling. Haining (1990a) discusses this in more detail, and gives references that suggest that systematic random sampling is often slightly better than stratified random sampling.

5.7.2 Sample Size Considerations

A basic question associated with sampling concerns the size of the sample. As we have seen in Section 5.2, the size of the sample will determine the accuracy of our estimate. Intuitively, larger samples will be required for more accurate estimates of means and proportions.

When estimating a mean, recall that the width of the confidence interval is, for a large sample, $\pm z_\alpha s / \sqrt{n}$. For 95% confidence intervals, this is $\pm 1.96 s / \sqrt{n}$. We can control the magnitude of this width by choosing the sample size. We do this by using the notation W for the width of the confidence interval, and then carry out two algebraic steps to solve for n:

$$W = \frac{z_\alpha s}{\sqrt{n}}$$

$$W^2 = \frac{z_\alpha^2 s^2}{n} \tag{5.56}$$

$$n = \frac{z_\alpha^2 s^2}{W^2}$$

For example, consider the case where we would like to estimate the average commuting distance for a community within $W = \pm 1.5$ km, with 95% confidence. To use Equation 5.56, we also need to know the variance, s^2. This is a problem – we haven't taken the sample yet, and don't even know the mean, so it seems unreasonable to know the variance! In practice, this is solved by taking a small pilot sample. Suppose that this small sample yields a sample variance of 58 km. Then Equation 5.56 implies that our sample size should be $n = 1.96^2(58)/(1.5^2) = 99$.

The approach to determining sample size when estimating proportions is similar to the one just described for means. From Section 5.2, the width of confidence intervals around sample proportions is $W = \pm z_\alpha \sqrt{p(1-p)/n}$. If we decide on W, we can solve for n:

$$W = \pm z_\alpha \sqrt{\frac{p(1-p)}{n}}$$

$$W^2 = \frac{z_\alpha^2 p(1-p)}{n} \tag{5.57}$$

$$n = \frac{z_\alpha^2 p(1-p)}{W^2}$$

A problem with using Equation 5.57 to determine the appropriate sample size is that we don't know p – after all, that is why we are taking the sample! However, notice from Table 5.5 that, for many possible values of p, the quantity $p(1-p)$ will be close to 0.25. We can therefore replace $p(1-p)$ in Equation 5.57 with 0.25 to obtain the equation for finding sample sizes when estimating proportions:

$$n = \frac{.25 z_\alpha^2}{W^2} = \frac{z_\alpha^2}{4W^2} \tag{5.58}$$

Table 5.5 Values of $p(1-p)$ for various values of p

p	$p(1-p)$
0.1	0.09
0.2	0.16
0.3	0.21
0.4	0.24
0.5	0.25
0.6	0.24
0.7	0.21
0.8	0.16
0.9	0.09

Note that when we desire a 95% confidence interval, this is equivalent to:

$$n = \frac{1.96^2}{4W^2} \approx \frac{1}{W^2} \tag{5.59}$$

Example

Determine the sample size required to estimate the proportion of households with two or more cars within $W = \pm 0.05$. Find the sample sizes required to achieve this objective with both 90% and 95% confidence.

Solution: Use Equation 5.58 to find a sample size of $n = 1.645^2/[4(0.05^2)] = 271$ to have 90% confidence that the true proportion is within 0.05 of the sample proportion. Use a sample size of $1.96^2/[4(0.05)^2] = 384$ to have 95% confidence. An approximate solution for the case of 95% confidence is $1/W^2 = 1/0.05^2 = 400$.

When the estimated proportion is thought to be close to zero or one, this approach is conservative in the sense that calculated sample sizes are larger than they have to be. For example, if we were estimating the proportion of people who took mass transit to work, we might expect the proportion to be no higher than 0.1. The quantity $p(1-p)$ would be equal to $0.1(0.9) = 0.09$, and if this is used in Equation 5.57 to estimate the true proportion within 0.05 with 95% confidence, we would have $n = 1.96^2 (0.09)/0.05^2 = 138$. This is substantially less than the sample size of $n = 384$ that would have been calculated using $p(1-p) = 0.25$.

Therefore, reasonable guidelines for determining the sample size when estimating proportions could be given as follows:

- First, make an educated guess regarding the true proportion.
- If the guess is within the range between 0.3 and 0.7, use $p(1-p) = 0.25$ and Equation 5.59.
- If the guess is less than 0.3, decide that the proportion could be no higher than $p*$; if the guess is greater than 0.7, decide that the proportion could be no lower than $p*$. Use $p*(1 - p*)$ in Equation 5.57 to find the sample size.

It is interesting to note that national polls do not necessarily require large sample sizes to estimate proportions accurately. If there is interest in estimating the proportion who will vote for a political candidate in a nation of many millions of people, accurate estimates can be obtained by randomly interviewing only several thousand people. To see this, use Equation 5.59 to realize that to estimate the proportion within ±0.03, the sample size should be about $1/0.03^2 = 1111$. To estimate the proportion within ±0.02, a sample size of approximately $1/0.02^2 = 2500$ is required. These calculations are independent of the national population size.

5.8 SOME TESTS FOR SPATIAL MEASURES OF CENTRAL TENDENCY AND VARIABILITY

Section 2.6 described spatial measures of central tendency and dispersion. Just as we have seen in this chapter that it is possible to construct confidence intervals around sample means, and to carry out statistical tests to test hypotheses about true or population means, it is also of interest to construct confidence intervals and carry out hypothesis tests for spatial measures. If we find that the mean center of a set of randomly chosen disease cases is in a particular location, how confident are we of our estimate of centrality? Could the mean center possibly be at some nearby location?

Statistical tests of hypotheses may be constructed using the descriptive spatial statistics from Section 2.6, together with knowledge of the expected values and variability of the quantities under particular null hypotheses.

To take a simple case, suppose the study area is square, and the coordinates are scaled so that each side has length equal to one. Suppose we wish to test the hypothesis that the true x-coordinate of the mean center is equal to 0.5 (of course, an identical test could also be carried out for the y-coordinate). A t-test may be employed:

$$t = (\bar{x} - 0.5) / (s / \sqrt{n}),\tag{5.60}$$

where x and s are the mean and standard deviation of the set of x-coordinates, respectively. This test has $n - 1$ degrees of freedom and assumes implicitly that the x-coordinates are independent and have a normal distribution. If the size of the sample is not too small the assumption of normality is not necessary and a z-test may be used.

Tests of spatial dispersion or clustering may also be carried out; these would be of interest, for example, if there was a hazardous site at the center of the study area, and it was of interest to know whether disease cases were more clustered around the center than would be expected if the disease cases were randomly located (this is clearly a simplified scenario – designed to illustrate the nature of hypothesis testing in a spatial setting, and not designed to account for other important factors, such as the distribution of the population around the hazardous site). Such tests may be implemented by using the following facts (Eilon et al. 1971):

- For a circle with radius R, the expected distance from the center to a randomly chosen point is $E[d] = 2R/3$.

- The variance of distances from the center to randomly chosen points is $V[d] = R^2/18$.

- For a square with side s, the expected distance from the center to a randomly chosen point is $E[d] = (s / 6)[\sqrt{2} + \ln(1 + \sqrt{2})] \approx 0.383s$.

- The variance of distances from the center of the square to randomly chosen points is $V[d] = 2s^2/12 - (0.383s)^2 \approx 0.02s^2$.

A simple z-test for either circular or square study areas may then be constructed as follows:

$$z = \frac{\bar{d} - E[d]}{\sqrt{V[d]/n}} \tag{5.61}$$

where $\bar{d}$ is the average distance of points from the center, and n is the number of points.

Example

For a square study area of 64 km², the average distance from the center to $n = 50$ observed locations is 2.2 km. Test the null hypothesis that points are distributed randomly around the center.

(Continued)

(Continued)

Solution: The side of the square is equal to $s = 8$ km. The expected distance from the center to a random point is $0.383(8) = 3.06$ km, and the variance of the distances from the center to points is $0.02(64) = 1.28$ km^2. The z-statistic is equal to $(2.2 - 3.06) / \sqrt{1.28 / 50} = -5.375$. Since this is less than the critical value of -1.96 (using a two-tailed test and $\alpha = 0.05$), we reject the null hypothesis, and conclude that points are closer to the center of the study area than we would have expected by chance.

Example

A circular study area has an area of 64 km^2. The average distance from the center to the 35 observed locations is 2.5 km. Test the null hypothesis that points are randomly distributed around the center.

Solution: Since the area of a circle is πr^2, we first use this to obtain the radius of the circular study area. Since $64 = \pi r^2$, $r^2 = 64/\pi$, and the radius of the circle is equal to $\sqrt{64 / \pi} = 4.514$ km. The expected distance from the center to a randomly chosen point is $2(4.514)/3 = 3.01$ km, and the variance of the distances from the center to points is equal to $4.514^2/18 = 1.132$ km^2. The z-statistic is equal to $(2.5 - 3.01) / \sqrt{1.132 / 35} = -2.84$. Again, we reject the null hypothesis that points are randomly distributed around the center.

5.9 ONE-SAMPLE TESTS OF MEANS IN *SPSS 21 FOR WINDOWS*

To test the null hypothesis that the true mean is equal to a specified value, enter the sample data in a column of the spreadsheet. Then click on `Analyze`, and then on `Compare Means`. Next, choose `One-Sample t-test`, move the name of the column variable over to the box on the right that is labeled `Test Variable`. Enter the hypothesized mean in the little box labeled `Test Value`. Then click `OK`. For example, using the data in Table 5.1, suppose we wish to test whether the true mean swimming frequency among individuals in the central city is equal to 55 days per year. Our sample data reveal a sample mean of 48.63, but we have only interviewed eight individuals and of course do not know the true mean. We want to know how unusual it would be to obtain a sample mean of 48.63, if the true mean is 55. We enter the eight central city values in the first column, and follow the instructions above; this results in the output seen in Table 5.6.

Table 5.6 *t*-test

One-sample statistics

	N	Mean	Std. deviation	Std. error mean
VAR00001	8	48.6250	19.87775	7.02785

One-sample test

	Test Value = 55					
			Sig.	Mean	95% confidence interval for the difference	
	t	df	(two-tailed)	difference	Lower	Upper
VAR00001	−.907	7	.395	−6.37500	−22.9932	10.2432

5.9.1 Interpretation

The output shows that the *t*-statistic in this example is −0.907. A key piece of the output is the *p*-value, which is given in the column headed Sig. (for significance). In this case, it is equal to 0.395; since this is greater than an alpha value of 0.05, we fail to reject the null hypothesis. A 95% confidence interval for the difference between the sample mean and the hypothesized mean is also provided. Although we observed a difference of 48.63 − 55 = −6.37, and this is our best estimate of how far away we may be from our initially hypothesized mean of 55, we are now 95% sure that the actual mean is between a value that is 22.99 less than the hypothesized value (55 − 22.99 = 32.01) and a value that is 10.24 higher than the hypothesized value (55 + 10.24 = 65.24). Note that the hypothesized mean of 55 is contained within the confidence interval of (32.01, 65.24). When the lower limit is negative and the upper limit is positive in the *SPSS* output, this implies that the hypothesized value is contained within the confidence interval, and we should not reject the null hypothesis.

5.10 TWO-SAMPLE *T*-TESTS IN *SPSS 21 FOR WINDOWS*

5.10.1 Data Entry

Suppose we wish to enter the data from Table 5.1 into *SPSS* and conduct a two-sample *t*-test of the null hypothesis that the mean annual swimming frequency among residents of the central city is equal to the mean annual swimming frequency among residents of the suburbs.

We begin by entering the data. In *SPSS*, this entails entering all of the swimming frequencies into one column. Another column contains a numeric value indicating which region the corresponding swimming frequency belongs to. For our two-region example, we would have

Swim	Location
38	1
42	1
50	1
57	1
80	1
70	1
32	1
20	1
58	2
66	2
80	2
62	2
73	2
39	2
73	2
58	2

The variable names 'Swim' and 'Location' are defined by right-clicking at the head of each column on the heading var that appears in the *SPSS* data editor. Then, under Define Variable, variable names may be assigned.

Note that here the first eight rows correspond to the data from the central city; location 1 refers to the central city. Similarly, the last eight rows contain the value '2' in the second column, and these correspond to the observations from the suburbs. In general, if there are n_1 observations on one variable and n_2 observations on the other, then there will be $n_1 + n_2$ rows and two columns once data have been entered into *SPSS*.

5.10.2 Running the *t*-Test

To run the analysis within *SPSS*, click on Analyze (Statistics in earlier versions of *SPSS for Windows*), then on Compare Means, and then on Independent Samples t-test. A box will open, and the variable Swim should then be

Table 5.7a Results of two-sample *t*-test

		N	Mean	Std. deviation	Std. error mean
Swimfreq	1.00	8	48.6250	19.8778	7.0278
	2.00	8	63.6250	12.6597	4.4759

Table 5.7b Independent samples test

		Levene's test for equality of variances		t-test for equality of means					95% confidence interval of the difference	
		F	Sig.	t	df	Sig. (two-tailed)	Mean difference	Std. error difference	Lower	Upper
SWIMFREQ	Equal variance assumed	1.776	.204	-1.800	14	.093	-15.0000	8.3321	-32.8706	2.8076
	Equal variance not assumed			-1.800	11.876	.097	-15.0000	8.3321	-33.1751	3.1751

highlighted and moved to the `Test Variable` box via the arrow tab. The variable Location is moved to the box headed `Grouping Variable` (since we are testing the variable Swim for differences by Location). Under the Grouping Variable box, click on `Define Groups`, and enter 1 for Group 1 and 2 for Group 2; these are the numeric values that *SPSS* will use from the second column of data to distinguish between groups. Then click `Continue`. Under options, the percentage associated with the confidence interval may be assigned if desired (the default is 95%). Finally, click `OK`.

An example of the output from a two-sample *t*-test is shown in Table 5.7a, which depicts the results of the test of equality of swimming frequencies in central city and suburbs using *SPSS 21 for Windows*.

First, the swimming frequencies in each region are summarized; location 1 (central city) has a mean response of 48.625 days and a standard deviation of 19.8778, while those in the suburbs apparently swim more often – the responses there have a mean of 63.625 and a standard deviation of 12.6597.

Below this are the results of the analysis. First note that there is a test of the assumption that the variances of the two groups are indeed equal. This test, Levene's test, is based upon an *F*-statistic. The key piece of output is the column headed 'Sig.', since this tells us whether to reject the null hypothesis that the two variances are equal. Since this value (which is also known as a *p*-value) is greater than 0.05, we fail to reject the null hypothesis, and conclude that the variances may be assumed equal.

The results of the *t*-test are given for both instances – one where the variances are assumed equal, and one where they are not. In both cases, the *t*-statistic is 1.8, and in both cases we do not reject the null hypothesis since the 'Sig.' column indicates a value higher than 0.05. Note that when equal variances are assumed, we come slightly closer to rejecting the null hypothesis (the *p*-value in that case is 0.093, compared with 0.097 when the variances are not assumed equal). The *p*-values differ despite identical *t*-statistics because the degrees of freedom differ.

Finally, note that the 95% confidence interval for the difference in means includes zero, indicating that the true difference between city and suburbs could be zero.

5.11 TWO-SAMPLE *T*-TESTS IN *EXCEL*

Unlike *SPSS*, here the data are entered in the more intuitive manner, where each variable represents a column. For the sample data in Table 5.1, two columns are created, each with eight observations; the first column represents the swimming frequencies for the central city, and the second column contains the frequencies for the suburbs.

Select Tools, and then Data Analysis; the user then has a choice of selecting either `t-test: Two-Sample Assuming Equal Variances`, or `t-test: Two-Sample`

Assuming Unequal Variances. With either choice, the user is then prompted in the Dialog Box to input the cell range associated with each variable. The user is also asked for the hypothesized difference. This will usually be zero, since we are most often interested in the hypothesis of no difference. Choosing the unequal variance option with the data of Table 5.1 results in the output shown in Table 5.8 (in comparison with the expedient method described earlier in this chapter, the degrees of freedom in this table have been calculated more precisely within Excel using a method beyond the scope of this text). Choosing the equal variance option results in the output shown in Table 5.9.

Table 5.8 *t*-test: two-sample assuming unequal variances

	Variable 1	Variable 2
Mean	48.625	63.625
Variance	395.125	160.2679
Observations	8	8
Hypothesized Mean Difference	0	
Df	12	
t Stat	−1.80026	
$P(T<t)$ one-tail	0.048495	
t Critical one-tail	1.782287	
$P(T>t)$ two-tail	0.096989	
t Critical two-tail	2.178813	

Table 5.9 *t*-test: two-sample assuming equal variances

	Variable 1	Variable 2
Mean	48.625	63.625
Variance	395.125	160.2678571
Observations	8	8
Hypothesized Mean Difference	0	
Df	14	
t Stat	−1.80026	
$P(T<t)$ one-tail	0.046698	
t Critical one-tail	1.761309	
$P(T>t)$ two-tail	0.093397	
t Critical two-tail	2.144787	

SOLVED EXERCISES

1. A researcher wants to estimate mean commuting distance within plus or minus two miles with 90% confidence. A small pilot study shows the standard deviation to be six miles. How big should the sample size be?

Solution. The width of a confidence interval for the mean is equal to $\pm\, z_\alpha s\sqrt{n}$. If we would like this quantity to be equal to two miles, and we know $s = 6$ and $z_\alpha = 1.645$ (associated with the 90% confidence interval), then we can solve for the sample size, n:

$$z_\alpha s / \sqrt{n} = 2 \Rightarrow \sqrt{n} = \frac{z_\alpha s}{2} \Rightarrow n = \left(\frac{z_\alpha s}{2}\right)^2 = \left(\frac{1.645(6)}{2}\right)^2 = 24.35$$

2. 24% of individuals in a sample of 100 indicate that they plan to move next year. Construct a 95% confidence interval around the sample proportion.

Solution. The 95% confidence interval is found directly from Equation 5.9:

$$0.24 \pm 1.96\sqrt{\frac{0.24(1-0.24)}{100}} = 0.24 \pm 0.837$$

3. The mean pH value of soil in an area is found to be 6.5, and this is based on a sample of 36. The sample variance was 9.0.

    (a) Find a 92% confidence interval around the sample mean.

Solution. We know that the z-value associated with a 95% confidence interval is 1.96; this is the value of z that leaves 2.5% of the area under the standard normal curve in each of the two tails. For a 92% confidence interval, we first seek the z-value that leaves 4% of the area in each tail. A check of the normal table reveals that the value of z is 1.75. Based upon Equation 5.9, the 92% confidence interval for the mean pH value is therefore:

$$6.5 \pm 1.75(3) / \sqrt{36} = 6.5 \pm 0.875$$

where we have used a standard deviation of 3 (equal to the square root of the sample variance). Note that it is not usual to compute 92% confidence intervals; 90%, 95%, and 99% confidence intervals are much more common. The present example is given primarily to reinforce concepts.

Now suppose that the statewide average pH value is 7.6.

    (b) Using the same sample data, test the null hypothesis that the true mean value in the area where the test was conducted is no different than the statewide value. State the null and alternative hypotheses, use $\alpha = 0.05$, find the test statistic and state your conclusion. Draw a diagram to show the results, including the test statistic and the critical value. What is the p-value?

Solution.

$$H_0 : \mu = 7.6$$
$$H_1 : \mu \neq 7.6$$

Test statistic

$$z = \frac{6.5 - 7.6}{3 / \sqrt{36}} = \frac{-1.1}{0.5} = -2.2$$

The critical values are equal to −1.96 and +1.96; therefore the null hypothesis is rejected. The p-value is found by looking up z = 2.2 in the normal table; the value of 0.0139 is multiplied by two (since this is a two-tailed test), to arrive at a p-value of 0.0278. Note that p<; this is consistent with rejecting the null hypothesis − the low p-value indicates that it is not very likely that a result this extreme would have been observed if the null hypothesis were indeed true. Only 2.78% of the time could we expect a sample this extreme or more extreme, if the null hypothesis was true.

(c) repeat part (b), this time assuming that the researcher has reason to believe, before the data are collected, that the mean pH in the area is lower than the statewide average.

Solution. Now we have H_1: $\mu < 7.6$. The critical value for the test statistic is z = −1.645, since this leaves 5% of the area in the left tail of the distribution. The null hypothesis is again rejected, because the test statistic of −2.2 is less than −1.645. The p-value is equal to 0.0139.

4. A sample of 14 sand grains reveals a mean diameter of 0.58 mm, and a standard deviation of 0.22 mm.

   (a) Find an 86% confidence interval for the mean sand grain size.

   (b) Test the null hypothesis that the true mean sand grain size is 0.65 mm. Find the p-value.

Solution.

(a) We use the t-distribution because the sample size is small. An 86% confidence interval has t-values that are associated with 7% of the area in each tail. Using Table A.4, with n − 1 = 13 degrees of freedom, we find that the appropriate t-value is between 1.5 and 1.6. (This t-table shows the cumulative probability, so to find the t-value associated with a tail probability of 0.07, we look for 0.93 in the body of the table.) A quick and rough interpolation reveals that t-value is a little more than halfway from 1.5 to 1.6; we could use a t-value of 1.56. The desired confidence interval is therefore $0.58 \pm 1.56(0.22) / \sqrt{14} = 0.58 \pm 0.0917$.
A more precise interpolation of the t-value could be found by asking precisely where 0.93 is situated between the t = 1.5 entry (0.92125) and the t = 1.6 entry

(Continued)

(Continued)

(0.93320). To do this, note that the difference between the two entries is 0.93320 − 0.92125 = 0.01195. The difference between 0.93 and the first entry (0.92125) is equal to 0.00875; this is about 73% of the way from 1.5 to 1.6 (since 0.00875/0.01195 is approximately 0.73). Thus, a more precise value of t would be 1.573.

(b) Since there is no indication that we expect our sample mean to be either higher or lower than the hypothesized mean, we use a two-tailed test. The test statistic is $t = (0.58 - 0.65) / (0.22) / \sqrt{14}) = -1.19$. The critical values, from Table A.3, using 13 degrees of freedom and $\alpha = 0.05$, are ± 2.15 (since this is a two-tailed test, we use the 0.025 column of the table). We therefore fail to reject the null hypothesis. Using Table A.4, again with 13 degrees of freedom, we will here first round the absolute value of the observed test statistic to $t = 1.2$, and find the entry of about 0.87. This implies that the fraction of area in the tail is 1 − 0.87, or 0.13. Finally, we double this (because we have a two-tailed test), to find $p = 0.26$. Again, a more precise result could be found by interpolation.

5. $n = 13$ samples are taken from a stream; the mean pollutant level is 16.7 mg/l, and the standard deviation is 5.1. Test the hypothesis that the true mean is no different than the 'normal' level of 14.2 mg/l. State the null and alternative hypotheses, and use a Type I error level of 0.05. Find the test statistic and its critical value, and state your conclusion. Also give the *p*-value.

Solution.

$$H_0 : \mu = 14.2$$
$$H_1 : \mu \neq 14.2$$

The alternative is two-sided, since there is no wording in the question to suggest that there is an a priori expectation that the sample mean will be lower or higher than the hypothesized mean.

Test statistic: $t = (16.7 - 14.2) / (5.1 / \sqrt{13}) = 1.767$

Critical values for this two-tailed test are found using the 0.025 column of Table A.2, with 12 degrees of freedom; this yields critical values of −2.179 and +2.179. Since the observed test statistic falls within the range of the critical values, we fail to reject the null hypothesis. The *p*-value is found by using the 12 df column of Table A.4 and roughly interpolating between the entries for $t = 1.7$ and $t = 1.8$. This yields a result of about 0.948, implying a tail area of 1 − 0.948 = 0.052. The *p*-value is twice this value (since it is a two-tailed test), so $p = 0.104$. Note that $p > \alpha$, which is consistent with failing to reject the null hypothesis.

6. A local planner believes that the average household size in a community may be significantly higher than the countywide average of 2.23. Based on a random

sample of 81 households, the planner finds an average household size of 2.61. The standard deviation of the sample data is $s = 2.20$. Using a significance level of $\alpha = 0.05$, test the hypothesis that the community's mean household size does not differ from the countywide average. State the null hypothesis, the alternative hypothesis, and the critical value of the test statistic. What is the planner's conclusion? Finally, give the p-value associated with the test statistic.

Solution.

$$H_0 : \mu = 2.23$$
$$H_1 : \mu > 2.23$$

This is a one-sided alternative hypothesis, since before the sample is taken, the planner believes that the sample mean will be higher than the countywide average.

The critical value of z is 1.645 (since this is a one-sided test). The test statistic is

$$z = (2.61 - 2.23) / (2.2 / \sqrt{81}) = 1.55$$

Decision: fail to reject the null hypothesis, since the observed value of the test statistic is less than the critical value and falls outside of the rejection region. The p-value is the area to the right of 1.55 under the standard normal curve. The normal table shows this area to be equal to 0.0606. Note that this is greater than 0.05; the test statistic is not unusual enough (when the null hypothesis is true) to reject the null hypothesis.

7. A researcher wishes to estimate the proportion of people in favor of a new highway within plus or minus 0.04, at a 95% level of confidence. How large should the sample be?

Solution. The width of a confidence interval for a proportion is $\pm z_\alpha \sqrt{p(1-p)/n}$. Setting this equal to 0.04, we can solve for the sample size, n:

$$z_\alpha \sqrt{p(1-p)/n} = 0.04$$

For a 95% confidence interval, $z_\alpha = 1.96$. We do not know the value of p, but we can still derive a conservative estimate of the sample size by realizing that the quantity $p(1 - p)$ can be no larger than 0.25. Thus:

$$1.96\sqrt{0.25/n} = 0.04$$

Squaring both sides yields 3.84 (0.25)/n = 0.0016. Then

$n = 3.84(0.25)/.0016 = 600$

Note that this is a conservative estimate in the sense that if $p<0.5$, the quantity $p(1- p)$ will be less than 0.25, yielding a sample size smaller than 600.

(Continued)

(Continued)

8. The statewide percentage of households classified as recent movers is 21%. A survey of 64 households in a neighborhood reveals that 14% may be classified as recent movers. Should one reject the null hypothesis that the neighborhood percentage is no different from the statewide percentage? State the null and alternative hypotheses, use a Type I error of 0.10, and give the *p*-value.

Solution.

$$H_0 : \rho = 0.21$$
$$H_1 : \rho \neq 0.21$$

The alternative hypothesis is set up so that it is two-sided; if there is nothing in the wording of the question to suggest that it should be one-sided, use a two-sided alternative.

Since $\alpha = 0.10$, the critical values of *z* are −1.645 and +1.645; this is a two-tailed test, with an area of 0.05 in each tail. The observed value of the test statistic is

$$z = (0.14 - 0.21) / \sqrt{0.21(1 - 0.21) / 64} = -1.375$$

Decision: Fail to reject the null hypothesis, since the test statistic falls within the range spanned by the critical values.

p-value: This is found by looking up *z* = 1.375 in the table of the normal distribution. Round to *z* = 1.38, and find 0.0838 in the table. Finally, multiply this by 2 to obtain *p* = 0.1676.

9. A researcher wishes to estimate mean length of residence within plus or minus 0.5 years, with 95% confidence. A small pilot study shows that the standard deviation is 3.5 years. How large should the sample size be?

Solution. The width of the confidence interval is $\pm z_\alpha s \sqrt{n}$. We set this quantity equal to 0.5, use $z_\alpha = 1.96$, (since it is a 95% confidence interval), substitute *s* = 3.5, and solve for *n*:

$$0.5 = 1.96(3.5) / \sqrt{n}$$

This leads to $\sqrt{n} = 1.96(3.5) / 0.5$; squaring both sides yields

$$n = \{1.96(3.5) / 0.5\}^2 = 188.$$

10. There are two alternative routes between a residential area and a downtown employment center. A transportation planner wants to know whether the mean number of cars using each route during the morning rush hour is the same. A survey over the period of $n_1 = n_2 = 30$ days reveals that the mean number of cars on Route

A is 940, and the mean number of cars on Route B is 1030. Using the results of the survey, the analyst finds $s_1 = 180$ and $s_2 = 90$. Set up the null and alternative hypotheses, use $\alpha = 0.05$, find the test statistic, and indicate whether the null hypothesis should be rejected. Also give the p-value. Assume that the variances of the two groups are the same, and use a pooled estimate of the variance.

Solution.

$$H_0 : \mu_A - \mu_B = 0$$
$$H_1 : \mu_A - \mu_B \neq 0$$

The critical values of the test statistic are $z = -1.96$ and $+1.96$, since it is a two-tailed test, with area equal to 0.025 in each tail.

$$\text{Pooled estimate of the variance: } s^2 = \frac{29(180^2) + 29(90^2)}{30 + 30 - 2} = 20{,}250$$

$$\text{Test Statistic: } z = (1{,}030 - 940) / \sqrt{\frac{20{,}250}{30} + \frac{20{,}250}{30}} = 90 / 36.74 = 2.45$$

Decision: Since the test statistic exceeds the critical value of $+1.96$, we reject the null hypothesis.

To find the p-value, we look up $z = 2.45$ in the table of the normal distribution, and find an area of 0.0071. Since this is a two-tailed test, we double this, to find $p = 0.0142$.

11. A survey of 40 individuals reveals that the proportion of people in a residential area who patronize a particular supermarket is 0.24; a repeat of the survey in the following year, using 50 individuals, finds that the proportion is 0.36. Test the null hypothesis that the proportions are equal, against the alternative that the proportion has increased over time. Set up the null and alternative hypotheses, show the rejection region (using $\alpha = 0.10$), find the test statistic, and indicate whether the null hypothesis should be rejected. Also give the p-value associated with the test statistic.

Solution. Here we have a one-sided alternative hypothesis, so that:

$$H_0 : p_1 = p_2$$
$$H_1 : p_2 > p_1$$

where p_1 is defined as the proportion in the first year, and p_2 is the proportion in the second year.

The critical value of the test statistic is $z = 1.28$, using the normal table and recognizing that we have a one-tailed test with $\alpha = 0.10$.

(Continued)

(Continued)

The pooled proportion is equal to $\{(40 \times 0.24) + (50 \times 0.36)\} / (40 + 50) = 27.6/90 = 0.3067$.

The observed test statistic is:

$$z = (0.36 - 0.24) / \sqrt{\frac{0.3067(1-.3067)}{40} + \frac{0.3067(1-.3067)}{50}}$$

and this is equal to $0.12/0.0978 = 1.227$.

Since the observed test statistic is less than the critical value, we fail to reject the null hypothesis that the proportions are equal. The p-value is found by looking up the observed value of $z = 1.23$ (after rounding) in the normal table. The resulting area gives $p = 0.1093$. This is just slightly higher than the value of α, and is consistent with failing to reject the null hypothesis.

We also could have used $p_1 - p_2$ in the numerator in the calculation of the observed test statistic. We would have found $z = -1.227$. This would be compared with a critical value of -1.28 (since we would want to place the critical region on the left side of the distribution, so that large negative values of z (i.e., cases where p_1 was much less than p_2) would lead to rejection of the null hypothesis.

12. The pollutant level is tested in two streams. The results are as follows:

$$\bar{x}_1 = 25.1 \text{ mg/l}; \ \bar{x}_2 = 15.7 \text{ mg/l}$$

$$s_1 = 14.0 \text{ mg/l}; \ s_2 = 12.2 \text{ mg/l}$$

$$n_1 = 10; n_2 = 25$$

Test the null hypothesis that the two means are equal, making sure to note the critical value. Show the rejection regions. Do not assume that the variances are equal. What is the p-value?

Solution.

$$H_0 : \mu_1 = \mu_2$$
$$H_1 : \mu_1 \neq \mu_2$$

The critical value of the test statistic, using $\alpha = 0.05$ (with areas of 0.025 in each tail), is found from a t-table, with min $(n_1, n_2) - 1 = 10 - 1 = 9$ degrees of freedom. This yields $t_{\text{crit}} = 2.262$.

The test statistic is:

$$t = \frac{25.1 - 15.7}{\sqrt{\dfrac{14.0^2}{10} + \dfrac{12.2^2}{25}}} = 1.8595$$

Note that we have to square the standard deviation, since we are given s, and need s^2. Since this observed statistic is less than the critical value, we fail to reject the null hypothesis of equal pollutant levels. The p-value is found by looking down the 9 df column of Table A.4 until you get to the rows labeled $t = 1.8$ (where the entry is 0.947) and $t = 1.9$ (where the entry is 0.955). Since these are cumulative areas, the tail areas are found by subtracting these quantities from one: $1 - 0.947 = 0.053$, and $1 - 0.955 = 0.045$. Since this is a two-tailed test, we multiply each of these by two, and conclude that the p-value must be between 0.09 and 0.106; both of these are greater than α, indicating that the observed result would not be all that unusual if the null hypothesis were true.

13. A survey of car ownership is carried out in two neighborhoods. In community 1, it is found that 22 out of 39 households own more than one car; in community 2, 18 out of 48 own more than one car. Test the null hypothesis that the communities have identical rates of multiple car ownership. Show the rejection regions. Use $\alpha = 0.10$, and give the p-value.

Solution.

$$H_0 : \rho_1 = \rho_2$$
$$H_1 : \rho_1 \neq \rho_2$$

$$p_1 = 22/39 = 0.5641; \, p_2 = 18/48 = 0.375$$

Pooled proportion: $(39 \times 0.5641 + 48 \times 0.375) / (39+48) = (18+22)/(39+48) = 0.4598$

Test statistic:

$$z = (0.5641 - 0.375) / \sqrt{\frac{0.4598(1 - 0.4598)}{39} + \frac{0.4598(1 - 0.4598)}{48}}$$

$$= 0.1891 / 0.1074 = 1.761$$

Since this observed value of z is greater than the critical value of $z = 1.645$ (based on a two-tailed test, with an area of 0.05 in each tail), we reject the null hypothesis that the true sample proportions are the same in the two neighborhoods. The p-value is found by looking up the observed value of $z = 1.76$ in the normal table; the resulting area (0.0392) is multiplied by two to yield a p-value of 0.0784.

EXERCISES

1. A planner wishes to estimate average household size for a community within 0.2. The planner desires a 95% confidence level. A small survey indicates that the standard deviation of household size is 2.0. How large should the sample be?

2. A transportation planner wishes to estimate the proportion of commuters who drive to work within 0.05. How large should the sample be, if the planner desires a confidence level of 90%?

3. The tolerable level of a certain pollutant is 16 mg/l. A researcher takes a sample of size $n = 50$, and finds that the mean level of the pollutant is 18.5 mg/l, with a standard deviation of 7 mg/l. Construct a 95% confidence interval around the sample mean, and determine whether the tolerable level is within this interval.

4. An analyst is interested in knowing whether the proportion of childless couples in a community deviates significantly from the statewide average of 0.12. The analyst suspects a priori that the proportion of childless couples in her community is higher than the statewide average. A sample of 100 couples is interviewed; the mean proportion of childless couples in the sample is found to be 0.16. Test the null hypothesis that the 'true' proportion could be 0.12 and give the p-value. Use $\alpha = 0.06$.

5. The proportion of people changing residence in the USA each year is 0.165. A researcher believes that the proportion may be different in the town of Amherst. She surveys 50 individuals in the town of Amherst and finds that the proportion who moved last year is 0.24. Is there evidence to conclude that the town has a mobility rate that is different from the national average? Use $\alpha = 0.05$ and find a 90% confidence interval around the sample proportion, and state your conclusion.

6. A political geographer is interested in the spatial voting pattern during the last presidential election. She suspects that university professors in her state were more likely than the statewide population to vote for candidate A. The statewide percentage of the population voting for candidate A was 0.38. She takes a random sample of 45 professors in the state, and finds that 20 voted for candidate A. Is there sufficient evidence to support her hypothesis? Use $\alpha = 0.05$. What is the p-value?

7. A survey of the white and nonwhite population in a local area reveals the following annual trip frequencies to the nearest state park:

$$\bar{x}_1 = 4.1, \qquad s_1^2 = 14.3, \qquad n_1 = 20$$
$$\bar{x}_2 = 3.1, \qquad s_2^2 = 12.0, \qquad n_2 = 16$$

where the subscript '1' denotes the white population and the subscript '2' denotes the nonwhite population.

    (a) Assume that the variances are equal, and test the null hypothesis that there is no difference between the park-going frequencies of whites and nonwhites.

(b) Repeat the exercise, assuming that the variances are unequal.

(c) Find the p-value associated with the tests in parts (a) and (b).

(d) Associated with the test in part (a), find a 95% confidence interval for the difference in means.

(e) Repeat parts (a)–(d), assuming sample sizes of $n_1 = 24$ and $n_2 = 12$.

8. Test the hypothesis that two communities have equal support for a political candidate using the following data:

Community A: $p_A = 0.33$ $n_A = 54$

Community B: $p_B = 0.18$ $n_B = 38$

In addition to testing the hypothesis, find the p-value.

9. A researcher suspects that the level of a particular stream's pollutant is higher than the allowable limit of 4.2 mg/l. A sample of $n = 17$ reveals a mean pollutant level of $\bar{x} = 6.4$ mg/l, with a standard deviation of 4.4 mg/l. Is there sufficient evidence that the stream's pollutant level exceeds the allowable limit? What is the p-value?

10. Information is collected by a researcher from 14 individuals on their use of rapid transit. Seven individuals were from suburb A and seven were from suburb B. The following data are the number of times per year the individual used rapid transit:

Individual	Suburb A	Suburb B	Pooled data
1	5	67	
2	12	56	
3	14	44	
4	54	22	
5	34	16	
6	14	61	
7	23	37	
Mean	22.29	43.29	32.79
Std. dev.	16.76	19.47	

Do the suburbs differ with respect to the mean number of rapid transit trips taken by individuals? Use the two-sample t-test with $\alpha = 0.05$, assuming that the variances are equal. Give the critical value of t, recalling that the degrees of freedom are equal to $n_1 + n_2 - 2$. Use $a = 0.05$. What is the p-value associated with this test?

(Continued)

(Continued)

11. The contour lines of the map below represent elevation.

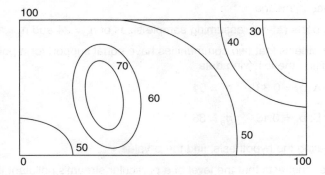

(a) Take a random spatial sample of $n = 18$ points and estimate the mean elevation of the study area.

(b) Divide the study region into a set of $3 \times 3 = 9$ strata of equal size. Take a stratified sample of size 18 by randomly choosing two points from within each strata. Estimate the mean.

(c) Using the same $3 \times 3 = 9$ strata from (b), choose a systematic random sample by first randomly selecting two points from within any individual stratum. Then use the configuration of points within that stratum to select points within the other strata (see Figure 5.11c). Estimate the mean elevation from the resulting 18 points.

Note: If answers from the entire class are pooled together, it will usually (but not always!) be the case that the means found in part (a) will display greater variability than those found in parts (b) and (c).

12. (a) A two-tailed test of a one-sample hypothesis of a mean yields a test statistic of $z = 1.47$. What is the p-value?

(b) A one-tailed test of a two-sample hypothesis involving the difference of sample means yields $t = 1.85$, with 12 degrees of freedom. What is the p-value?

13. Suppose we want to know whether the mean length of unemployment differs among the residents of two local communities. Sample information is as follows:

Community A: sample mean = 3.4 months

$s = 1.1$ month

$n = 52$

Community B: sample mean = 2.8 months

$$s = 0.8 \text{ month}$$

$$n = 62$$

Set up the null and alternative hypotheses. Use $\alpha = 0.05$. Choose a particular test, and show the rejection regions on a diagram. Calculate the test statistic, and decide whether to reject the null hypothesis. (Do not assume that the two standard deviations are equal to one another – therefore a pooled estimate of s should not be found.)

14. The mean number of children among a sample of 15 low-income households is 2.8. The mean number of children among a sample of 19 high-income households is 2.4. The standard deviations for low- and high-income households are found to be 1.6 and 1.7, respectively. Test the hypothesis of no difference against the alternative that high-income households have fewer children. Use $\alpha = 0.05$ and a pooled estimate of the variance.

15. Find the 90% and 95% confidence intervals for the following mean stream link lengths:

    100, 426, 322, 466, 112, 155, 388, 1155, 234, 324, 556, 221, 18, 133, 177, 441.

16. A researcher surveys 50 individuals in Smithville and 40 in Amherst, finding that 30% of Smithville residents moved last year, while only 22% of Amherst residents did. Is there enough evidence to conclude that mobility rates in the two communities differ? Use a two-tailed alternative, and $\alpha = 0.10$. Again, find the p-value and a 90% confidence interval for the difference in proportions.

17. A survey of two towns is carried out to see whether there are differences in levels of education. Town A has a mean of 12.4 years of education among its residents; Town B has a mean of 14.4 years. Fifteen residents were surveyed in each town. The sample standard deviation was 3.0 in Town A, and 4.0 in Town B. Is there a significant difference in education between the two towns?

    (a) Assume the variances are equal.

    (b) Assume the variances are not equal.

In each case, state the null and alternative hypotheses, and test the null hypothesis, using $\alpha = 0.05$. Find the p-values and a 95% confidence interval for the difference.

18. A sample of 45 households reveals that 23% have traveled overseas. Is this significantly different from a hypothesized value of 30%, which characterizes the entire country? Use $\alpha = 0.05$. Also, find the p-value, and a confidence interval for the observed proportion.

19. The Times of London reported on 4 January 2002 that Polish mathematicians Gliszczyński and Zawadowski recorded 140 heads in 250 spins (56%) of the

(Continued)

(Continued)

Belgian one euro coin, suggesting that it was unbalanced; this touched off an international media frenzy. The story is reported in the 11.02 issue of *Chance News*, available through www.dartmouth.edu/~chance

(a) Test the null hypothesis that the true probability of heads is equal to 0.5.

(b) Place a 95% confidence interval around the sample proportion.

20. A survey of $n = 50$ people reveals that the proportion of residents in a community who take the bus to work is 0.15. Is this significantly different from the statewide average of 0.10? Use a Type I error probability of 0.05.

On the Companion Website

For Chapter 5, the website's 'Further Resources' section contains links to other material which supplements and expands upon the text. The website further provides links to the journal papers and chapters referenced in this chapter.

6

ANALYSIS OF VARIANCE

6.1 INTRODUCTION

The two-sample difference of means test may be generalized to handle cases with more than two samples. In general, we may wish to test the null hypothesis that a set of k population means are all equal:

$$H_0 : \mu_1 = \mu_2 ... = \mu_k \tag{6.1}$$

where $k \geq 2$.

Null hypotheses of this type may possibly concern variation in means over time or space. For example, we may wish to know if traffic counts vary by month, or whether the number of weekly shopping trips made by households varies among the central city, suburban, and rural portions of a county. In the former case, we would have $k = 12$ categories, one for each month of the year. In the latter case, we would have $k = 3$, with each category representing a geographic region.

When testing the null hypothesis of equality of k population means, the data are typically given in a table such as Table 6.1, with the categories constituting the columns. The first subscript on the X variable represents the row, and the second represents the column. For example, X_{32} refers to the observation in row 3 and column 2, corresponding to the value of the 3rd observation for group 2. Note that $\bar{X}_j$ designates the mean of column j and $\bar{X}$ denotes the mean of all observations, summed over rows and columns.

Analysis of variance (ANOVA) represents a conceptual extension of the two–sample t-test for differences of means. It involves the introduction of some new ideas, though the underlying assumptions of the test are similar to those used in the two–sample t-test.

The assumptions of analysis of variance may be stated as follows:

1. Observations between and within samples are random and independent.

2. The observations in each category are normally distributed.

3. The population variances are assumed equal for each category:

$$\sigma_1^2 = \sigma_2^2 \ldots = \sigma_k^2 = \sigma^2 \tag{6.2}$$

The assumed equality of variances across categories is referred to as the assumption of *homoscedasticity* (sometimes written as *homoskedasticity*).

The idea behind the test is to compare the variation *within* columns to the variation *between* column means. If the variation between group means is much greater than the variation of observations within columns, we will be inclined to reject the null hypothesis. If, however, the variation between group means is not very large relative to the variation within columns, this

Table 6.1 Arrangement of data for analysis of variance

	Category 1	Category 2	... Category k
Obs. 1	X_{11}	X_{12}	X_{1k}
Obs. 2	X_{21}	X_{22}	X_{2k}
Obs. 3	X_{31}	X_{32}	X_{3k}
.	.	.	.
.	.	.	.
.	.	.	.
Obs. i	X_{i1}	X_{i2}	X_{ik}
.	.	.	.
.	.	.	.
.	.	.	.
No. of obs.	n_1	n_2	n_k
Mean	$\bar{X}_1$	$\bar{X}_2$	$\bar{X}_k$
Standard deviation	s_1	s_2	s_k

Overall Mean: $\bar{x}$

Overall standard deviation: s

Total sample size: $n = n_1 + \ldots + n_k$

suggests that any differences in group means may be due to sampling fluctuation, and hence we are more inclined not to reject H_0. For example, in Table 6.2, there is variability within columns; different individuals within each subregion have differing levels of participation. There is also variability between columns; the sample means in each region are different. If the between-column variability is high relative to the within-column variability, we will reject the null hypothesis and conclude that the true column means are not equal.

Although the analysis of variance test is one that tests for the equality of group means, the test itself is carried out using two independent estimates of the common variance, σ^2. One estimate of the variance is a pooled estimate of the within-group variances. The other estimate of the variance is a between-group variance. To be more specific, we may define the total sum of squares as the sum of the squared deviations of all observations from the overall mean. This total sum of squares (TSS) may be partitioned into a 'between sum of squares' (BSS) and a 'within sum of squares' (WSS). The between sum of squares is equal to the sum of squared deviations of the column means from the overall mean, where each element in the sum is weighted by the number of observations in the category. The within sum of squares is the sum of squared deviations of observations from their respective column means, and these sums are summed over all categories.

This comparison of between-column variation to within-column variation leads to an F-statistic. The partitioning of the sum of squares is as follows:

$$\text{TSS} = \sum_{j=1}^{k} \sum_{i=1}^{n_j} (X_{ij} - \bar{X})^2 = (n-1)s^2$$

$$\text{BSS} = \sum_{j=1}^{k} n_j (\bar{X}_j - \bar{X})^2$$

$$\text{WSS} = \sum_{j=1}^{k} \sum_{i=1}^{n_j} (X_{ij} - \bar{X}_j)^2 = \sum_{j=1}^{k} (n_j - 1)s_j^2$$

(6.3)

Table 6.2   Annual swimming frequencies for three regions

	Annual swimming frequencies		
	Central city	Suburbs	Rural
	38	58	80
	42	66	70
	50	80	60
	57	62	55
	80	73	72
	70	39	73
	32	73	81
	20	58	50
	48.63	63.63	67.63
Mean $\bar{X}$ = 59.96			
Standard deviation s = 16.69	19.88	12.66	11.43

The F-statistic is constructed by first dividing the between and the within sum of squares by their respective degrees of freedom (where the between sum of squares has $k - 1$ degrees of freedom, and the within sum of squares has $n - k$ degrees of freedom), and then forming a ratio of the two variables:

$$F = \frac{\text{BSS} / (k - 1)}{\text{WSS} / (n - k)}. \tag{6.4}$$

When the null hypothesis is true, this statistic has a F-distribution, with $k - 1$ degrees of freedom and $n - k$ degrees of freedom associated with the numerator and denominator, respectively. Observed values of F that are larger than the critical value lead to rejection of the null hypothesis, since ratios of between to within variation that are this large would be unusual and unexpected by chance, if the null hypothesis was true.

6.1.1 A Note on the Use of F-Tables

As we have seen in the previous section, F-statistics are based on ratios. There are degrees of freedom associated with both the numerator and the denominator. F-tables are typically arranged so that the columns correspond to particular degrees of freedom associated with the numerator, and rows correspond to particular degrees of freedom associated with the denominator. Entries in the table give the critical F-values, and the entire table is associated with a given significance level, α. Many texts give separate tables for $\alpha = 0.01, 0.05$, and 0.10; these are provided, for example, in Table A.5 in Appendix A. Because F-tables are displayed in this way, it is often difficult to state the p-value associated with the test. Stating the p-value would require a very complete set of F-tables for many more values of α. Software for statistical analysis is often useful in this regard, since p-values are usually provided in the results.

6.2 ILLUSTRATIONS

6.2.1 Hypothetical Swimming Frequency Data

We now extend the previous example on swimming frequencies (Table 5.1) to include residents of the outlying rural region, using the data in Table 6.2. We formulate the null hypothesis of no difference in the mean annual swimming frequency between the three regions:

$$\mu_{SUB} = \mu_{CC} = \mu_{R} \tag{6.5}$$

With $\alpha = 0.05$, the critical value of F is found by using $k - 1 = 2$ degrees of freedom for the numerator, and $n - k = 24 - 3 = 21$ degrees of freedom for the denominator. Table A.5 reveals that the critical value is equal to 3.47; a common notation for this is $F_{.05,2,21}$ =3.47. The observed F-statistic along with its components is given below:

$$\text{Total Sum of Squares} = \text{TSS} = 6406.96$$
$$\text{Between Sum of Squares} = \text{BSS} = 1605.33$$
$$\text{Within Sum of Squares} = \text{WSS} = 4801.63$$

$$F = \frac{1605.33 \ / \ (3-1)}{4801.63 \ / \ (24-3)} = 3.51 \tag{6.6}$$

Since the observed F value exceeds the critical value of 3.47 (found from Table A.5, using 2 and 21 degrees of freedom for the numerator and denominator, respectively), the null hypothesis is rejected.

These results are usually presented in an ANOVA table, such as the one produced by *SPSS* and shown in Table 6.3. The first column of the panel presents the sum of squares,

Table 6.3   *SPSS* output for ANOVA

Descriptives

SWIMFREQ

	N	Mean	Std. deviation	Std. error	95% confidence interval for mean Lower bound	Upper bound	minimum	maximum
1.00	8	48.6250	19.8778	7.0278	32.0068	65.2432	20.00	80.00
2.00	8	63.6250	12.6597	4.4759	53.0412	74.2088	39.00	80.00
3.00	8	67.6250	11.4260	4.0397	58.0726	77.1774	50.00	81.00
Total	24	59.9583	16.6902	3.4069	52.9107	67.0060	20.00	81.00

Test of Homogeneity of Variances

SWIMFREQ

Levene statistic	df1	df2	Sig.
1.509	2	21	.244

ANOVA

SWIMFREQ

	Sum of squares	df	Mean square	F	Sig.
Between groups	1605.333	2	802.667	3.510	.048
Within groups	4801.625	21	228.649		
Total	6406.958	23			

and the second column gives the degrees of freedom. The third column shows the results of dividing column one by column two (that is, dividing each sum of squares by its degrees of freedom); this is called the mean square. The next column gives the observed F-statistic, derived as the ratio of the two mean squares. Although the critical value of 3.47 does not appear in this table, the p-value is provided in the last column (under the heading of Sig.); since the p-value for this test is less than 0.05, we conclude that the observed F-statistic is unlikely to have occurred if the null hypothesis of equal means is indeed true.

How are the sums of squares most easily calculated? One approach is to recognize that the total sum of squares is equal to the overall variance multiplied by $n - 1$, where n is equal to the total number of observations (i.e., $n = n_1 + \cdots + n_k$). Thus, $(16.69)^2(23) = 6406.79$; the actual result of 6406.96 that appears in software output is a slightly more accurate result that is achieved by not rounding the overall variance to 16.69. Similarly, the within sum of squares for a particular column of data is equal to the variance of the observations in the column, multiplied by one less than the number of observations in the group. The within sum of squares is simply found by finding the total of these column-specific within sums of squares. Thus, $7 \times (19.88^2 + 12.66^2 + 11.43^2) = 4802.94$, which, had the individual variances not been rounded to two decimal places, would have yielded 4801.63 in statistical software. The between sum of squares is then derived as the difference between the total and within sum of squares: BSS $= 6406.96 - 4801.63 = 1605.33$.

6.2.2 Diurnal Variation in Precipitation

The effects of urban areas on temperature are well known – temperatures are generally higher in cities than in the surrounding countryside (this is known as the urban 'heat island' effect). But what about the effects of urban areas on precipitation?

One possibility is that the particulate matter ejected by urban factories forms the condensation nuclei necessary for precipitation. If this is correct, one might expect to see diurnal variation in precipitation, since factories are generally idle on the weekends. If there is no lag, precipitation would be lightest on the weekends, and heaviest during the week.

I collected the data in Table 6.4 while I was an undergraduate, in conjunction with an assignment in my geography statistics class! For each day of the week, the data are lumped into six-month categories. One consequence of this lumping is to make the assumption of normality more plausible (since sums of variables from any type of distribution tend to be normally distributed).

A look at the data reveals that the largest amount of precipitation occurs on Fridays and Sundays, and the least on Mondays and Tuesdays. Perhaps there is a roughly two-day lag between the buildup of particulate matter during the week, and the precipitation events.

Table 6.4 Precipitation data for LaGuardia airport, New York

Year	Precipitation at LaGuardia airport (inches)						
	Sat	Sun	Mon	Tue	Wed	Thur	Fri
1971 II	2.30	6.84	4.47	3.40	0.94	1.71	8.30
1972 I	5.56	6.81	1.97	2.26	3.03	4.42	5.08
1972 II	5.31	1.50	1.74	3.00	5.89	4.16	2.88
1973 I	2.15	4.39	3.96	1.17	4.35	4.78	7.09
1973 II	1.71	4.12	2.87	0.79	3.90	3.11	5.68
1974 I	2.60	2.50	1.68	1.36	0.45	4.03	5.27
Mean	3.27	4.36	2.78	2.00	3.09	3.70	5.72
Std. dev.	1.70	2.18	1.20	1.06	2.08	1.12	1.86

Overall Mean: 3.56				Overall std. dev.: 1.90			
	Sum of squares			d.f.		Variance	
Between:	51.97			6		8.663	
Within:	96.34			35		2.753	

$F = 3.15$

$F_{0.05,6,35} = 2.37; F_{0.01,6,35} = 3.37$

The null hypothesis is that mean precipitation in each six-month period does not vary with day of the week. The results of the analysis of variance, shown in the table, reveal that the null hypothesis is rejected using $\alpha = 0.05$, since the observed F-statistic of 3.15 is greater than the critical value of 2.37. The critical value is found from the table by using 6 degrees of freedom for the numerator and 35 degrees of freedom for the denominator. (Note that the null hypothesis would not be rejected with $\alpha = 0.01$, since the observed value of 3.15 is less than the critical value of 3.37; consequently, the p-value is between 0.01 and 0.05.) In the exercises at the end of the chapter, you will be asked to repeat this analysis for Boston and Pittsburgh.

6.3 ANALYSIS OF VARIANCE WITH TWO CATEGORIES

The analysis of variance with two categories ($k = 2$) gives the same results as the two-sample t-test. To illustrate, consider once again the first two columns of the swimming frequency data (i.e., the central city and suburbs). Analysis of variance yields the following:

$$BSS = 900$$

$$WSS = 3888$$

$$F = \frac{(900 / 1)}{(3888 / 14)} = 3.24$$

$$F_{.05,1.14} = 4.6$$

$$F_{0.10,1,14} = 3.10.$$

(6.7)

The null hypothesis of no difference is therefore rejected using $\alpha = 0.10$, and not rejected using $\alpha = 0.05$. The p-value must be close to, but less than, 0.10. The result here is in fact the same as that found in the last chapter using the two-sample t-test, under the assumption of equal variances. You may also note that the observed F-value (3.24) turns out to be equal to the square of the observed t-statistic (1.8); this is a relationship that always holds between the tests. When there are two categories, either the F-test or the t-test may be used; the results will be identical.

6.4 TESTING THE ASSUMPTIONS

Since the analysis of variance depends upon a number of assumptions, it is important to know whether these assumptions are satisfied. One way of testing the assumption of homoscedasticity is to use Levene's test. There are also a number of ways to test normality. Two common methods are the Kolmogorov–Smirnov test and the Shapiro–Wilk test. Although a detailed discussion of all of these tests is beyond the scope of this text, most statistical software packages do provide these tests to allow researchers to test the underlying assumptions.

6.5 CONSEQUENCES OF FAILURE TO MEET ASSUMPTIONS

What do we do if the assumptions are not satisfied? What do we do if we are not sure that the assumptions are satisfied? One option is to proceed with the analysis of variance anyway, and 'hope' that we get a conclusion that we can have confidence in. Fortunately, that is often not a bad way to proceed. The F-test is said to be relatively 'robust' with respect to deviations from the assumptions of normality and homoscedasticity. This means that the results of the F-test may still be used effectively, if the assumptions are at least 'reasonably close' to being satisfied. If either (a) the assumptions are close to being satisfied, or (b) the F-statistic yields a 'clear' conclusion (say, for example, a p-value much less than say 0.01, or greater than 0.20), the conclusion will generally be acceptable. In the next section, we investigate these assumptions in more detail.

6.6 IMPLICATIONS FOR HYPOTHESIS TESTS WHEN ASSUMPTIONS ARE NOT MET

In this section, we examine the implications for hypothesis testing when any of the three assumptions needed for ANOVA are not met. These implications are illustrated through the use of controlled numerical experiments. The experiments are conducted by making up data that are consistent with both the null hypothesis and two of the three assumptions. In this way, we know the null hypothesis to be true, and we can isolate the specific effects of each assumption, because we know the data are consistent with the other two assumptions.

6.6.1 Normality

With $k = 4$ groups and $n = 15$ observations in each group, observations for each group were taken from exponential distributions, all with mean 0.5, to assess the implication of deviations from the normality assumption. Analysis of variance was carried out on the data, and this was repeated 100,000 times. The resulting 100,000 F-values were ordered; the 95th percentile was 2.67. If we had proceeded in the usual way, the critical value of $F = 2.77$ (from the F-table, based on $k - 1 = 3$ and $N - k = 56$ degrees of freedom for numerator and denominator, respectively, using $\alpha = 0.05$) would have been exceeded 4.4% of the time. This numerical experiment was repeated for an exponential distribution with mean 0.1; the 95th percentile was 2.69, and the critical value of 2.77 from the F-table was exceeded 4.8% of the time.

In general, ANOVA is relatively robust with respect to departures from normality; in this example, the presence of non-normality leads to slightly conservative decision-making, in the sense that Type I errors will be just a little bit less likely than the chosen value of 5%.

6.6.2 Homoscedasticity

With $k = 4$ and $n = 15$, observations were chosen for each group from normal distributions with means of zero. For three of these groups, the normal distribution had variance equal to one; the fourth group had variance equal to $v > 1$; results are shown in Table 6.5. We see that heteroscedasticity has led to liberal decisions; if the tabled critical value of F is used when heterogeneity is present, Type I errors will be committed a little more often

Table 6.5   **Effects of heteroscedasticity on ANOVA hypothesis testing when H_0 is true ($\alpha = 0.05$)**

v	95th percentile of simulated F-values	Percentage of null hypotheses rejected using tabled value of $F = 2.77$
2	2.99	0.0611
3	3.22	0.0725
4	3.30	0.0760
5	3.32	0.0765
10	3.37	0.0793

than the desired value of α. Ideally, larger critical values should be used when this assumption is not satisfied (these are shown in the middle column of Table 6.5); otherwise, too many true null hypotheses will be rejected. Note, however, that again there is a good deal of robustness, and the proportion of true null hypotheses that are rejected does not exceed α by very much.

6.6.3 Independence of Observations

One of the assumptions in ANOVA is that the observations within each category are independent. With spatial data, observations are often *dependent*, and some adjustment to the analysis should be made. The general effect of spatial dependence will be to render the effective number of observations smaller than the actual number of observations. With an effectively smaller number of observations, results are not as significant as they appear in the F-tests outlined in this chapter. With spatial data, therefore, it is possible that significant findings are due to the spatial dependence among the observations, and not to any real underlying differences in the means of the categories.

Again with $k = 4$ and $n = 15$, the observations for each column were generated to come from a normal distribution and have a mean of zero and a variance of one. After the first observation was chosen in a column, successive observations in the column were chosen to be dependent upon the previous observation according to:

$$x_{i+1} = \rho x_i + \varepsilon \qquad (6.8)$$

where ε is normally distributed with mean zero and variance $(1 - \rho^2)$. Table 6.6 shows how dependence amongst the observations affects hypothesis testing. Note that greater amounts of dependence lead to higher probabilities of Type I errors; i.e., the null hypothesis may too often be rejected when it is actually true. Of the three assumptions we have examined, this one is the most sensitive. Rejection of the null hypothesis therefore may be due to dependence in the data. Dependence in spatial data is common since nearby observations

Table 6.6 Effects of dependence on ANOVA hypothesis testing when H_0 is true ($\alpha = 0.05$)

ρ	95th percentile of F	Percentage of null hypotheses rejected using tabled value of $F = 2.77$
0.1	3.32	0.0916
0.2	4.19	0.1524
0.3	5.12	0.2338
0.4	6.83	0.3341
0.5	8.82	0.4633

tend to exhibit similar values, and the possibility that rejection of a null hypothesis has occurred because of dependence of observations (instead of a false null hypothesis) should be considered.

Griffith (1978) has proposed a spatially adjusted ANOVA model. The details of his model are beyond the scope of this text. Griffith's paper may also be of interest since it contains citations to other studies in geography that also use analysis of variance.

6.7 THE NONPARAMETRIC KRUSKAL-WALLIS TEST

If the data deviate drastically from the assumptions, or if the p-value is close to α, then an alternative test that does not rely on the assumptions might be considered. Tests that do not make assumptions regarding how the underlying data are distributed are called *nonparametric* tests. The nonparametric test for testing the equality of means for two or more categories is the Kruskal–Wallis test.

There is another set of circumstances for which the Kruskal–Wallis test is useful for testing hypotheses about a set of means – namely when only ranked (i.e., ordinal) data are available. In such situations, there is insufficient information to use ANOVA, which requires interval or ratio level data. (Recall that with interval and ratio data the magnitude of the difference between the observations is meaningful.)

The application of the Kruskal–Wallis test begins by ranking the entire pooled set of n observations, from lowest to highest. That is, the lowest observation is assigned a rank of 1, and the highest observation is assigned a rank of n. The idea behind the test is that, if the null hypothesis is true, then the sum of the ranks in each column should be about the same. Again, no assumptions about normality and homoscedasticity are required (although the assumption of independent observations is retained). The test statistic is:

$$H = \left(\frac{12}{n(n+1)} \sum_{i=1}^{k} \frac{R_i^2}{n_i} \right) - 3(n+1), \tag{6.9}$$

where R_i is the sum of the ranks in category i, and n_i is the number of observations in category i. Under the null hypothesis of no difference in category means, the statistic H has a chi-square distribution, with $k - 1$ degrees of freedom. Table A.6 contains critical values for the chi-square distribution.

6.7.1 Illustration: Diurnal Variation in Precipitation

The LaGuardia airport precipitation data are ranked and displayed in Table 6.7. Also shown is the sum of the ranks, for each column. Employing Equation 6.9 yields a value of $H = 13.17$. This is just slightly higher than the critical value of 12.59, and so the null hypothesis of no variation in precipitation by day of the week is rejected at the $\alpha = 0.05$ significance

Table 6.7　Ranked precipitation data for LaGuardia airport

Ranks of observations (1 = lowest; 42 = highest)

Year	Sat	Sun	Mon	Tue	Wed	Thur	Fri
1971 II	14	40	31	22	3	8	42
1972 I	36	39	11	13	20	30	33
1972 II	35	6	10	19	38	27	18
1973 I	12	29	24	4	28	32	41
1973 II	9	26	17	2	23	21	37
1974 I	16	15	7	5	1	25	34
SUM	122	155	100	65	113	143	205

Kruskal–Wallis statistic: $H = 13.17$

Critical value: $\chi^2_{0.05,6} = 12.59$; $\chi^2_{0.01,6} = 16.81$

level. Note that the hypothesis would not have been rejected using $\alpha = 0.01$. The p-value associated with the test is approximately 0.04, meaning that if the null hypothesis were true, a test statistic this high would be observed only 4% of the time. The reader should compare this with the p-value associated with the ANOVA results. The ANOVA results yielded a p-value just slightly higher than 0.01 (we know this since the observed F-value is just slightly less than the critical F value of 3.37, using $\alpha = 0.01$). This result is a typical one – the Kruskal–Wallis test, though not relying on as many assumptions as the analysis of variance, is not as powerful. That is, it is harder to reject false hypotheses. Thus, we would have rejected H_0 with ANOVA using say $\alpha = 0.02$ or above, whereas we would only have rejected H_0 using the Kruskal–Wallis test had we chosen $\alpha = 0.04$ or above.

6.7.2 More on the Kruskal-Wallis Test

If there are values for which the ranks are tied, an adjustment is made to the value of H. Suppose that we have $n = 10$ original observations, ranked from lowest to highest: 3.2, 4.1, 4.1, 4.6, 5.1, 5.2, 5.2, 5.2, 6.1, and 7.0. There are two sets of tied observations. When the data are assigned ranks, the tied values are each assigned the average rank. Thus, the ranks of these ten observations are: 1, 2.5, 2.5, 4, 5, 7, 7, 7, 9, 10; the second and third items on the list each are assigned 2.5, and the 6th, 7th, and 8th items on the list are each assigned a rank of 7, which is the average of 6, 7, and 8. In instances where tied ranks exist, the usual value of H is divided by the quantity:

$$1 - \frac{\sum_i (t_i^3 - t_i)}{n^3 - n},\qquad (6.10)$$

where t_i is the number of observations tied at a given rank, and the sum is over all sets of tied ranks. In our present example containing ten observations, the adjustment is:

$$1 - \frac{(2^3 - 2) + (3^3 - 3)}{10^3 - 10} = 1 - \frac{30}{990} = \frac{32}{33} \tag{6.11}$$

The effect of this adjustment is to make H slightly larger, and therefore to give the Kruskal–Wallis test slightly higher power, since it is then easier to reject false hypotheses.

6.8 THE NONPARAMETRIC MEDIAN TEST

An alternative to the Kruskal–Wallis test is the median test. Like the Kruskal-Wallis test, the median test is a nonparametric test that makes no assumptions about the distribution of the data (and in the case of ANOVA, this means that it is not necessary to assume that the data in each category come from a normal distribution).

The median test is straightforward, and may be described as follows. First, 5 pool all of the data together, and rank the data from lowest to highest. For each of the k groups, categorize the observations into those that fall below the median, and those that are at or above the median. This results in a $2 \times k$ table. The results observed in this table are then compared with those expected when the null hypothesis is true. The expected values for each cell in the $2 \times k$ table are found by dividing the product of the column sum and the row sum in the observed table by the total number of observations. Then the chi-square statistic is calculated:

$$\chi^2 = \sum_{\text{all cells}} \frac{(f_{obs} - f_{exp})^2}{f_{exp}} \tag{6.12}$$

where the subscripts refer to observed and expected frequencies occupying the cells of the observed and expected $2 \times k$ tables, respectively.

When the null hypothesis is true, this statistic has a chi-square distribution with $k - 1$ degrees of freedom. An observed value of the chi-square statistic that exceeds the critical value found from the table implies that the null hypothesis should be rejected, since there are significant differences between the observed and expected frequencies. This in turn means that some of the k groups have significantly more observations that are less than the median than would be expected, while other groups have significantly fewer observations that are less than the median, in comparison with what would be expected if the null hypothesis were true.

6.8.1 Illustration

The data in Table 6.8 show the observed and expected 2×7 tables that result from categorizing the LaGuardia precipitation data into observations that are below the median, and those that are at or above the median. The observed table is constructed by starting with the data in Table 6.4 and finding the median. The median is 3.255, midway between the 21st observation (3.11) and the 22nd observation (3.4). A simple count of observations at or above the median and those below the median (i.e., with ranks of 21 or less) is carried out. For example, from Table 6.4, Saturdays have four observations less than the median, and two observations greater than or equal to the median. This results in the upper part of Table 6.8.

Table 6.8 Illustration of median test using LaGuardia precipitation data

Observed	Sat	Sun	Mon	Tue	Wed	Thu	Fri	Total
≥ median	2	4	2	1	3	4	5	21
< median	4	2	4	5	3	2	1	21
Total	6	6	6	6	6	6	6	42
Expected	Sat	Sun	Mon	Tue	Wed	Thu	Fri	Total
≥ median	3	3	3	3	3	3	3	21
< median	3	3	3	3	3	3	3	21
Total	6	6	6	6	6	6	6	42

Each element in the expected table (i.e., the lower part of Table 6.8) is found by multiplying the row sum (which in this example is always equal to 21) by the column sum (always 6 in this example), and then dividing the result by the total (42).

The data in Table 6.8 are then used to compute the chi-square statistic set out in Equation 6.13. Each element in the expected table is first subtracted from its corresponding element in the observed table, and the results are squared. Then this result is divided by the expected value. The results for each cell are added to yield the chi-square statistic:

$$\chi^2 = 4\left(\frac{(2-3)^2}{3}\right) + 4\left(\frac{(4-3)^2}{3}\right)$$

$$+ 2\left(\frac{(1-3)^2}{3}\right) + 2\left(\frac{(5-3)^2}{3}\right) + 2\left(\frac{(3-3)^2}{3}\right) = 8 \tag{6.13}$$

(There are four cells with observations of two, two cells with observations of three, four cells with observations of four, two cells with observations of one, and two cells with observations of five.)

The observed statistic of 8 is less than the critical value of 12.59 (found from using $\alpha = 0.05$ and $k - 1 = 6$ degrees of freedom in the chi-square table). Thus, we do not reject the null hypothesis. Note that we were able to reject the null hypothesis using the Kruskal–Wallis test. One could expect the median test to be less powerful than the Kruskal–Wallis test (i.e., it will be less able to reject false null hypotheses), because it uses less information. This reduction in statistical power associated with the median test is sometimes balanced by the fact that it is relatively quicker and simpler to carry out.

6.9 CONTRASTS

The analysis of variance, as a test for the equality of means, can sometimes leave the analyst with a sense of unfulfillment. In particular, if the null hypothesis is rejected, what have we learned? We've learned that there is significant evidence to conclude that the means are not equal, but we do not know *which* means differ from one another. We might look at the data and get a feel for which means seem high and which seem low, but it would be nice to have a way of testing to see whether particular combinations of categories had significantly different means. Suppose we want to know, in an example involving five categories, whether the difference between categories 2 and 5 (that is, $\mu_2 - \mu_5$) differs significantly from zero. Differences that are of interest may involve more than two separate means. For instance, with the precipitation data, we may wish to contrast weekends with weekdays. In that case, we would want to contrast the mean of the first two categories (Saturday and Sunday) with the mean of the last five weekday categories. This could be represented as:

$$\frac{\mu_{SAT} + \mu_{SUN}}{2} - \frac{\mu_{MON} + \ldots + \mu_{FRI}}{5}. \tag{6.14}$$

Scheffé (1959) described a formal procedure for contrasting sets of means with one another. A *contrast*, ψ, is defined as a combination of the population means. Usually, one defines linear combinations, so:

$$\psi = \sum_{i=1}^{k} c_i \mu_i, \tag{6.15}$$

where the values of c_i are specified by the analyst, in a manner that is consistent with the contrast of interest. In our first example, categories 2 and 5 would be contrasted with each other using the values $c_1 = 0$, $c_2 = 1$, $c_3 = 0$, $c_4 = 0$, and $c_5 = -1$. This choice of coefficients arises from writing:

$$\mu_2 - \mu_5 = 0\mu_1 + 1\mu_2 + 0\mu_3 + 0\mu_4 + (-1)\mu_5. \tag{6.16}$$

In the precipitation example, the coefficients for the weekend days would each be equal to 1/2, and the coefficients for the weekdays would each be −1/5. Why? This combination arises from realizing that Equation 6.15 can be written as:

$$\frac{\mu_{SAT} + \mu_{SUN}}{2} - \frac{\mu_{MON} + \cdots + \mu_{FRI}}{5} =$$

$$\frac{1}{2}\mu_{SAT} + \frac{1}{2}\mu_{SUN} - \frac{1}{5}\mu_{MON} - \cdots - \frac{1}{5}\mu_{FRI}. \tag{6.17}$$

Note that the sum of the coefficients (i.e., the c_i) is always equal to zero. If the original null hypothesis of no difference among the k means is rejected, then there is at least one contrast that is significantly different from zero. If the original null hypothesis of no difference among the k means is *not* rejected, then there are no contrasts that will be found significant.

6.9.1 A Priori Contrasts

Contrasts such as those described above can be either a posteriori (or post hoc) contrasts, or a priori contrasts, since they may be set up either after or before the analysis of variance test. For example, with the swimming data, we may find that the analysis of variance gives significant results, and after looking at the data, we may suspect that swimming frequencies among rural residents differ from those for everyone else. The contrast between these two groups (rural vs. remainder of population) would be a post hoc contrast. With the precipitation data, *before* conducting the analysis of variance, we may form the null hypothesis that weekend and weekday magnitudes do not differ (thinking that they indeed *might* differ). This would be an example of an a priori contrast.

When contrasts are specified prior to the analysis of variance, confidence intervals are narrower in comparison with the width of a posteriori confidence intervals. This allows more null hypotheses associated with contrasts to be rejected. Thus, it is preferable, if possible, to set up contrasts before the analysis of variance, to improve the possibility of rejecting the null hypothesis of no significant contrast.

An example illustrating the use of contrasts is given in Section 6.10.

6.10 ONE-WAY ANOVA IN *SPSS 21 FOR WINDOWS*

6.10.1 Data Entry

We will now see how the data in Table 6.2 can be analyzed using *SPSS*. As for the two-sample difference of means test, there is a separate row for each observation. Since we now have a total of 24 observations, we will have 24 rows and two columns. Again, the second column designates the group number, and now we have added a third value to correspond with the rural region. The data are entered into the Data Editor of *SPSS* as follows:

Swim	Location
38	1
42	1
50	1
57	1
80	1
70	1
32	1
20	1
58	2
66	2
80	2
62	2
73	2
39	2
73	2
58	2
80	3
70	3
60	3
55	3
72	3
73	3
81	3
50	3

6.10.2 Data Analysis and Interpretation

The analysis-of-variance proceeds in *SPSS 21 for Windows* by clicking on Analyze, then on Compare Means, and then on One-Way ANOVA. Swim (or whatever name is given to the variable in the first column) is then highlighted and moved over into the dialog box entitled Dependent List, and Location is highlighted and moved over into the dialog box entitled Factor. At this point, one can simply click OK to proceed with the analysis, but here we will also click on Options, and then check the boxes entitled Descriptive and Homogeneity-of-Variance. Also, post hoc contrasts can be made by simply clicking on Post Hoc, and then clicking on the box labeled Scheffé. A priori contrasts are chosen by clicking on the box labeled Contrasts. Suppose we wish to contrast swimming frequency in the central city with the average swimming frequency in the other two regions. After choosing Contrasts, click on Polynomial, and leave Linear as the selected polynomial. We then need to specify the contrast coefficients (the c's in Equation 6.15). Here, we could use either $c_1 = 1$, $c_2 = -0.5$, and $c_3 = -0.5$ (representing the quantity $\bar{x}_1 - (\bar{x}_2 + \bar{x}_3)/2$), or $c_1 = -1$, $c_2 = 0.5$, and $c_3 = 0.5$ (representing the quantity $(\bar{x}_2 + \bar{x}_3)/2 - \bar{x}_1$). Enter the coefficients one at a time, clicking on Add after each entry. Finally, choose Continue, and then OK. The output that results is shown in Tables 6.3 and 6.9.

The first box in Table 6.3 provides descriptive information on the variable in each region. Note that the mean frequency among respondents in the rural region is higher (67.625) than that in other regions, and its standard deviation is lower (11.426).

The second box gives us the results of a test of the assumption of homoscedasticity. Levene's test supports the null hypothesis that the variances of the three region's responses could be equal (since the column headed 'Sig.' has an entry greater than 0.05), and that we have merely observed sampling variation. Had the p-value associated with this test been less than 0.05, we would have had to take the results of the analysis of variance more cautiously, since one of the underlying assumptions would have been violated.

The next box displays the results of the analysis of variance. The table gives the sums of squares, the mean squares, the degrees of freedom, and the F-statistic. Note that these match the results discussed previously (as they should!). Importantly, the output also includes the p-value associated with the test under the column labeled 'Sig'. Since this value is less than 0.05, we reject the null hypothesis, and conclude that there are significant differences in swimming frequencies among the residents of these three regions, and these differences cannot be attributed to sampling variation alone (unless we just happened to get a fairly unusual sample).

The results of the a priori contrasts are shown in Table 6.9 and they indicate that there is indeed a significant difference between the swimming frequencies in the central city and other areas. The value of the contrast is 17, which is the mean difference in swimming frequencies (65.63 − 48.63). The significance or p-value is indicated in the last column, and this is less than 0.05 when variances are assumed equal (and equal to 0.051 when variances are not assumed equal). Results of the post hoc contrasts indicate that there is one paired difference that is close to significant – that between the central city (region 1) and rural region (region 3). This is indicated by the p-value (in the column headed 'Sig.') of 0.063. Confidence intervals for the difference in swimming frequencies associated with each paired comparison are also given. Note that all of these confidence intervals include zero, implying insignificance.

6.11 ONE-WAY ANOVA IN *EXCEL*

As is the case with the two–sample *t*-test, the input format is more intuitive with *Excel* than it is with *SPSS*; one simply enters the observations for each variable in a separate column. For example, the data in Table 6.2 would be entered in three columns, each with eight entries, and with the completed spreadsheet appearing exactly as it does in Table 6.2. Once the data are entered, select `Tools: Data Analysis`, and `Anova: Single Factor`. Next, the cells used for input are specified in the `Input Range` box (for example, with the data in Table 6.2 entered into a spreadsheet, one would type A1:C8, reflecting the fact that the data have been entered into the first eight rows and the first three columns of the spreadsheet).

Table 6.9 SPSS output for contrast tests

Contrast tests

		Contrast	Value of contrast	Std. error	*t*	df	Sig. (2-tailed)
VAR00001	Assume equal variances	1	17.0000	6.5476	2.596	21	.017
	Does not assume equal variances	1	17.0000	7.6471	2.223	9.648	.051

Post Hoc Tests

Multiple comparisons

Dependent Variable: VAR00001

Scheffé

(I) VAR00002	(J) VAR00002	Mean difference (I–J)	Std. error	Sig.	95% confidence interval Lower bound	95% confidence interval Upper bound
1.00	2.00	−15.0000	7.5606	.165	−34.9083	4.9083
	3.00	−19.0000	7.5606	.063	−38.9083	.9083
2.00	1.00	15.0000	7.5606	.165	−4.9083	34.9083
	3.00	−4.0000	7.5606	.870	−23.9083	15.9083
3.00	1.00	19.0000	7.5606	.063	−.9083	38.9083
	2.00	4.0000	7.5606	.870	−15.9083	23.9083

Table 6.10 shows the output format used for reporting the results.

Table 6.10   Excel output for ANOVA

Anova: single factor

SUMMARY

Groups	Count	Sum	Average	Variance
Column 1	8	389	48.625	395.125
Column 2	8	509	63.625	160.2679
Column 3	8	541	67.625	130.5536

ANOVA

Source of variation	SS	df	MS	F	P-value	F crit
Between groups	1605.333	2	802.6666667	3.510478	0.04838732	3.46679485
Within groups	4801.625	21	228.6488095			
Total	6406.958	23				

SOLVED EXERCISES

1. Data are collected from four separate communities on the distance (in miles) that individuals live from their parents. The results are as follows:

Community:	1	2	3	4	
Sample size	15	15	20	20	Total: 70
Mean	30	40	20	20	overall mean: 26.42
Std. dev.	20	10	10	20	overall std dev.: 17.49

Complete the analysis of variance (ANOVA) table, find the critical value, and decide whether to reject the null hypothesis that there is no difference in mean distance across the four communities.

Solution. The total sum of squares is based upon the overall variance and the sample size, and is equal to $(n - 1)s^2 = 69(17.49^2) = 21,107$. The within sum of squares is equal to this same quantity as determined for each category, and then summed over categories; thus, the within sum of squares is equal to $(15 - 1)(20^2) +$ $(15 - 1)(10^2) + (20 - 1)(20^2) + (20 - 1)(10^2) = 16,500$. The between sum of squares is equal to the difference between the total and within sum of squares, and is therefore equal to $21,107 - 16,500 = 4,607$. Note that the between sum of squares may also be found by using Equation 6.3: $15(30-26.42)^2 + 15(40-26.42)^2 + 20(20-26.42)^2 + 20(20-26.42)^2 = 4,607$. The degrees of freedom is equal to the number of categories (4), minus 1, for the between sum of squares, and is equal to the total sample size (70) minus one, for the total sum of squares. The within sum of squares has degrees of freedom equal to the sample size (70), minus the number of categories (4). Mean squares are found by dividing the sum of squares by the appropriate degrees of freedom; thus the between mean square is $4,607/3 = 1,535.7$, and the within mean square is equal to $16,500/66 = 250$. Finally, the F-statistic is the ratio of these two results: $F = 1,535.7/250 = 6.14$.

	Sum of squares	df	Mean square	F
Between	4,607	3	1,535.7	6.14
Within	16,500	66	250	
Total	21,107	69		

From the F-table, the critical value, using $\alpha = 0.05$ with 3 and 66 degrees of freedom for the numerator and denominator, respectively, is about 2.75 (from Table A.5). Since the observed statistic (6.14) is greater than the critical value, we reject the null hypothesis and conclude that there are differences in the mean distance across the four communities.

2. A one-way ANOVA test for the equality of means is carried out. There are six categories. There are 48 total observations, and the standard deviation of these observations is 50. The between sum of squares is 25,000. Fill in the ANOVA table. Find and state the critical value of F from the table. Do you reject the null hypothesis that the means of the categories are equal?

Solution. Since the variance is equal to the total sum of squares, divided by $n-1$, the total sum of squares is equal to the variance, multiplied by $n-1$. Thus the total sum of squares is equal to $50^2(48-1) = 117,500$. Since we are given that the between sum of squares is equal to 25,000, the within sum of squares must be equal to $117,500 - 25,000 = 92,500$. Degrees of freedom for the total sum of squares are always equal to $n-1$. For the between sum of squares, the degrees of freedom are equal to $k-1$, where k is the number of categories. Thus, here we have df = $6-1 = 5$. For the within sum of squares, degrees of freedom are equal to $N-k$, or $48-6 = 42$. The mean squares are found by dividing the respective sums of squares by their degrees of freedom. Thus, $25,000/5 = 5000$, and $92,500/42 = 2202.38$. Finally, the observed F-statistic is found by dividing the between mean square by the within mean square. Thus $F = 5000/2202.38 = 2.27$.

(Continued)

(Continued)

	Sum of squares	df	Mean square	F
Between	25,000	5	5,000	2.27
Within	92,500	42	2,202.38	
Total	117,500	47		

The critical value is found from Table A.5. This table is arranged using different values of α; here we will use the $\alpha = 0.05$ table. We use the column headed by df = 5 (these are the degrees of freedom for the numerator) and we then go down to the row headed 'df = 40', since it is closest to the desired value of df = 42. We find the value to be 2.45; the actual critical value is a little less than this, since note that the entry for df = 60 is 2.37. If desired, we could interpolate between the two values to obtain a more precise value. Since our observed statistic is less than the critical value, we fail to reject the null hypothesis that all means are equal.

EXERCISES

1. Using the following data: Precipitation at Boston Airport (Inches)

Year	Sat	Sun	Mon	Tue	Wed	Thur	Fri
1971 II	0.83	3.14	4.20	1.28	1.16	4.25	2.08
1972 I	4.66	4.15	3.40	1.74	3.91	5.15	5.06
1972 II	3.03	5.80	2.29	3.17	3.50	3.40	3.04
1973 I	3.69	3.72	4.29	2.06	3.04	2.30	4.26
1973 II	2.35	3.62	3.56	2.27	4.46	2.52	3.36
1974 I	3.18	3.28	1.82	3.75	2.07	3.54	2.27

(a) Find the mean and standard deviation for each day of the week.

(b) Use *SPSS* to carry out Levene's test to determine whether the assumption of homoscedasticity is justified.

(c) Perform an analysis of variance to test the null hypothesis that precipitation does not vary by day of the week. Show the between and within sum of squares, the observed *F*-statistic, and the critical *F*-value.

(d) Repeat the analysis using data for Pittsburgh:

Precipitation at Pittsburgh Airport (Inches)

Year	Sat	Sun	Mon	Tue	Wed	Thur	Fri
1971 II	1.64	5.55	3.19	2.45	1.44	1.07	1.66
1972 I	2.20	3.37	0.78	2.63	2.32	5.57	2.80
1972 II	2.75	1.72	2.34	3.40	3.68	3.48	2.50
1973 I	2.23	4.31	2.02	1.83	4.35	4.07	2.66
1973 II	3.65	2.66	3.95	2.31	1.85	2.63	1.11
1974 I	4.96	3.00	2.61	1.75	2.70	2.45	4.06

2. Assume that an analysis of variance is conducted for a study where there are $N = 50$ observations and $k = 5$ categories. Fill in the blanks in the following ANOVA table:

	Sums of squares	Degrees of freedom	Mean square	F
Between	_____	_____	116.3	_____
Within	2000	_____	_____	
Total	_____	_____		

With $\alpha = 0.05$, what is your conclusion regarding the null hypothesis that the means of the categories are equal?

3. What are the assumptions of analysis of variance? What does it mean to say that analysis of variance is relatively robust with respect to deviations from the assumptions? What does it mean to say that the Kruskal–Wallis test is not as powerful as ANOVA?

4. Fill in the blanks in the following analysis of variance table. Then compare the F-value with the critical value, using $\alpha = 0.05$.

	Sum of squares	df	Mean square	F
Between SS	34.23	2	_____	_____
Within SS	_____	__	_____	
Total SS	217.34	35		

(Continued)

(Continued)

5. Using

$$\text{WSS} = \sum_{i=1}^{k} (n_i - 1)s_i^2$$

(a) Find the within sum of squares for the following data:

Toxin levels in shellfish (mg)

Observation	Long Island Sound	Great South Bay	Shinnecock Bay
1	32	54	15
2	23	27	18
3	14	18	19
4	42	11	21
5	13	10	28
6	22	34	9
Mean	24.33	25.67	18.33
Std. dev.	11.08	16.69	6.31
Overall mean: 22.78		Overall std. dev.: 11.85	

(b) Find the value of the ANOVA test statistic and compare it with the critical value. Use $\alpha = 0.05$.

(c) Rank the data (1 = lowest), using the average of the ranks for any set of tied observations. Find the Kruskal–Wallis statistic

$$H = \left(\frac{12}{n(n+1)} \sum_{i=1}^{n} \frac{R_i^2}{n_i} \right) - 3(n+1),$$

and then adjust the value of H by dividing it by

$$1 - \frac{\sum_i (t_i^3 - t_i)}{n^3 - n},$$

where t_i is the number of observations that are tied for a given set of ranks. Compare this test statistic with the critical value of chi-square (with $\alpha = 0.05$), which has $k - 1$ degrees of freedom to decide whether to reject the null hypothesis.

6. Is there a significant difference between the distances moved by low- and high-income individuals? Twelve respondents in each of the income categories are interviewed, with the following results for the distances associated with residential moves:

Respondent	Low income	High income
1	5	25
2	7	24
3	9	8
4	11	2
5	13	11
6	8	10
7	10	10
8	34	66
9	17	113
10	50	1
11	17	3
12	25	5
Mean	17.17	23.17
Std. dev.	13.25	33.45

Test the null hypothesis of homogeneity of variances by forming the ratio s_1^2 / s_2^2, which has an F-ratio with $n_1 - 1$ and $n_2 - 1$ degrees of freedom. Then use ANOVA (with $\alpha = 0.10$) to test whether there are differences in the two population means. Set up the null and alternative hypotheses, choose a value of α and a test statistic, and test the null hypothesis. What assumption of the test is likely not satisfied?

7. Are the confidence intervals associated with a priori contrasts in ANOVA narrower or wider than a posteriori contrasts? Why? Which would be more powerful in rejecting the null hypothesis that the contrast was equal to zero?

8. A study classifies 72 observations into nine groups, with eight observations in each group. The study finds that the variance among the 72 observations is 803. Complete the following ANOVA table:

	Sum of squares	df	Mean square	F
Between	6000	—		—
Within	___	—	___	
Total	___			

What do you conclude about the hypothesis that the means of all groups are equal? Assume $\alpha = 0.05$. What can you conclude about the p-value?

(Continued)

(Continued)

9. A sample is taken of incomes in three neighborhoods, yielding the following data:

	Neighborhood			Overall (combined sample)
	A	B	C	
N	12	10	8	30
Mean	43.2	34.3	27.2	35.97
Std. dev.	36.2	20.3	21.4	29.2

Use analysis of variance (with $\alpha = 0.05$) to test the null hypothesis that the means are equal.

10. Use the Kruskal–Wallis test (with $\alpha = 0.05$) to determine whether you should reject the null hypothesis that the means of the four columns of data are equal:

Col 1	Col 2	Col 3	Col 4
23.1	43.1	56.5	10002.3
13.3	10.2	32.1	54.4
15.6	16.2	43.3	8.7
1.2	0.2	24.4	54.4

11. A researcher is interested in differences in travel behavior for residents living in four different regions. From a sample of size 48 (12 in each region), she finds that the mean commuting distance is 5.2 miles, and that the standard deviation is 3.2 miles. What is the total sum of squares? Suppose that the standard deviations for each of the four regions are 2.8, 2.9, 3.3, and 3.4. What is the within sum of squares? Fill in the table:

	Sum of squares	df	Mean square	F
Between	_____	__	_____	_____
Within	_____	__	_____	
Total	_____			

12. A researcher wishes to know whether distance traveled to work varies by income. Eleven individuals in each of three income groups are surveyed. The resulting data are as follows (in commuting miles, one-way):

Observations	Income		
	Low	Medium	High
1	5	10	8
2	4	10	11
3	1	8	15
4	2	6	19
5	3	5	21
6	10	3	7
7	6	16	7
8	6	20	4
9	4	7	3
10	12	3	17
11	11	2	18

Use analysis of variance (with $\alpha = 0.05$) to test the hypothesis that commuting distances do not vary by income. Also evaluate (using, e.g., *SPSS* and the Levene test) the assumption of homoscedasticity. Finally, lump all of the data together and produce a histogram, and comment on whether the assumption of normality appears to be satisfied.

13. Data are collected on automobile ownership by surveying residents in central cities, suburbs and rural areas. The results are:

	Central cities	Suburbs	Rural areas
Number of observations	10	15	15
Mean	1.5	2.6	1.2
Std. dev.	1.0	1.1	1.2
Overall mean: 1.725			
Overall std. dev.: 1.2			

Test the null hypothesis that the means are equal in all three areas.

(Continued)

(Continued)

14. Using *SPSS*, with $\alpha = 0.05$, use the data from exercise 1 to:

    (a) carry out an a priori test of the hypothesis that weekend precipitation differs from weekday precipitation in Boston.

    (b) Repeat (a), using the data for Pittsburgh.

    (c) Repeat (a) and (b), but use a two-day lag (so that the null hypothesis is that precipitation on Monday and Tuesday is no different from precipitation for other days in the week).

15. Use the Hypothetical UK Housing Prices dataset to determine whether the floor area of a home varies with the era it was constructed (use the variables 1940s, 1950s, 1960s, 1970s, 1980s, and 1990s to define the categories).

On the Companion Website

The **Hypothetical UK Housing Prices** dataset required for the exercises of Chapter 6 is available within the 'Datasets' section of the website, along with information about its fields and format. Further links to other resources and texts are provided in the 'Further Resources' section.

7

CORRELATION

7.1 INTRODUCTION AND EXAMPLES OF CORRELATION

A common objective of researchers is to determine whether two variables are associated with one another. Does patronage of a public facility vary with income? Does interaction vary with distance? Do housing prices vary with accessibility to major highways? Researchers are interested in how variables *co-vary*. The concept of *covariance* is a straightforward extension of the concept of variance. Whereas the sample variance is based upon the squared deviation of observations of a single variable from their mean, the sample covariance is based upon the product of the two variables' (say X and Y) respective deviations from their means:

$$\text{Cov}(X, Y) = \frac{\sum\limits_{i=1}^{n}(x_i - \overline{x})(y_i - \overline{y})}{n - 1} \tag{7.1}$$

Note that the covariance of a variable with itself may be found by replacing the y's with x's, and this is simply the formula for the variance.

The covariance of X and Y may be negative or positive. In Figure 7.1, observations consisting of x,y pairs are plotted. The axes represent the mean of each variable, and these are used to define four quadrants. The covariance will be positive if most of the points lie in quadrants I and III, and will be negative when most of the points lie in quadrants II and IV. In Figure 7.1a,

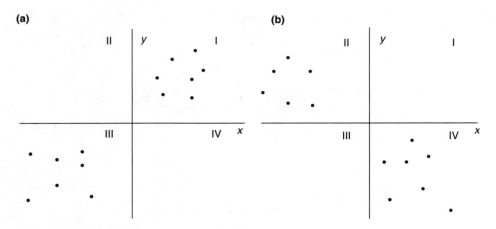

Figure 7.1 Scatterplots illustrating (a) positive and (b) negative correlation

the covariance is positive; x- and y-coordinates are either simultaneously above their means (quadrant I) or simultaneously below their means (quadrant III). Both quadrant I and quadrant III points make positive contributions to the numerator of Equation 7.1. In Figure 7.1b, the covariance is negative; when x is above its mean, y is below its mean (these are the quadrant IV points), and when x is below its mean, y is above its mean (these are the quadrant II points). Both quadrant II and quadrant IV points make negative contributions to the numerator of Equation 7.1, since a positive value is multiplied by a negative value.

The magnitude of the covariance will depend upon the units of measurement. The covariance may be standardized so that its values lie in the range from −1 to +1 by dividing by the product of the standard deviations. This standardized covariance is known as the *correlation coefficient*. The correlation coefficient provides a standardized measure of the linear association between two variables. The sample correlation coefficient, r, may be found from

$$r = \frac{\sum\limits_{i=1}^{n}(x_i - \bar{x})(y_i - \bar{y})}{(n-1)s_x s_y} \tag{7.2}$$

where s_x and s_y are the sample standard deviations of variables x and y, respectively. This is known as *Pearson's* correlation coefficient. Note that this is equivalent to

$$r = \frac{\sum\limits_{i=1}^{n} z_x z_y}{n-1}, \tag{7.3}$$

where z_x and z_y are the z-scores associated with x and y, respectively (the z-scores are created by subtracting the mean from each observation, and then dividing the result by the standard deviation).

It is important to note that the correlation coefficient is a measure of the strength of the *linear* association between variables. Points lying precisely along a straight line with positive slope will have a correlation of −1, while points lying precisely along a line with negative slope will have a correlation of +1. Points that are randomly scattered on the graph will have a correlation close to zero. However, as Figure 7.2 demonstrates, a correlation of zero does not necessarily mean that x and y are not related — it simply means that they are not related in a linear fashion. The figure reinforces the fact that it is possible to have a strong, *nonlinear* association between two variables, and yet have a correlation coefficient close to zero. One implication of this is that it is important to plot data (the term *scatterplot* is often used to refer to graphs such as Figures 7.1 and 7.2, where each observation is represented by a point in the plane, and where the two axes represent the levels of the two variables), since potential associations between the variables might be revealed in those cases where the value of r is low.

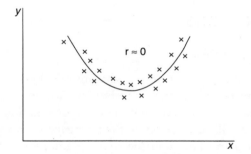

Figure 7.2   **Nonlinear relationship with r approximately equal to zero**

It is also important to realize that the existence of a strong linear association does not necessarily imply that there is a *causal* connection between the two variables. A strong correlation was once found between British coal production and the death rate of penguins in the Antarctic, but it would be a stretch of the imagination to connect the two in any direct way! Changes in both British coal production and the death rate of penguins in the Antarctic happened to go in the same direction over a period of time, but this does not necessarily imply a causal connection between the two. Another article once pointed out the strong connection between the annual number of tornadoes and the volume of automobile traffic in the United States. The claim was that both the number of tornadoes and the volume of automobile traffic had steadily increased in the United States throughout the 20th century. If for each year traffic was used as the x-variable, and the number of tornadoes was used for the y-variable, a very strong positive correlation would be observed; years with many tornadoes would coincide with years with a high volume of traffic. Strong linear relationships often prompt deep thought about possible explanations, and in this case an explanation (presumably made in jest) was offered. The correlation was deemed to be due to the fact that Americans drive on the right-hand side of the road! As cars pass one another, counterclockwise movements of air are generated, and we all know that counterclockwise movements of air are associated with low-pressure systems. Some of these low-pressure systems spawn tornadoes. Increasing traffic, then, would understandably lead to more tornadoes. Furthermore, since the British drive on the left, it should come as no surprise that there are not many tornadoes there! (Though I doubt one could claim that they have the great weather one would expect from the high-pressure systems created by the clockwise movement of traffic-generated air currents!) A better explanation of the relationship is that the two variables have both increased over time, but for very different reasons. The increase in the number of tornadoes is likely due to the simple fact that the weather observation network is better than it used to be. The search for a causal relationship is an important one, but the effort may sometimes be carried too far!

7.2 MORE ILLUSTRATIONS

7.2.1 Mobility and Cohort Size

Easterlin (1980) suggested that young adults who are members of a large cohort (like the baby boom) will face a more difficult time in labor and housing markets. For these cohorts, the supply of people is large, relative to the number of job and housing opportunities available. Consequently, there will be a tendency for mortgage rates and unemployment to be higher when large cohorts pass through their young adult years. Similarly, mortgage and unemployment rates will tend to be lower when small cohorts reach their twenties and thirties. Rogerson (1987) extended this argument to hypothesize that large cohorts of young adults will exhibit lower mobility rates, since the cohort's opportunities for changing residence will be limited by the relatively inferior state of the labor and housing markets. The mobility rate is measured as the percentage of individuals changing residence during a one-year period, and the size of the young adult cohort is measured by the fraction of the total population that is in a specified young adult age group. Annual data on these variables for the period 1948–84 are presented in Table 7.1.

Table 7.1 US mobility data, 1948–1984

	Mobility rate		Fraction of total population	
Year	20–24	25–29	20–24	25–29
1948–49	35.0	*	.0804	.0829
1949–50	34.0	*	.0784	.0821
1950–51	37.7	33.6	.0767	.0812

Year	Mobility rate		Fraction of total population	
	20–24	25–29	20–24	25–29
1951–52	37.8	31.6	.0746	.0794
1952–53	40.5	33.4	.0720	.0774
1953–54	38.1	30.5	.0757	.0762
1954–55	41.8	31.3	.0735	.0758
1955–56	44.5	32.3	.0713	.0750
1956–57	41.2	32.0	.0694	.0734
1957–58	42.6	34.6	.0671	.0715
1958–59	42.5	33.2	.0645	.0701
1959–60	41.2	32.1	.0617	.0682
1960–61	43.6	34.4	.0616	.0605
1961–62	43.2	33.0	.0625	.0592
1962–63	42.0	34.6	.0641	.0582
1963–64	43.4	35.2	.0672	.0580
1964–65	45.0	35.8	.0692	.0528
1965–66	42.4	35.5	.0708	.0584
1966–67	41.0	33.0	.0715	.0593
1967–68	41.5	33.2	.0767	.0609
1968–69	42.5	32.6	.0787	.0638
1969–70	41.8	32.6	.0813	.0658
1970–71	41.2	32.4	.0839	.0669
...				
1975–76	38.0	32.6	.0897	.0790
1980–81	36.8	30.1	.0943	.0859
1981–82	35.5	30.0	.0949	.0868
1982–83	33.7	29.8	.0939	.0892
1983–84	34.1	30.1	.0928	.0904

Note: Data unavailable for missing years.

Source: Rogerson (1987).

For 20–24-year-olds, $n = 28$, and the correlation coefficient between the mobility rate and cohort size is equal to -0.747; for 25–29 year olds, $r = -0.805$. Details of the calculation for 20–24-year-olds are as follows. The sum of the cross-products $(x_i - \overline{x})(y_i - \overline{y})$ (i.e., the numerator of the covariance) is $-.694$. Dividing by $n - 1 = 27$ yields a covariance of $-.0257$. The standard deviations for x and y are 3.371 and 0.0102, respectively. Dividing the covariance by the product of these standard deviations as in Equation 7.2 yields the

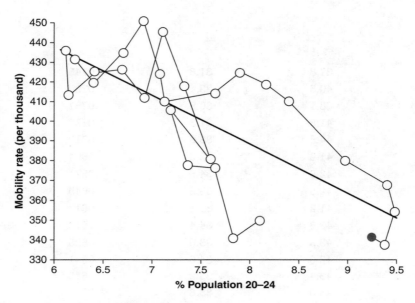

Figure 7.3 Correlation of mobility with cohort size
Source: Plane and Rogerson (1991)

correlation coefficient of −0.747. The data for 20–24-year-olds are graphed in Figure 7.3, where the negative relationship between the variables is apparent.

7.2.2 Statewide Infant Mortality Rates and Income

As part of a class assignment back in my graduate school days, I decided to investigate geographic variation in infant mortality rates in the United States. The data I collected were at the state level. I was interested in understanding whether infant mortality rates varied with factors such as educational attainment, income, access to health care, etc. As part of my analysis, I graphed the relationship between infant mortality rates and personal income for the white population. The graph is shown in Figure 7.4. Most of the states fall close to a line with negative slope, ranging from states such as Mississippi (MS) and Kentucky (KY), with low values of median personal income and high infant mortality rates, to states such as Connecticut (CT), where the statewide median personal income was high, and mortality rates were low. Pearson's correlation coefficient for the 50 states is equal to −0.28. Notice though the presence of six states that have infant mortality levels above the level expected, given the personal income in the state (TX, CO, AZ, WY, NM, and NV, corresponding to Texas, Colorado, Arizona, Wyoming, New Mexico, and Nevada). Cases such as these, that do not fit the general trend, are known as *outliers*. It is interesting to note that these states also form a compact geographic cluster. Additional inspection of the data had raised the possibility that these six states had a relatively low number of physicians per 100,000 residents. A two-tail *t*-test confirmed that indeed these six states had a significantly lower number of physicians per 100,000 residents, relative to the other states.

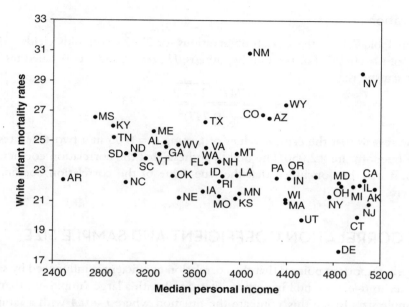

Figure 7.4 White infant mortality rates as a function of median personal income

The treatment of outliers depends upon the circumstances. A good underlying understanding of *why* particular points are outliers provides some rationale for removing those points from the analysis. In this case, we have a reasonably good explanation for the outliers, and we are justified in asking what the correlation would be without the outliers (it is $r = -0.64$, a much stronger negative relationship than was found with the original 50 observations). Of course, we would not want to get in the habit of plotting variables and arbitrarily eliminating those points that do not fall close to the line, just so we can report a high value of r. But it *is* good practice to plot the data and think carefully about the reasons for any outliers.

7.3 A SIGNIFICANCE TEST FOR R

To test the null hypothesis that the true correlation coefficient, ρ, is equal to zero, the data for each variable are assumed to come from normal distributions. In addition, it is assumed that the observations for each variable are independent; an observation of x does not affect, and is not affected by, other values of x. The same holds for y. If these assumptions are satisfied, the test may be carried out by forming the t-statistic:

$$t = \frac{r\sqrt{n-2}}{\sqrt{1-r^2}} \tag{7.4}$$

If the null hypothesis is true, this statistic has a t-distribution, with $n-2$ degrees of freedom.

7.3.1 Illustration

The data in Table 7.1 for the $n = 28$ observations for 20–24-year-olds yields a correlation coefficient of $r = -0.747$. For the null hypothesis, H_0: $\rho = 0$, and a two-tailed test with $\alpha = 0.05$, the t-statistic is:

$$t = \frac{(-0.747)\sqrt{28-2}}{\sqrt{1-(-0.747)^2}} = -5.73 \qquad (7.5)$$

A t-table reveals that the critical values of t, using $\alpha = 0.05$ in a two-tailed test with 26 degrees of freedom, are ± 2.056. The null hypothesis that the correlation coefficient is zero is rejected; it may be concluded that the true value of the correlation coefficient is not equal to zero.

7.4 THE CORRELATION COEFFICIENT AND SAMPLE SIZE

An extremely important point is that the correlation coefficient is influenced by sample size. It is far easier to reject the null hypothesis that $\rho = 0$ with a large sample size than it is with a small sample size. To see this, compare the situation where $r = 0.4$ with a sample size of 11, and the situation where $r = 0.4$ with a sample size of $n = 62$. In the former case, the observed t-statistic is $1.31 (= 0.4\sqrt{9} / \sqrt{1-0.4^2})$, which is less than the critical value $t_{0.05,9} = 2.262$, and the null hypothesis is not rejected. In the latter case, when $n = 62$, the t-statistic is $3.38 = 0.4\sqrt{60} / \sqrt{1-0.4^2}$, and the null hypothesis is rejected, since the t-statistic is greater than the critical value, $t_{.05,60} = 2.00$.

One of the implications of this is that there really should be no rules of thumb that are invoked to decide whether r is sufficiently high to make the researcher happy about the level of correlation. Such rules of thumb do seem to be popular – for instance, an r value of 0.7 or 0.8 is often taken as important or significant. But, as we have just seen, whether a correlation coefficient is truly significant depends upon the sample size. Thus, when the researcher is working with large datasets, a relatively low value of r should not be as disappointing as that same value of r when the sample size is smaller. A value of $r = 0.4$ could be quite meaningful if $n = 1000$, and the researcher should not necessarily throw the results out the window just because the r-value is noticeably less than 1 and perhaps less than some arbitrary, rule-of-thumb value, such as $r = 0.8$. Table 7.2 gives, for various values of n, the minimum absolute value of r needed to achieve significance. For example, with a sample size of $n = 50$, any value of r greater than 0.288 or less than -0.288 would be found significant using the t-test described above. The reader will note how even quite small values of r are significant when the sample size is only modestly large. For values of $n > 30$, the quantity $2/\sqrt{n}$ is equal to the approximate absolute value of r that is needed for significance using $\alpha = 0.05$. For example, if $n = 49$, a correlation coefficient with absolute value greater than $2 / \sqrt{49} = 0.286$ would be significant.

Table 7.2   Minimum values of r required for significance

Sample size, n	Minimum absolute value of r needed to attain significance (using $\alpha = 0.05$)
15	.514
20	.444
30	.361
50	.279
100	.197
250	.124

For large n, r_{crit} is approximately $2/\sqrt{n}$

While we have just argued that we should not too hastily discard the results of an analysis because of a seemingly low correlation (since with a large sample size that correlation may be significantly different from zero), there is also some concern about attaching too *much* importance to the results of a significance test. Meehl (1990) has noted that, with many datasets, there is a strong tendency to find that 'everything correlates to some extent with everything else' (p. 204). This is sometimes referred to as the 'crud factor'. If the sample size is large enough, we will often be able to reject the null hypothesis that any two particular variables are unrelated. For example, Standing et al. (1991) found that in a dataset containing 135 variables related to the educational and personal attributes of 2058 individuals, the typical variable exhibited a significant correlation with 41% of the other variables. The extreme case was the variable measuring Grade 5 mathematics scores – it was significantly correlated with 76% of the other variables, leading the authors to conclude that 'the number of statistically significant possible "causes" of mathematics achievement available to the unbridled theorizer will almost be as large' (p. 125). With a large enough sample size, the confidence interval around the correlation coefficient becomes so narrow that we can almost always conclude confidently that the true correlation coefficient is not *exactly* equal to zero. It is possible however that the true correlation, while not precisely equal to zero, is so small that it has no substantive meaning.

7.5 SPEARMAN'S RANK CORRELATION COEFFICIENT

In situations where only ranked data are available, or where the assumption of normality required for the test $H_0: \rho = 0$ is not satisfied, it is appropriate to use a nonparametric test based upon Spearman's rank correlation coefficient. As the name implies, the rank correlation coefficient is a measure of correlation that is based only upon the ranks of the data. Two separate sets of ranks are developed, one for each variable. A rank of 1 is assigned to the lowest value for each variable and a rank of n to the highest observation in each column (where each column contains the data for one of the variables). Spearman's rank correlation coefficient is defined as Pearson's correlation coefficient, after the data have been converted to ranks. When there are no tied ranks in either column of data, Spearman's rank correlation coefficient, r_s, can be simplified to:

$$r_s = 1 - \frac{6 \sum\limits_{i=1}^{n} d_i^2}{n^3 - n} \tag{7.6}$$

where d_i^2 is the squared difference between the ranks for observation i, and n is the sample size.

When there are tied ranks, Equation 7.6 should not be used; instead one calculates Spearman's correlation coefficient by calculating Pearson's correlation coefficient (Equation 7.2), using

Table 7.3 US mobility data, 1948–84, with ranks

Year	Mobility rate 20–24	Rank	Fraction of total population 20–24	Rank	d_i
1950–51	37.7	5	.0768	18	−13
1951–52	37.8	6	.0746	15	−9
1952–53	40.5	9	.0720	13	−4
1953–54	38.1	8	.0757	16	−8
1954–55	41.8	15.5	.0735	14	1.5
1955–56	44.5	25	.0713	11	14
1956–57	41.2	12	.0694	9	3
1957–58	42.6	21	.0671	6	15
1958–59	42.5	19.5	.0645	5	14.5
1959–60	41.2	12	.0617	2	10
1960–61	43.6	24	.0616	1	23
1961–62	43.2	22	.0625	3	19
1962–63	42.0	17	.0641	4	13
1963–64	43.4	23	.0672	7	16
1964–65	45.0	26	.0692	8	18
1965–66	42.4	18	.0708	10	8
1966–67	41.0	10	.0715	12	−2
1967–68	41.5	14	.0767	17	−3
1968–69	42.5	19.5	.0787	19	0.5
1969–70	41.7	15.5	.0813	20	−4.5
1970–71	41.2	12	.0839	21	−9

...

Year	Mobility rate 20–24	Rank	Fraction of total population 20–24	Rank	d_i
1975–76	38.0	7	.0897	22	−15
1980–81	36.8	4	.0943	25	−21
1981–82	35.5	3	.0949	26	−23
1982–83	33.7	1	.0939	24	−23
1983–84	34.1	2	.0928	23	−21

Source: **Plane and Rogerson (1991).**

the ranks as observations. Note that tied ranks are treated by replacing them with the average of the tied ranks.

The mobility data for 20–24-year-olds, for the period 1950–1984, are repeated in Table 7.3, with ranks adjacent to the mobility and cohort size variables. In this case, there are tied ranks in the mobility data and we should therefore calculate Pearson's correlation using the ranked data. Using Equation 7.2 on these ranks results in a Spearman correlation of $r_s = -0.730$. If Equation 7.6 *had* been used, the result would not have been much different. The differences in ranks, d_i, are given in the last column of the table. For these data, r_s would have turned out to be $1 - (6\ (5055))/(26^3 - 26) = -0.728$.

To test hypotheses for cases where n is greater than about 10, we may use the fact that the quantity $r_s \sqrt{n-1}$ has, approximately, a standard normal distribution. In our example, the observed value of t is therefore $-0.737\sqrt{25} = -3.69$. This is less than the critical value, $t_{0.05,25} = -1.96$, and so the null hypothesis of no association is rejected.

7.6 ADDITIONAL TOPICS

7.6.1 The Effect of Spatial Dependence on Significance Tests for Correlation Coefficients

The tests of significance outlined for both Pearson's r and Spearman's r_s assume that the observations of x are independent and that the values of y are also independent. When the x and y variables come from spatial locations, this assumption of independence may not be satisfied. Indeed, one of the most important points in this book is that *spatial data often exhibit dependence* – the value of x in one location is often related to the value of x in nearby locations. Thus, two nearby locations do not necessarily provide two independent pieces of information. In the extreme case of perfect dependence, one observation could be predicted from the other, and we would effectively have just one observation instead of two. Thus, our sample size effectively does not consist of n independent pieces of information.

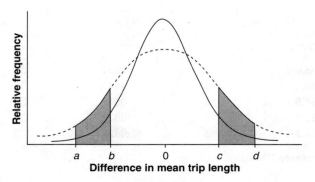

Figure 7.5   Assumed (———) and actual (- - - - - - -) variability in *r* when spatial dependence is present but not accounted for

It effectively contains less than *n* independent pieces of information; if we proceed as if we have *n* observations, we will be too prone to reject true null hypotheses. Spatial dependence affects the outcome of statistical tests, and this point should always be borne in mind when interpreting statistical results.

When spatial dependence is present and not accounted for, the variance of the correlation coefficient under the null hypothesis of no correlation is overestimated (since the variability in an effectively smaller sample will be larger). Alternatively stated, when repeated samples are taken from spatially dependent data, and the *x* and *y* values follow the null hypothesis of no correlation between *x* and *y*, the frequency distribution of *r* values will look like the dotted line in Figure 7.5. The dotted line has a wider frequency distribution than the solid line. The solid line corresponds to the variability in *r* that is calculated as present when standard significance tests such as Equation 7.4 are applied. The critical values associated with the standard statistical tests (*b* and *c*) are lower in absolute value than those that *should* be used (*a* and *d*). When sample values of *r* fall in the shaded regions, the standard statistical tests will incorrectly imply that the correlation coefficient is significantly different from zero. Correlation coefficients falling in the shaded regions are likely *not* significant, and may simply be the result of the underlying spatial dependence exhibited by the *x* and *y* variables. Haining (1990a) states this as follows:

> The important issue here is not to use conventional procedures to test for the significance of the correlation coefficient, and to recognize that a large *r* (or *r*ₛ) value may be due to spatial correlation [i.e. dependence] effects … The risks of inferring association between variables that is nothing other than the products of the spatial characteristics of the system are real and call for caution on the part of the user. (p. 321)

To see the effects of spatial dependence on correlation tests, consider the following model for the value of a variable a, at location x_1, y_1, applied to the interior cells of a region that has been subdivided into a grid of square cells:

$$a(x_1, y_1) = \mu + 4\rho(\overline{a} - \mu) + \varepsilon(x_1, y_1) \qquad (7.7)$$

Table 7.4   Type I error probabilities with spatial dependence

ρ_1	ρ_2	Type I error probability
0	0	0.0566
0	0.2	0.0500
0	0.24	0.0400
0.1	0.1	0.0700
0.1	0.225	0.1000
0.15	0.15	0.1000
0.15	0.225	0.1500
0.2	0.2	0.1900
0.2	0.24	0.3100
0.225	0.225	0.3366
0.24	0.24	0.5000

where μ is the overall mean, $\overline{a}$ is the mean value of the variable in the four surrounding cells, and $\varepsilon(x_1,y_1)$ is a normally distributed error term with mean zero. If $\rho = 0$, the values of the variable $a(x_1,y_1)$ are independent of the values at other sites. In this case, the values of a are simply equal to the overall mean and a normally distributed error term. When $\rho > 0$, the value in a particular interior cell depends upon the values of the four surrounding cells. The value of ρ measures the amount of dependence, and it can range up to 0.25. Note that when $\rho = 0.25$, the value of a at a location is precisely equal to the average of the values in surrounding locations, plus an error term. Clifford and Richardson (1985) use Equation 7.7 to simulate two spatial variables that are not correlated with one another (thus they are conducting a controlled experiment of sorts, where they know the null hypothesis to be true). Next, they find r and use Equation 7.4 and $\alpha = 0.05$ to see if r is significant. One would expect to find significant values 5% of the time (since 0.05 is the Type I error probability).

Table 7.4, as reported by Haining (1990a), displays the findings. ρ_1 and ρ_2 represent the amount of spatial dependence used in generating the two variables.

When ρ_1 and ρ_2 are zero, the Type I error probability is near its expected value of 0.05. Note that when one variable has no spatial dependence ($\rho_1 = 0$), the other variable can exhibit strong spatial dependence (e.g., $\rho_2 = 0.24$), and there is still no effect on the test for correlation, since the Type I error probability is still near 0.05. But when both variables exhibit strong spatial dependence, the Type I error probabilities – where one incorrectly finds significant correlation coefficients – rise dramatically. With strong spatial dependence and no corrective action, one will too often reject true null hypotheses.

7.6.2 The Modifiable Area Unit Problem and Spatial Aggregation

Gehlke and Biehl (1934) noted that correlation coefficients tend to increase with the level of geographic aggregation when census data are analyzed. A smaller number of large geographic units tends to give a larger correlation coefficient than does an analysis with a larger number of small geographic units. In a classic study, Robinson (1950) noted that the correlation between race and illiteracy rose with the level of geographic aggregation. It is important to keep in mind the fact that the size and configuration of spatial units may affect the outcome of the analysis. What is significant at one spatial scale may not be significant at another.

7.7 CORRELATION IN *SPSS 21 FOR WINDOWS*

Each variable should be represented by a column of data. Then click on `Analyze` and `Correlate`. Next, click on `Bivariate`, and move the variables you wish to correlate from the list on the left to the box on the right. You may move more than two variables into the box, if you wish to see a table of correlations among a number of variables. If desired, check the box to have `Spearman's correlation coefficient` calculated (Pearson's correlation is calculated by default).

7.7.1 Illustration

The data in Table 7.5 are a random sample of 29 census block groups (areas of about 1000 people) from Erie County, New York, in 1990. For each block group, there are data on population (totpop90), median household income (medhsinc), the median age of heads of households (medage), the standard deviation of the householder's age (sage), which is a measure of age mixing in the block group, the median age of housing (mage), and the percentage of housing that is owner-occupied (pctown). Rogerson and Plane (1998) discuss the age structure of householders in residential neighborhoods, and develop a model showing how age structure is related to variables such as age of the housing in the neighborhood, mobility, and homeownership.

Table 7.6 shows the bivariate correlation coefficients (Pearson and Spearman) among four of the variables. The median age of householders in block groups has a significant correlation with homeownership; as one might expect the association is positive, and

Table 7.5   1990 census data for a random sample of census tracts in Erie County, New York

AREANAME	BLOCKGROUP#	TOTPOP90	MEDHSINC	MEDAGE	SAGE	MAGE	PCTOWN
Tract 0010	BG 4	999	20862	49.24	19.08	60	.510
Tract 0016	BG 2	477	17804	50.70	17.49	60	.354
Tract 0026	BG 1	647	10545	51.24	16.92	58	.500
Tract 0028	BG 2	856	14602	50.66	18.29	60	.479
Tract 0045	BG 5	994	33603	45.12	15.36	60	.683
Tract 0057	BG 4	1083	24440	52.31	18.05	60	.516
Tract 0060	BG 1	879	15964	39.43	15.83	60	.416
Tract 0068	BG 2	374	33750	45.07	16.92	60	.202
Tract 0068	BG 4	806	14597	42.93	18.30	60	.346
Tract 0073.02	BG 9	2194	39779	52.49	16.58	39	.906
Tract 0076	BG 4	1150	29250	54.65	17.97	46	.884
Tract 0079.01	BG 3	1720	44205	53.63	13.83	41	.978
Tract 0079.02	BG 5	540	34625	60.55	14.06	44	.967
Tract 0079.02	BG 8	1128	32439	58.49	14.18	44	.899
Tract 0085	BG 1	434	39375	54.07	17.89	54	.967
Tract 0087	BG 3	1415	29513	51.42	17.77	60	.706
Tract 0097.01	BG 2	1639	39104	53.35	12.48	33	.992
Tract 0100.02	BG 1	3072	26174	54.41	16.36	30	.886
Tract 0100.02	BG 5	1715	28477	49.28	15.38	42	.638
Tract 0101.01	BG 5	755	35000	56.58	15.48	54	1000
Tract 0101.02	BG 2	731	17647	43.49	16.66	45	.239
Tract 0111	BG 4	544	28438	57.78	17.04	47	.796
Tract 0115	BG 1	885	33214	51.68	19.54	47	.862
Tract 0117	BG 2	633	36346	48.22	16.09	47	.856
Tract 0120.02	BG 1	851	28500	59.56	17.39	42	.876
Tract 0142.05	BG 1	681	38125	46.44	18.23	19	.885
Tract 0150.03	BG 2	1270	25515	51.64	16.78	60	.580
Tract 0152.02	BG 9	1334	29554	49.15	17.89	37	.740
Tract 0153.02	BG 1	634	47083	52.38	15.21	46	.860

median age is higher in areas of higher homeownership. The variability of ages in a neighborhood (sage) is negatively related to income (high-income neighborhoods are more homogeneous with respect to age), and negatively related to homeownership (where ownership is high, the ages of the householders are more homogeneous). Finally, note the similarity between Pearson's and Spearman's coefficients.

Table 7.6 Bivariate correlations among four neighborhood variables

		MEDHSINC	SAGE	MEDAGE	PCTOWN
MEDHSINC	Pearson Correlation	1.000	−.428*	.362	.739**
	Sig. (two-tailed)	.	.021	.053	.000
	N	29	29	29	29
SAGE	Pearson Correlation	−.428*	1.000	−.242	−.370*
	Sig. (two-tailed)	.021	.	.207	.048
	N	29	29	29	29
MEDAGE	Pearson Correlation	.362	−.242	1.000	.684**
	Sig. (two-tailed)	.053	.207	.	.000
	N	29	29	29	29
PCTOWN	Pearson Correlation	.739**	−.370*	.684**	1.000
	Sig. (two-tailed)	.000	.048	.000	.
	N	29	29	29	29

* Correlation is significant at the 0.05 level (two-tailed).

** Correlation is significant at the 0.01 level (two-tailed).

Nonparametric correlations

			MEDHSINC	SAGE	MEDAGE	PCTOWN
Spearman's rho	MEDHSINC	Correlation Coefficient	1.000	−.433*	.347	.742**
		Sig. (two-tailed)	.	.019	.065	.000
		N	29	29	29	29
	SAGE	Correlation Coefficient	−.433*	1.000	−.261	−.403*
		Sig. (two-tailed)	.019	.	.172	.030
		N	29	29	29	29
	MEDAGE	Correlation Coefficient	.347	−.261	1.000	.745**
		Sig. (two-tailed)	.065	.172	.	.000
		N	29	29	29	29
	PCTOWN	Correlation Coefficient	.742**	−.403*	.745**	1.000
		Sig. (two-tailed)	.000	.030	.000	.
		N	29	29	29	29

* Correlation is significant at the 0.05 level (two-tailed).

** Correlation is significant at the 0.01 level (two-tailed).

7.8 CORRELATION IN *EXCEL*

Each variable is represented by a column in the spreadsheet; rows of the spreadsheet represent observations. To find Pearson's bivariate correlation, click on Tools, then Data Analysis, and then Correlation. The dialog box that opens requires you to specify the cells containing the variables in the box titled Input Range. For example, if there are 29 observations, and the variables to be correlated are in columns A and B, one would enter A1:B29 in this box. There is also an option to indicate whether the observations are grouped by rows or columns; since the column button is checked by default, there is no need to make any change in this selection.

SOLVED EXERCISES

1. Find Spearman's rank correlation coefficient for the following data:

Obs.	X	Y
1	21	22
2	12	12
3	11	8
4	26	4
5	25	7777
6	24	11

Solution. Use Equation 7.6. First, we rank each column, using 1 for the lowest value, and 6 for the highest, and we add a column for the difference in the ranks (although it doesn't matter which way we subtract, here we use 'rank of X' minus the 'rank of Y'):

Obs.	X	Y	difference in ranks (d)
1	3	5	−2
2	2	4	−2
3	1	2	−1
4	6	1	5
5	5	6	−1
6	4	3	1

The quantity $\sum d_i^2$ is equal to $(-2)^2 + (-2)^2 + \dots + 1 = 36$. Now, using the formula, Spearman's rank correlation coefficient is:

$$1 - \frac{6(36)}{6^3 - 6} = 1 - \frac{216}{210} = -0.0286$$

(Continued)

(Continued)

2. A sample of size 102 yields a correlation of 0.24. Would one reject the null hypothesis that the true correlation was zero?

Solution. A test of the null hypothesis is carried out using Equation 7.4:

$$t = \frac{r\sqrt{n-2}}{\sqrt{1-r^2}} = \frac{0.24\sqrt{102-2}}{\sqrt{1-0.24^2}} = \frac{2.4}{0.9708} = 2.47$$

We reject the null hypothesis, since this observed value of t is greater than the critical value of about 1.96, which would be used in a two-tailed test with large sample size.

EXERCISES

1. (a) Find the correlation coefficient, r, for the following sample data on income and education:

Observation	Income ($ × 1000)	Education (Years)
1	30	12
2	28	13
3	52	18
4	40	16
5	35	17

   (b) Test the null hypothesis $\rho = 0$.

   (c) Find Spearman's rank correlation coefficient for these data.

   (d) Test whether the observed value of r_s from part (c) is significantly different from zero.

2. (a) Draw a graph depicting a situation where the correlation coefficient is close to zero, but there is a clear relationship between two variables.

   (b) Draw a graph depicting a situation where there is a strong positive relationship between two variables, but where the presence of a small number of outliers makes the strength of the relationship less strong.

3. The distribution of the t-statistic for testing the significance of a correlation coefficient has $n - 2$ degrees of freedom. If the sample size is 36 and $\alpha = 0.05$, what is the smallest absolute value a correlation coefficient must have to be significant? What if the sample size is 80?

4. Find the correlation coefficient for the following data:

Obs.	X	Y
1	2	6
2	8	6
3	9	10
4	7	4

5. (a) Why is a 'rule of thumb' for the significance of a correlation coefficient (e.g., r^2 above 0.7 is significant) not a good idea?

   (b) Why is a very large sample a 'problem' in the interpretation of significance tests of the correlation coefficient?

6. Find the correlation coefficient between median annual income in the United States and the number of horse races won by the leading jockey, for the period 1984–1995:

Year	Median income	Number of races won by leading jockey
1984	35,165	399
1985	35,778	469
1986	37,027	429
1987	37,256	450
1988	37,512	474
1989	37,997	598
1990	37,343	364
1991	36,054	430
1992	35,593	433
1993	35,241	410
1994	35,486	317

Test the hypothesis that the true correlation coefficient is equal to zero. Interpret your results.

7. For the following ranked data, find Spearman's r, and then test the null hypothesis (using a Type I error probability of 0.10) that the true correlation is equal to zero.

(Continued)

(Continued)

Observation	Rank of x	Rank of y
1	1	8
2	2	4
3	5	12
4	6	3
5	11	10
6	12	7

8. Find Pearson's r for the following data, and then test the null hypothesis that the correlation coefficient is equal to zero. Use a Type I error probability of 0.05.

Observation	x	y
1	3.2	6.2
2	2.4	7.3
3	1.6	8.1
4	8.3	2.6
5	7.2	6.3
6	5.1	4.3

9. Using *SPSS* or *Excel* and the Milwaukee dataset described in Section 1.9.2, find the correlation between number of bedrooms and lot size. Also do this without the aid of a computer using the first seven observations in the dataset.

10. Using *SPSS* or *Excel* and the Hypothetical UK Housing Prices dataset described in Section 1.9.4, find the correlation between floor area and number of bedrooms.

On the Companion Website

Chapter 7 makes use of a number of resources made available on the website: the **Home Sales in Milwaukee, Wisconsin** dataset and the **Hypothetical UK Housing Prices** dataset required for the Exercises as well as the **1990 Census Data for Erie County, New York** dataset used in the text are available within the 'Datasets' section. The website also includes information about their fields and formats, and the 'Further Resources' section provides links to other websites and videos that also explain the concepts described here.

8

INTRODUCTION TO
REGRESSION ANALYSIS

8.1 INTRODUCTION

Whereas correlation is used to measure the strength of the linear association between variables, regression analysis refers to the more complete process of studying the relationship between a dependent variable and a set of independent, explanatory variables. Linear regression analysis begins by assuming that a linear relationship exists between the *dependent* or *response* variable (y) and the *independent* or *explanatory* variables (x), proceeds by fitting a straight line to the set of observed data, and is then concerned with the interpretation and analysis of the effects of the x variables on y, and with the nature of the fit. An important outcome of regression analysis is an equation that allows us to predict values of y from values of x.

As discussed in Chapter 1, the process of description often leads one to suspect that two or more variables are related. Once we specify *how* the variables are related, we have a model, which may be thought of as a simplification of reality. Regression analysis provides us with (a) a simplified view of the relationship between variables, (b) a way of fitting the model with our data, and (c) a means for evaluating the importance of the variables and the correctness of the model.

For example, we may wish to know whether the distance that adults live away from their parents is dependent upon education, whether snowfall is dependent upon elevation, whether infant mortality is related to income, or whether park attendance is related to the income of the population living within a certain distance of the park. In each case, a good place to begin is to plot the data on a graph, and to measure the correlation between variables. Linear regression analysis takes this a step further by finding the best-fitting straight line through the set of points.

When there is just one independent, explanatory variable, as is the case in the examples above, we wish to fit a straight line through the set of data points; the equation of this line is

$$\hat{y} = a + bx \, , \qquad\qquad (8.1)$$

where $\hat{y}$ is the predicted value of the dependent variable, x is the observed value of the independent variable, a is the intercept (or point where the line intersects the vertical axis), and b is the slope of the line. The quantities a and b represent *parameters* describing the line, and these will be estimated from the data. This case, with one independent variable, is known as *simple regression* or *bivariate regression*, and it is depicted in Figure 8.1. We shall in this chapter confine our attention to this special case. More generally, *multiple regression* (treated in the next chapter) refers to the case where there is more than one independent variable.

The slope of the line, b, may be interpreted as the change in the dependent variable expected from a unit change in the independent variable. For example, suppose a regression of housing sale prices on the square footage of houses yielded the following equation:

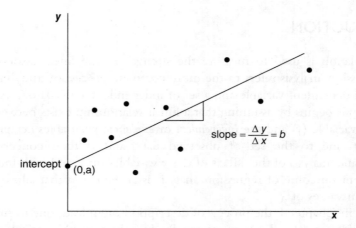

Figure 8.1   Regression line through a set of points

$$\hat{p} = 30,000 + 70s \tag{8.2}$$

where $\hat{p}$ is the predicted housing sales price, and s represents square footage. The slope in this equation is 70, and this means that an increase of one square foot leads, on average, to a \$70 increase in sales price. The price predicted for a house with 2000 square feet is $30,000 + 70 (2000) = \$170,000$. The intercept is the predicted value of the dependent variable when the independent variable is set equal to zero. In this example, a house with 0 square feet would sell at a predicted price of \$30,000! The intercept of \$30,000 could in this case be interpreted as the value of the land on which the house is built. More generally, the intercept does not always have a realistic interpretation, since a zero value for the independent variable may lie well outside the range of observed values.

In studying the linear relationship between variables, each observation (i) of the dependent variable, y_i, may be expressed as the sum of a predicted value and a residual term:

$$y_i = a + bx_i + e_i = \hat{y}_i + e_i, \tag{8.3}$$

where $\hat{y}_i = a + bx_i$ is the predicted value, and e_i is termed the *residual*. The value $\hat{y}_i$ represents the value of the dependent variable predicted by the regression line. Note that the residual is equal to the difference between observed and predicted values:

$$e_i = y_i - \hat{y}_i \tag{8.4}$$

In keeping with the distinction between sample and population, note that a and b are estimates of some 'true', unknown regression line. The slope and intercept of this true regression line could, in theory, be determined by taking a complete, 100% sample of the population. As usual, we use Greek letters to denote the population values of the parameters:

$$y_i = \alpha + \beta x_i + \varepsilon_i, \tag{8.5}$$

where α and β are the intercept and slope of the true regression line, respectively. Each observation, y_i, may be viewed as the sum of a component that predicts the value of y_i on the basis of the value of x_i (using the true coefficients α and β) and some random error (ε_i). The error term reflects the fact that we do not expect the model to work 'perfectly'; inevitably there will be other variables that also influence y, though we hope their influence is relatively minor. Observations on the dependent variable may be expressed as the sum of the predicted value and a 'true' population error, ε_i, where:

$$\varepsilon_i = y_i - \tilde{y}_i \tag{8.6}$$

is the difference between the observed value (y_i) and that predicted by the true regression line ($\tilde{y}_i$). The latter quantity is:

$$\tilde{y}_i = \alpha + \beta x_i \qquad (8.7)$$

In Figure 8.2, both the true regression line (Equation 8.7) and the best-fitting line based on the sample of points (Equation 8.1) are shown. It is important to keep in mind that had a different sample been collected, the regression line based on the sample would be different, but the true regression line would remain the same.

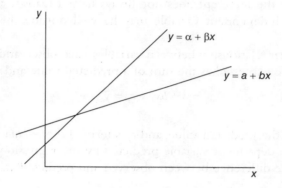

Figure 8.2   **True and sample regression lines**

8.2 FITTING A REGRESSION LINE TO A SET OF BIVARIATE DATA

Figure 8.3 shows a straight line fit through a set of four points plotted in a two-dimensional, x–y space. In regression analysis, the objective is to find the slope and intercept of a best-fitting line that runs through the observed set of data points. But what is meant by best-fitting? There are certainly many ways to fit a line through a set of points. One way would be to fit the line so that the sum of the minimum distances of the observations to the line was a minimum. In this case, the distances are represented in Figure 8.3 geometrically as (dashed) lines that run from the observations to the regression line, with these lies being perpendicular to the regression line.

In linear regression analysis, the sum of squared vertical distances from the observed points to the line (i.e., the solid lines in Figure 8.3) is minimized. The fact that vertical distances are used is consistent with the idea that the dependent variable, which is always portrayed on the vertical axis, is being predicted from the independent variable (portrayed on the horizontal axis). In fact, the vertical distance is identical to the value of the residual,

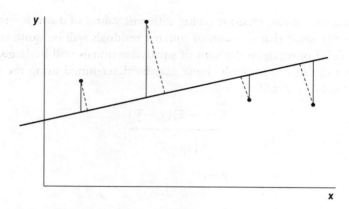

Figure 8.3   Alternative measures of distance from points to regression line

which, as we have indicated, is the difference between the observed and predicted values of the dependent variable. Thus regression analysis minimizes the sum of squared residuals. The sum of *squared* residuals is used partly for reasons of mathematical convenience – expressions for finding the values of *a* and *b* from the data are much easier to derive and express. Thus, the objective is to find values of *a* and *b* that minimize the sum of squared residuals:

$$\min_{a,b} \sum_{i=1}^{n} (y_i - \hat{y})^2 = \min_{a,b} \sum_{i=1}^{n} (y_i - a - bx_i)^2 \tag{8.8}$$

Geometrically, this problem corresponds to finding the minimum of a three-dimensional parabolic cone, where *a* and *b* are coordinates in the two-dimensional plane, and the sum of squared residuals is the vertical axis (Figure 8.4).

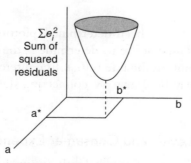

Figure 8.4   Minimization of sum of squared residuals

Viewing the figure, we can imagine trying different values of a and b – some will work relatively well in the sense that the sum of squared residuals will be quite small, and other combinations will be poor, since the sum of squared residuals will be large. The values of a and b at the bottom of the parabolic cone may be determined using the data as follows (see box below for more detail):

$$b = \frac{\sum_{i=1}^{n}(x_i - \bar{x})(y_i - \bar{y})}{\sum_{i=1}^{n}(x_i - \bar{x})^2};$$

$$a = \bar{y} - b\bar{x}.$$

(8.9)

The solution of Equation 8.8 to find the values of the slope and intercept is the solution of a calculus problem. Solving the calculus problem amounts to finding the a, b combination that leads to the smallest sum of squared residuals, at the bottom of the parabolic bowl. For those with a calculus background, we proceed by taking the derivatives of Equation 8.8 with respect to a and b, setting them each to zero (corresponding to the fact that the line that is tangent to the bottom of the bowl has a slope of zero), and solving for the two unknowns a and b. The result is the set of equations in 8.9.

The reader should note that the numerator of the expression for the slope, b, is identical to that for the correlation coefficient, r. In fact, for bivariate regression, the slope may be written in terms of the correlation coefficient:

$$b = r\frac{s_y}{s_x}$$

(8.10)

Once the values of a and b have been determined, plotting the line is straightforward. As indicated in the equation used above to determine a, the regression line goes through the mean $(\bar{x}, \bar{y})$. Another point on the line is $(0, a)$ (the intercept). After plotting these two points, the regression line may be drawn by connecting the two points together with a straight line.

8.2.1 Illustration: Income Levels and Consumer Expenditure

A supermarket is interested in how income levels (x) may affect the amount of money spent per week by its customers (y). The null hypothesis is that income levels do not affect the amount of money spent per week by customers, and the alternative hypothesis is that higher incomes are associated with greater spending. Table 8.1 depicts the data collected from ten survey respondents.

To fit the regression line, we can first compute the following quantities:

$$\bar{x} = 52.3; \qquad s_x = 20.20$$

$$\bar{y} = 79.1; \qquad s_y = 28.34$$

$$\sum_{i=1}^{n}(x_i - \bar{x})(y_i - \bar{y}) = 4301.7$$

$$r = \frac{\sum_{i=1}^{n}(x_i - \bar{x})(y_i - \bar{y})}{(n-1)s_x s_y} = \frac{4301.7}{9(20.2)(28.34)} = 0.835$$

(8.11)

Table 8.1 Data on spending and income

Amount spent/week (y)	Income (× 000) (x)
$120	65
$68	35
$35	30
$60	44
$100	80
$91	77
$44	32
$71	39
$89	44
$113	77

Income (x) is given in thousands of dollars, and weekly supermarket spending is given in dollars. From these, we may find the slope from either of the following:

$$b = r\frac{s_y}{s_x} = 0.835 \frac{28.34}{20.2} = 1.171$$

$$b = \frac{\sum_{i=1}^{n}(x_i - \bar{x})(y_i - \bar{y})}{\sum_{i=1}^{n}(x_i - \bar{x})^2} = \frac{\sum_{i=1}^{n}(x_i - \bar{x})(y_i - \bar{y})}{(n-1)s_x^2}$$

(8.12)

$$= \frac{4301.7}{(9)202^2} = 1.171$$

The intercept is:

$$a = \bar{y} - b\bar{x} = 79.1 - 1.171(52.3) = 17.8 \tag{8.13}$$

Our regression line may be expressed as:

$$\hat{y} = \$17.8 + 1.171x \tag{8.14}$$

implying that every increase of $1000 of income leads to an increase in $1.17 spent at the supermarket each week.

For each observation, we can compute a predicted value and a residual, using the regression line, along with the values of x. For example, for the first observation, income is $65,000 and the predicted value of the dependent variable is therefore:

$$\hat{y} = \$17.8 + (1.171)(\$65) = \$93.90 \tag{8.15}$$

The residual for this observation is the observed value of y ($120) minus the predicted value:

$$e = \$120 - \$93.90 = \$26.10 \tag{8.16}$$

Table 8.2 depicts the results, including residuals and predicted values for all observations. The sum of the residuals (subject to a bit of rounding error) is equal to zero – the amount by which the positive residuals lie above the regression line is equal to the amount by which the negative residuals lie below the regression line.

Table 8.2 Predicted values and residuals associated with the spending and income data

Amount spent/week (y)	Income (× 000) (x)	Predicted ŷ	Residual e
$120	65	93.9	26.1
$68	35	58.8	9.2
$35	30	52.9	−17.9
$60	44	69.3	−9.3
$100	80	111.5	−11.5
$91	77	108.0	−17.0
$44	32	55.3	−11.3
$71	39	63.5	7.5
$89	44	69.3	19.7
$113	77	108.0	5.0

8.3 REGRESSION IN TERMS OF EXPLAINED AND UNEXPLAINED SUMS OF SQUARES

Another way to understand regression is to recognize that it provides a way to partition the variation in the observed values of a dependent (y) variable. In particular, the variability one observes in y may be decomposed into (a) a part that is explained by the regression line (via the assumption that y is linearly related to one or more independent variables) and (b) a part that remains unexplained.

More specifically, the variability in y may be measured by the sum of squared deviations of the y values from their mean. Some of this variability is explained by the regression line – the values of y vary partly because of the assumed relationship with the x variables. The partitioning of the total sum of squares into explained and unexplained components is analogous to the analysis of variance, where the sum of squares is divided into between and within column components. Here we have:

$$\sum_{i=1}^{n}(y_i - \overline{y})^2 = \sum_{i=1}^{n}(\hat{y}_i - \overline{y})^2 + \sum_{i=1}^{n}(y_i - y)^2 \qquad (8.17)$$

The left-hand side is the *total sum of squares*; the deviation between the observed value and the mean is represented geometrically by the distance A in Figure 8.5. In the figure, the points clearly vary about the horizontal line representing the mean value of y. Some of this total variability may be attributed to the regression line; we expect points on the right-hand side of the diagram to be above the mean, and points on the left-hand side of the diagram to be below the mean.

The first term on the right-hand side is the *regression* (or *explained*) *sum of squares* – it is the sum of squared differences between predicted values and the mean value of y. These differences are the deviations of the regression line from the mean and constitute the 'explained' portion of the variability in y. The distance between the predicted value and the mean is represented by 'C' in Figure 8.5. Finally, the second term on the right-hand side is the *unexplained*, or *residual sum of squares*. This is the sum of squared differences between the observed and predicted values. It is this quantity that is minimized when the coefficients are estimated. The distance between the observation and the regression line (i.e., the residual) is represented by 'B' in the figure.

The proportion of the total variability in y explained by the regression is sometimes called the *coefficient of determination*, and it is equal to the square of the correlation coefficient:

$$r^2 = \frac{\text{regression sum of squares}}{\text{total sum of squares}} = \frac{\sum_{i=1}^{n}(\hat{y}_i - \overline{y})^2}{\sum_{i=1}^{n}(y_i - \overline{y})^2} \qquad (8.18)$$

$$= 1 - \frac{\text{residual sum of squares}}{\text{total sum of squares}} = 1 - \frac{\sum_{i=1}^{n} e_i^2}{(n-1)s_y^2}$$

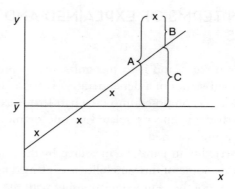

Figure 8.5 Partitioning the variability in y

where e is the value of the residual. Note that r^2 is equal to the regression sum of squares divided by the total sum of squares. It is also equal to one minus the ratio of the residual sum of squares to the total sum of squares.

The value of r^2 varies from 0 to 1; a value of zero would indicate that none of the variability in y has been explained by the x-variable, whereas a value of one would imply that all of the residuals are zero, and that the regression line fits perfectly, through all of the observed points.

One way to determine whether the regression has been successful at explaining a significant portion of the variation in y is to test the null hypothesis that the proportion of the variability in y explained by x is equal to zero:

$$H_0 : \rho^2 = 0 \qquad (8.19)$$

This is carried out with an F-test, analogous to the F-test used in the analysis of variance. Specifically, for simple regression, the null hypothesis that $\rho^2 = 0$ is tested with the F-statistic:

$$F = \frac{r^2(n-2)}{1-r^2} \qquad (8.20)$$

which has an F-distribution with 1 and $n - 2$ degrees of freedom for the numerator and denominator, respectively, when the null hypothesis is true. By looking back to the previous chapter on correlation, you will notice that this F-statistic is the square of the t-statistic (equation 7.4) used to test the hypothesis that the correlation coefficient, ρ, is equal to zero. The tests are identical in the sense that they will always yield identical conclusions and p-values.

The origins of this F-test lie in the partitioning of the sums of squares just described. We can create an analysis of variance table, for a hypothetical example, with 12 observations, as follows:

	Sum of squares	df	Mean square	F
Regression (explained)	578	1	578	13.7
Residual (unexplained)	422	$n - 2 = 10$	42.2	
Total	1000	$n - 1 = 11$		

The total sum of squares has $n - 1$ degrees of freedom associated with it. In simple regression, where there is one independent variable, there is always 1 degree of freedom associated with the regression sum of squares, leaving $n - 2$ degrees of freedom associated with the residual sum of squares. As was the case with ANOVA, the F-ratio is found by forming a ratio of sums of squares, where each has been divided by its associated degrees of freedom. Here the F-ratio is found by taking the ratio of (a) the regression sum of squares, divided by its degrees of freedom, to (b) the residual sum of squares, divided by its associated degrees of freedom. The quantities defined in (a) and (b) are sometimes termed 'mean squares'.

$$F = \frac{\text{Regression SS} / 1}{\text{Residual SS} / (n - 2)} = \frac{578 / 1}{422 / 10} = 13.7 \tag{8.21}$$

Recalling the definition of r^2, this ANOVA-like expression for F can be seen to be equivalent to Equation 8.20, since:

$$F = \frac{\text{Regression SS} / 1}{\text{Residual SS} / (n - 2)} = \frac{r^2(n - 2)}{1 - r^2} \tag{8.22}$$

8.3.1 Illustration

The analysis of variance table associated with the regression carried out in Section 8.2.1, using the data in Table 8.1, is shown in Table 8.3.

This table can be constructed by recalling that, for bivariate regression, the degrees of freedom associated with the explained sum of squares is equal to one, and the degrees of freedom associated with the total sum of squares is equal to $n - 1$. Also recall that the mean square is simply the sum of squares divided by the degrees of freedom, and the F-statistic is the ratio of the mean squares. The first column may be completed by recognizing that the total sum of squares is equal to the variance of y, multiplied by $n - 1$:

Table 8.3   Analysis of variance table for regression

	Sum of squares	df	Mean square	F
Regression (explained)	5039.2	1	5039.2	18.4
Residual (unexplained)	2189.7	$n - 2 = 8$	273.7	
Total	7228.9	$n - 1 = 9$		

$$\text{Total sum of squares} = \sum_{i=1}^{n}(y_i - \bar{y})^2 = (n-1)s_y^2 \qquad (8.23)$$

Since $r^2 = 0.835^2$ is the proportion of the total sum of squares explained by the regression, the regression sum of squares is equal to $7228.9(0.835^2) = 5040.2$ (Table 8.3 shows a slightly more accurate result, where there is no rounding error). The residual sum of squares is simply the difference between the total and regression sum of squares.

The observed F-ratio of 18.4 in the table may be compared with the critical value of 5.32 found in the F-table, using $\alpha = 0.05$ and 1 and 8 degrees of freedom, respectively, for numerator and denominator. Since the observed value of F exceeds the critical value of 5.32, we reject the null hypothesis that the true correlation coefficient, ρ^2, is equal to zero, and conclude that income explains a significant amount of the variability in supermarket spending.

8.4 ASSUMPTIONS OF REGRESSION

The assumptions of regression analysis for simple regression are:

1. The relationship between y and x is linear; that is, there is an equation, $y = \alpha + \beta x + \varepsilon$ that constitutes the population model.

2. The errors (i.e., the residuals) have mean zero, and constant variance; that is, $E[\varepsilon] = 0$ and $V[\varepsilon] = \sigma^2$. The errors about the regression line do not vary with x; that is, $V[\varepsilon \mid x] = \sigma^2_x = \sigma^2$.

3. The residuals are independent; the value of one error is not affected by the value of another error.

4. For each value of x, the errors have a normal distribution about the regression line. This normal distribution is centered on the regression line. This assumption may be written $\varepsilon \sim N(0, \sigma^2)$.

Multiple regression, treated in the next chapter, adds another assumption – namely, that the x variables have no multicollinearity (that is, the independent variables are not significantly correlated with one another).

8.5 STANDARD ERROR OF THE ESTIMATE

The standard error of the estimate is another name for the standard deviation of the residuals; for the case of simple regression, it is estimated by:

$$s_e = \sqrt{\frac{\sum_{i=1}^{n}(y_i - \hat{y}_i)^2}{n-2}} \qquad (8.24)$$

Note that the standard error of the estimate is equal to the square root of the residual mean square (where the residual mean square is the residual sum of squares divided by its associated degrees of freedom). For the data in Table 8.1, the standard error of the estimate is $\sqrt{2737} = 16.54$. The standard error of the estimate may be interpreted loosely as the approximate magnitude of a typical residual.

8.6 TESTS FOR β

We are often interested in testing the null hypothesis that the true value of the slope is equal to zero, i.e., $H_0: \beta = 0$. We are of course usually interested in the prospect of rejecting this null hypothesis, thereby accumulating some evidence that the variable x is important in understanding y. This can be done via a t-test:

$$t = \frac{b - \beta}{s_b} = \frac{b}{s_b} \tag{8.25}$$

where s_b is the standard deviation of the slope; this is given by:

$$s_b = \sqrt{\frac{s_e^2}{(n-1)s_x^2}} \tag{8.26}$$

When the null hypothesis is true, t has a t-distribution with $n - 2$ degrees of freedom.

8.6.1 Illustration

We can test the hypothesis that the true regression coefficient is equal to zero by using Equations 8.24–8.26. This first requires finding the variance of the residuals (s_e^2), and the standard error of the estimate (s_e):

$$s_e^2 = \frac{\sum_{i=1}^{n} e_i^2}{n-2} = \frac{2191.03}{8} = 273.71; \tag{8.27}$$

$$s_e = \sqrt{273.71} = 16.54$$

Next, the standard deviation of the estimate of the slope is given by:

$$s_b = \sqrt{\frac{s_e^2}{(n-1)s_x^2}} = \sqrt{\frac{273.71}{9(20.2^2)}} = \sqrt{.0745} = 0.273 \tag{8.28}$$

The test of the null hypothesis, $H_0: \beta = 0$ is then carried out with the t-statistic:

$$t = \frac{b - \beta}{s_b} = \frac{1.171}{0.273} = 4.29 \tag{8.29}$$

Since the observed value of t exceeds the critical value of 2.306 found from the table using a two-tailed test with $\alpha = 0.05$ and $n - 2 = 8$ degrees of freedom, we reject the null hypothesis that the true regression coefficient is equal to zero. Note that the observed value of t (4.29) is equal to the square root of the observed value of the F-statistic associated with the analysis of variance table for the regression (which is shown in Table 8.3). This relationship between t and F-statistics was also pointed out in Section 6.3. In bivariate regression, the t-test and the F-test will always give consistent results.

8.7 ILLUSTRATION: STATE AID TO SECONDARY SCHOOLS

In New York state, aid is given to schools by the state government. The aid formula is based upon a number of factors, but one principle is that districts with residents that have relatively higher household incomes should receive relatively less in state aid.

A graph of state aid per pupil versus average household income for 27 public school districts in western New York reveals that there is in fact a downward trend in aid with increasing income (see Figure 8.6). Regression analysis of the 27 pairs of points confirms this:

$$\hat{y} = 6106.93 - 0.0818x \tag{8.30}$$

where $\hat{y}$ is the predicted amount of school aid per pupil, in dollars, and x is the average household income. The slope implies that every increase of \$1 in household income brings with it a decline in state aid of \$0.08 per pupil. Equivalently, every increase of \$1000 in household income leads to a decline of \$81.80 in state aid, per pupil (although this estimate is also capturing the effects of other variables in the state aid formula that have been omitted here; see Sections 9.1.2 and 9.2). The value of r^2 is 0.257, and the slope is significantly

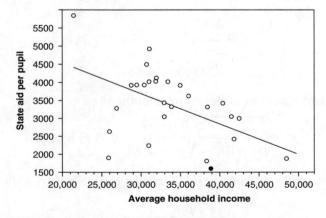

Figure 8.6   New York state school aid per pupil vs average household income

negative (a t-test yields $t = -2.94$, which is more extreme than the critical value, and implies a p-value of 0.007).

Of particular interest to me was the darkened circle, which represents the Amherst school district where my children attended school! Note that the state aid per pupil is significantly lower than the expected value, given the average household income in the district. A partial explanation of this has to do with the way in which the state collects information on income. Each taxpayer must note their school district on their tax return, using a three-digit code taken from a list at the back of the tax form instruction booklet. The problem is that the Amherst school district lies in the town of Amherst, but so too do a number of other districts, including the Williamsville and Sweet Home school districts. Residents of the Williamsville and Sweet Home districts dutifully go to the back of the instruction booklet, begin to scan the list, and quickly come across Amherst, since it is near the beginning of the alphabet. Since they live in the town of Amherst, they (incorrectly) copy the Amherst school district code onto their tax form. State officials then tally up the income in each district, and find (incorrectly) that residents of the Amherst school district make a lot of money, and consequently should receive less in school aid! The average household income data used in Figure 8.6 represent the more accurate household income data taken from the US census. Even though it may be out of date, since it was collected in 1990 and the analysis was done in 1997, the census data give a truer picture than do the tax return data of how deserving of aid each district is.

If the aid received by the Amherst school district from the state was in line with expectations, the black dot would be raised vertically, from its present location to the regression line. This would represent an increase of more than $1100 per pupil, which is approximately a 70% increase over the present figure of about $1600 per pupil. Since the district contains over 3000 students, this would represent an increase of over three million dollars (roughly 10% of the district's budget). Unfortunately for the residents of the Amherst district, it is difficult to correct the imbalance, and this is true despite the potential of geographic information systems to attribute each resident's address to the correct local school district. There is a legal clause that limits the degree to which the problem can be corrected, since any change that gave more to the Amherst District would give less to surrounding districts. So, despite the fact that residents in the Williamsville and Sweet Home districts have less income attributed to their district than they should (and consequently have higher state aid), the imbalance has been only partially corrected.

One solution I suggested at a meeting of the Amherst School Board was that we change the name of our district! Since the problem is caused by the fact that the school district has the same name as the town, perhaps we could simply change the name to something else. This seemingly creative suggestion met with difficulty too, since it turns out that it is difficult, if not impossible, to legally change the school district's name. Another, less savory, approach would be to have a campaign to encourage residents of the Amherst school district to put the school codes of Sweet Home or Williamsville on their tax forms. In any event, this example serves to illustrate how regression analysis can be used to both estimate the magnitude of effects of one variable on another (the term 'effect' in the current example refers to the effect of income on state aid) and to interpret unusual observations. The example also illustrates that, because of the quirky nature of data, potential pitfalls abound when one attempts to establish relationships between one variable (income) and another (state aid).

8.8 LINEAR VERSUS NONLINEAR MODELS

It should be understood that the word 'linear' refers to the fact that, in linear regression analysis, the relationship is one that is *linear in the parameters*. The parameters are the intercept and slope coefficients, *a* and *b*. Thus, linear regression could be used to study the effects of the square of some variable x on the value of y. Similarly, the equations:

$$y = a + b\sqrt{x} \tag{8.31}$$
$$y = a + b(\ln x)^2$$

may also be studied using the methods of linear regression, since the parameters a and b appear linearly (that is, they are raised to the power 1). An example of an equation that is *not* linear in its parameters is:

$$y = e^{a + bx^2 + \varepsilon} \tag{8.32}$$

In many instances, a nonlinear relationship may still be analyzed using linear regression, since the nonlinear curve may be transformed into a linear one. For example, a common finding in geographic research is that there is a distance-decay effect for many kinds of

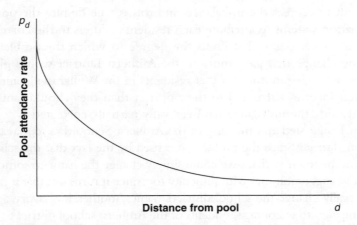

Figure 8.7   Negative exponential decline in pool attendance rate with distance

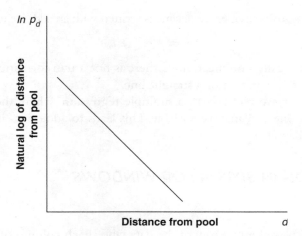

Figure 8.8 Linear decline in log of pool attendance rate with distance

interaction. Furthermore, this effect is often well-modeled with a negative exponential curve. Attendance at a local swimming pool, for instance, may appear to decline exponentially with the distance that individuals reside from the pool (see Figure 8.7). Negative exponential decay in this example could be modeled with an equation of the form:

$$p_d = p_0 e^{-bd} \tag{8.33}$$

where p_d is the pool attendance rate among residents residing a distance d from the pool, and p_0 is the attendance rate among those living as close as possible to the pool. Furthermore, e is the constant 2.718 ..., and b is the rate at which attendance declines with distance (that is, it is a measure of the steepness of the exponential decline; higher values of b imply a greater effect of distance on attendance). In this example, we can transform the curvilinear relationship seen in Figure 8.7 into a linear one. One reason for wanting to do this is to be able to make use of the well-developed methods of linear regression analysis. The transformation is brought about by taking the logarithms of both sides of Equation 8.33:

$$\ln p_d = \ln p_0 - bd \tag{8.34}$$

This is the equation of a straight line; if the relation is linear, when $\ln p_d$ is plotted against distance (d), the result will be a straight line, with slope equal to b, and intercept equal to $\ln p_0$ (see Figure 8.8).

Note also that models with negative exponential decline have to be initially written with a multiplicative error term, since that allows them to be linearized:

$$p_d = p_0 e^{-bd} \varepsilon \Rightarrow \ln p_d = \ln p_0 - pd + \varepsilon \tag{8.35}$$

If a model with negative exponential decline is written with an additive error term, as follows:

$$p_d = p_0 e^{-bd} + \varepsilon \qquad (8.36)$$

it is said to be intrinsically nonlinear, since there is not a transformation that can convert Equation 8.36 into the equation of a straight line.

In the next chapter, we will focus on multiple regression, where the regression model includes more than one explanatory variable. This leads to additional issues, and these are also discussed in Chapter 9.

8.9 REGRESSION IN *SPSS 21 FOR WINDOWS*

8.9.1 Data Input

Each observation is placed into a row of the data table. Each column of the data table corresponds to a variable. It is often convenient, but not necessary, to place the value of the dependent variable in the first column, and the value of the independent variable in the second column.

8.9.2 Analysis

To carry out a regression analysis, first click on Analyze, then Regression, and then Linear; this will open a box. Within the box, select the dependent variable from the list of variables on the left, and use the arrow key to move it into the box titled Dependent. Likewise, move the independent variables from the left to the box on the right labeled Independent(s). Clicking on OK will then carry out a regression analysis. This produces information on r, s_e, an ANOVA-type summary table, and information on the coefficients (i.e., slopes and intercept) and their significance. Options for additional output are discussed below.

8.9.3 Options

Clicking on alternatives in the categories Statistics, Plots, and Save can produce additional output. It is common, for example, to want to save additional information. Under Save, one can click on boxes to save, among other items, predicted values, residuals, and confidence intervals associated with both the mean predicted value of y given x, and individual values of y. New columns containing the desired information are attached to the right-hand side of the data table.

8.9.4 Output

An example of output is shown in Table 8.4. This output corresponds to the results of the regression associated with the data in Table 8.1. Note that the value of r^2 is 0.697.

8.10 REGRESSION IN *EXCEL*

8.10.1 Data Input

Each observation is placed in a row of the spreadsheet. Each column corresponds to a variable.

8.10.2 Analysis

Once the data have been entered, click on `Tools`, then `Data Analysis`, and then `Regression`. A dialog box then pops up, and this requires entries for `Input Y range` and `Input X range`. Options are also available in this dialog box to print residuals (and predicted values are supplied as well), and to plot both the regression line and the residuals. When the ten y observations of Table 8.1 are entered into column A and the ten x observations are entered into column B, one enters A1:A10 for the y-variable input and B1:B10 for the x-variable. Output is shown in Table 8.5.

Table 8.4 *SPSS* output for regression of amount spent per week vs income

Variables entered(/)removed[a]

Model	Variables entered	Variables removed	Method
1	INCOME[b]	.	Enter

[a] Dependent variable: AMTWEEK.

[b] All requested variables entered.

Model Summary

Model	R	R square	Adjusted R square	Std. error of the estimate
1	.835[a]	.697	.659	16.5441

[a] Predictors: (Constant), INCOME

ANOVA[a]

Model		Sum of squares	df	Mean squares	F	Sig.
1	Regression	5039.248	1	5039.248	18.411	.003[b]
	Residual	2189.652	8	273.706		
	Total	7228.900	9			

[a] Dependent variable: AMTWEEK.

[b] Predictors: (Constant), INCOME.

Coefficients[a]

Model		Unstandardized coefficients		Standardized coefficients	t	Sig.
		B	Std. error	Beta		
1	(Constant)	17.833	15.207		1.173	.275
	INCOME	1.171	.273	.835	4.291	.003

[a] Dependent variable: AMTWEEK.

Table 8.5 *Excel* output for regression of amount spent per week vs income

SUMMARY OUTPUT

Regression statistics	
Multiple R	0.834924
R square	0.697097
Adjusted R square	0.659235
Standard error	16.54408
Observations	10

ANOVA

	df	SS	MS	F	Significance F
Regression	1	5039.248	5039.248	18.41114	0.002648
Residual	8	2189.652	273.7065		
Total	9	7228.9			

	Coefficients	Standard error	t Stat	P-value	Lower 95%	Upper 95%	Lower 95.0%	Upper 95.0%
Intercept	17.8329	15.20692	1.172684	0.274656	−17.2343	52.90014	−17.2343	52.90014
X Variable 1	1.171455	0.273014	4.29082	0.002648	0.541883	1.801027	0.541883	1.801027

Separate functions may also be used to find the slope and intercept. Use the function = SLOPE(known_y's, known_x's) to obtain the slope. For example, when the ten *y* observations of Table 8.1 are entered into column A and the ten *x* observations are entered into column B, the function is = SLOPE(A1:A10,B1:B10); *Excel* reports a slope of 1.171455.

The intercept is found similarly, by replacing the word SLOPE with the word INTERCEPT. This yields a result of 17.8329 for the intercept. *Excel* reports more digits than are usual or necessary.

SOLVED EXERCISES

1. Find the slope and intercept of the regression line from the following information:

	Dependent variable	Independent variable
Mean	19.4	10.6
Std. dev.	8.2	6.4

You are also given that the correlation between the dependent and independent variables is $r = 0.31$. In addition, construct the ANOVA table, assuming that there are 25 observations. Indicate whether the independent variable has a significant effect on the dependent variable, and justify your answer.

Solution. We first use the relationship between the slope and the correlation coefficient (Equation 8.10):

$$b = r\frac{s_y}{s_y} => b = 0.31(8.2)/6.4 = 0.3972$$

We can find the intercept by recognizing that the regression line goes through the mean. Using the second equation in (8.9), we have that the intercept, a, is equal to

$$a = \bar{y} - b\bar{x}; \ a = 19.4 - 0.3972(10.6) = 15.19$$

For the ANOVA table, we can first recognize that the total sum of squares refers to the variability in the dependent (y) variable. The variance is equal to the total sum of squares, divided by $n - 1$. Thus, the (total) sum of squares is equal to the variance of y, multiplied by $n - 1$: $(25-1)(8.2^2) = 1613.76$.

Since r^2 is equal to the proportion of the total variability in y explained by the regression, r^2 is also equal to the ratio of the regression sum of squares to the total sum of squares. Thus, we have $0.31^2 =$ (regression sum of squares)/1613.76. This implies that the regression sum of squares is equal to $1613.76(0.31^2) = 155.08$. Since regression and residual sum of squares must add to the total sum of squares, the residual sum of squares is equal to $1613.76 - 155.08 = 1458.68$.

The degrees of freedom associated with the regression sum of squares is equal to the number of independent variables (1); the degrees of freedom for the total sum of squares is equal to $n - 1$ ($= 24$), and since the degrees of freedom for regression and residual sums of squares must total to 24, we see that the degrees of freedom for the residual sum of squares must be $n - 2$, or 23. The mean squares are found by dividing the sums of squares by the respective degrees of freedom; thus, $155.08/1 = 155.08$, and

(Continued)

(Continued)

1458.68/23 = 63.42. Finally, the observed *F*-statistic is found by dividing the regression mean square by the residual mean square: 155.08/63.42 = 2.445. We fail to reject the null hypothesis, since this is less than the critical value of *F* (of 4.28, found in Table A.5, using $\alpha = 0.05$, and 1 degree of freedom for the numerator (which is associated with the first column of the table), and 23 degrees of freedom for the denominator.

	Sum of squares	df	Mean square	F
Regression	155.08	1	155.08	2.445
Residual	1458.68	23	63.42	
Total	1613.76	24		

2. A regression of income (in thousands of dollars) on education (in years) is performed on data collected from 27 individuals. The results are summarized below:

   Intercept   31.8

   Slope   6.1

	Sum of squares	df	Mean square	F
Regression	2000			
Residual	20,000			
Total				

   (a) Fill in the blanks in the ANOVA table.

   (b) What is the predicted income for someone with 11 years of education?

   (c) In words, what is the meaning of the coefficient 6.1?

   (d) Is the regression coefficient significantly different from zero? How do you know?

   (e) What is the value of the correlation coefficient?

   (f) What is the standard deviation of the residuals?

Solution.

   (a) The regression and residual sums of squares must add to the total sum of squares. The degrees of freedom associated with the regression sum of squares is always equal to the number of independent, explanatory variables. The degrees of freedom associated with the total sum of squares is equal to the number of observations, minus one. The mean squares are found by dividing the sums of squares by their respective degrees of freedom: 2000/1 = 2000, and 18,000/25 = 720. The F-ratio is the ratio of the regression mean square to the residual mean square.

	Sum of squares	df	Mean square	F
Regression	2000	1	2000	2.78
Residual	18,000	25	720	
Total	20,000	26		

(b) The predicted income for someone with 11 years of education is found by using the regression equation, $y = a + bx$, and substituting the intercept for a, the slope for b, and 11 for x: 31.8 + 6.1(11) = 98.9. Thus, the predicted income is $98,900.

(c) For every increase in education equal to one year, the predicted income for the individual will rise by $6100.

(d) The observed F-value of 2.78 can be compared with the critical value of F found from Table A.5. Using the 0.05 F-table (with 1 and 25 degrees of freedom), we see that the critical value is approximately 4.26 (found using the first column, and going down to 24, since there is no entry for 25). Since the observed value of F is less than the critical value, we conclude that the regression coefficient (i.e., the slope) is not significantly different from zero.

(e) The value of r^2 is equal to the regression sum of squares divided by the total sum of squares. Since r^2 = 2000/20,000 = 0.1, r = 0.316.

(f) The standard deviation of the residuals is equal to the square root of the residual mean square: $\sqrt{720}$ = 26.83 .

EXERCISES

1. A regression of weekly shopping trip frequency on annual income (data entered in thousands of dollars) is performed on data collected from 24 respondents. The results are summarized below:

Intercept   0.46

Slope   0.19

(Continued)

(Continued)

	Sum of squares	df	Mean square	F
Regression				
Residual	1.7			
Total	2.3			

(a) Fill in the blanks in the ANOVA table.

(b) What is the predicted number of weekly shopping trips for someone making $50,000/year?

(c) In words, what is the meaning of the coefficient 0.19?

(d) Is the regression coefficient significantly different from zero? How do you know?

(e) What is the value of the correlation coefficient?

2. Name four assumptions of simple, linear regression.

3. A regression of infant mortality rates (annual deaths per thousand births) on median annual household income (data entered in thousands of dollars) is performed on data collected from 34 counties. The results are summarized below:

Intercept   18.46

Slope   −0.14

	Sum of squares	df	Mean square	F
Regression				
Residual	1.8			
Total	3.4			

(a) Fill in the blanks in the ANOVA table.

(b) What is the predicted infant mortality rate in a county where the median annual household income is $40,000?

(c) In words, what is the meaning of the coefficient −0.14? Do *not* simply say that this is the slope or the regression coefficient; indicate what it means and how it can be interpreted.

(d) What is the standard error of the residuals?

    (e) What is the value of the correlation coefficient?

4. A simple regression of y vs x reveals, for $n = 22$ observations, that $r^2 = 0.73$. The standard deviation of x is 2.3. The regression sum of squares is 1324. What is the value of the standard deviation of y? What is the value of the slope, b?

5. Given a simple regression with slope $b = 3$, $s_y = 8$, $s_x = 2$, and $n = 30$, find the standard error of the estimate (i.e., the standard deviation of the residuals).

6. The following data are collected in an effort to determine whether snowfall is dependent upon elevation:

Snowfall(inches)	Elevation(feet)
36	400
78	800
11	200
45	675

Without the aid of a computer, show your work on exercises (a) through (g).

    (a) Find the regression coefficients (the intercept and the slope coefficient).

    (b) Estimate the standard error of the residuals about the regression line.

    (c) Test the hypothesis that the regression coefficient associated with the independent variables is equal to zero. Also place a 95% confidence interval on the regression coefficient.

    (d) Find the value of r^2.

    (e) Make a table of the observed values, predicted values, and the residuals.

    (f) Prepare an analysis of variance table portraying the regression results.

    (g) Graph the data and the regression line.

7. Confirm the results given in Section 8.2.1 by using a computer software program. You may find small differences in the results due to round-off error.

8. Use the following data on distance and interaction frequency:

	y Interaction frequency	x Distance
Mean	10.3	6.2
Std. dev.	4.1	2.8

(Continued)

(Continued)

Correlation between interaction frequency and distance: −0.682.

Sample size: 30

(a) Find the slope and intercept, where interaction frequency is the dependent variable, and distance is the independent variable.

(b) Construct an ANOVA table, and indicate whether the regression is significant.

(c) What is the predicted interaction frequency for someone at a distance of 5.4 miles?

(d) What is the standard error of the estimate (i.e., the standard deviation of the residuals)?

9. The following ANOVA table results from the regression of y on x for a particular dataset:

	Sum of Squares	df	Mean Square	F
Regression	5000	1	5000	29.0
Residual	5000	29	172.4	
Total	10,000			

(a) Find r^2.

(b) If the standard deviation of x is equal to 18, find the slope, b.

10. Use *SPSS* or *Excel* and the Milwaukee dataset (described in Section 1.9.2) to:

(a) perform a regression using sales price as the dependent variable and lot size as the independent variable.

11. Use *SPSS* or *Excel* and the Hypothetical UK Housing Prices dataset (described in Section 1.9.4) to:

(a) perform a regression using house price as the dependent variable, and number of bedrooms as the independent variable;

(b) repeat part (a), using the number of bathrooms as the independent variable, and comment on the results.

On the Companion Website

The website includes a number of resources utilised in Chapter 8. Both the **Home Sales in Milwaukee, Wisconsin** dataset and the **Hypothetical UK Housing Prices** dataset are available within the 'Datasets' section of the website, including information about their respective fields and formats. The website also provides links to other websites and videos that also explain the concepts described here, to supplement the text.

9

MORE ON REGRESSION

9.1 MULTIPLE REGRESSION

It is often the case that there is more than one variable that is thought to affect the dependent variable. For example, housing prices are affected by many characteristics of both the house and the neighborhood. The number of shopping trips generated by a residential neighborhood is affected by the income of its residents, but also by the number of automobiles its residents own, accessibility to shopping alternatives, and so on.

With p independent, explanatory variables, the regression equation is:

$$\hat{y} = a + b_1 x_1 + b_2 x_2 + \ldots + b_p x_p \qquad (9.1)$$

where $\hat{y}$ is the predicted value of the dependent variable. With a given set of observations on the dependent (y) and independent (x) variables, the problem is to find the values of the parameters a and $b_1, b_2, \ldots, b_p$. The solution is found by minimizing the sum of the squared residuals:

$$\min_{\{a, b_1 \ldots b_p\}} (y - a - b_1 x_1 - \ldots - b_p x_p)^2 \qquad (9.2)$$

The problem and solution are identical in concept to that of bivariate regression discussed in the previous chapter, except that there are now more parameters to estimate, and the geometric interpretation is carried out in a higher-dimensional space. If $p = 2$, we wish to find a, b_1, and b_2 by fitting a plane through the set of points plotted in a three-dimensional space (where the axes are represented by the y-variable and the two x-variables (see Figure 9.1). The intercept a is the point of the plane on the y-axis, when $x_1 = x_2 = 0$. The value of b_1 describes how much the value of y changes in the plane when x_1 increases by one unit, along any line where x_2 is constant. Similarly, the value of b_2 describes the change in y when x_2 changes by one unit, while x_1 is held constant. Although it is difficult (if not impossible!) to visualize, we wish to find the minimum of a four-dimensional parabolic cone, where the sum of squared residuals (i.e., the quantity being minimized in Equation 9.2) is represented along the vertical axis, and the values of a, b_1, and b_2 are represented along the other dimensions.

More generally, for p independent variables, we fit a p-dimensional hyperplane through the set of points that are plotted in a $p + 1$ dimensional space (one dimension for the y-variable and p additional dimensions, one for each of the independent variables). The coefficients a and $b_1...,b_p$ are found at the base of a $p + 2$ dimensional parabolic cone. Though it is of course not possible to actually picture this for high-dimensional spaces, this geometric description serves to reinforce what is actually being carried out in regression analysis.

When there is more than one independent variable, the value of r^2 may be interpreted in at least two ways. One interpretation is the usual one, as the proportion of the variance in the dependent variable that is 'explained' by the independent variables. In addition, it may also be thought of as the maximal correlation between the dependent variable and a weighted combination of the independent variables. (The weights happen to be the regression coefficients that result when the dependent and independent variables are put in their standardized, z-scores form.)

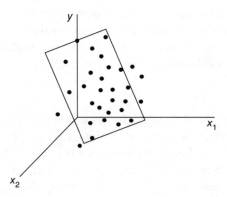

Figure 9.1 Fitting a plane through a set of points in three dimensions

9.1.1 Multicollinearity

In addition to the assumptions given for bivariate regression in the previous chapter, multiple regression analysis makes use of one additional assumption. In particular, it is assumed that there is no *multicollinearity* among the independent variables. This means that the correlation among the explanatory *x*-variables should not be high. In the extreme case where two variables are perfectly correlated, it is not possible to estimate the coefficients (and computer software will not provide results for this situation). In the more common case where multicollinearity is high, but not perfect, the estimates of the regression coefficients become very sensitive to individual observations; the addition or deletion of a few observations can change the coefficient estimates dramatically. Also, the variance of the coefficient estimates becomes inflated. Because the coefficients are less stable, it is not uncommon for insignificant independent, explanatory variables to appear significant.

9.1.2 Interpretation of Coefficients in Multiple Regression

Suppose that a regression of house prices in dollars (y) on lot size in square feet (x_1) and the number of bedrooms (x_2) results in the following equation:

$$\hat{y} = 4000 + 20x_1 + 10,000x_2 \qquad (9.3)$$

The coefficient on lot size means that every increase of one square foot adds an average of $20 to the house price, *holding constant* the number of bedrooms. Similarly, the coefficient on the number of bedrooms implies that an added bedroom will increase the value of the house by an estimated $10,000, for houses with identical lot sizes.

   As with simple regression, the coefficients tell us the effect on the dependent variable of an increase of one unit in the independent variable. In addition, they control for the effects of other variables in the equation. That is, to understand the effect of a particular explanatory variable on the dependent variable, it is not sufficient to simply include it in the right-hand side of a regression equation. Since other variables may also affect the dependent variable, they also have to be included, so the separate effects of each contributing variable may be estimated. If all relevant variables are not included, this may lead to misspecification error. This is explored further in Section 9.2.

9.1.3 Adjusted r^2

The calculated value of r^2 overestimates the true value, ρ^2. Note that if the number of observations is exactly equal to one more than the number of independent variables, r^2 will always equal one, even if the variables are unrelated (i.e., $\rho^2 = 0$). Thus a straight line fit on a graph with one independent variable using two observations will necessarily have a perfect fit (simply a straight line through the two points). Similarly, when there are three observations and two independent variables, a three-dimensional plane will fit perfectly – a plane can go through all three of the points when any three points are plotted in the three dimensional space based on y and the two independent x variables.

If $\rho^2 = 0$, then the expected value of r^2 is $p/(n-1)$, where p is the number of independent variables, and n is the number of observations. For example, if $p = 10$ and $n = 21$, then the expected value of r^2 is $10/20 = 0.5$, even when the individual variables are not truly correlated! This serves to emphasize the importance of having a large number of observations, relative to the number of variables. The *adjusted* r^2 represents a downward adjustment of r^2 that takes this difference between the sample and population values into account. We will see examples of this adjustment in the illustration given in Sections 9.4. You may also look back on the sample output associated with Sections 8.9 and 8.10 to see the adjusted r^2 values in the illustration where there was just one independent variable.

9.2 MISSPECIFICATION ERROR

Suppose that Equation 9.3 characterizes the 'true' relationship between house prices, lot size, and the number of bedrooms. We will examine the effects of a misspecified regression equation that has an omitted variable by first generating some sample data from the equation:

$$y = 4000 + 20x_1 + 10{,}000x_2 + \varepsilon, \tag{9.4}$$

where ε is a normal random variable with mean 0 and variance equal to 3000^2 (this implies that one can, with 95% confidence, predict house prices within about two standard deviations, or $6000). Table 9.1 displays the data associated with ten observations, where the data on lot size and the number of bedrooms are simply hypothetical, and then Equation 9.4 is used to generate housing prices.

Table 9.1   Observations of house price, lot size and number of bedrooms

House price	Lot size	Bedrooms
132,767	5000	3
134,689	5500	2
159,718	6000	4
164,937	6500	3
132,489	5200	2
125,766	5400	1
146,568	5700	3
168,932	6100	4
171,180	6300	4
187,921	6400	5

Now suppose that we incorrectly assume that house prices are a function of lot size only. A regression of house prices on lot size yields:

$$\hat{y} = -57,809.7 + 36.2x_1$$
$$(-1.7)\ (6.21)$$

(9.5)

where the t-values associated with each coefficient are given below the equation in parentheses. We see that the coefficient of lot size is significant (since its t-value is greater than the one-sided critical value of $t = 1.86$ with $n - 2 = 8$ degrees of freedom, using $= 0.05$), and is in the 'right' direction (i.e., larger lot sizes lead to higher housing prices, as we would expect), but it is quite a bit larger than the 'true' value of 20. We have overestimated the effect of lot size by omitting the number of bedrooms, which also affects housing price.

Similarly, if we had incorrectly assumed that house prices were a function of the number of bedrooms only, we would find that the regression equation, based on the observed data is:

$$\hat{y} = 103,361 + 15,580x_2$$
$$(12.05)\quad (6.1)$$

(9.6)

The effect of number of bedrooms on housing price is significant and in the expected direction, but we have again overestimated somewhat the 'true' effect of an added bedroom, which we know to be \$10,000.

In both Equations 9.5 and 9.6, the true magnitude of the variable's positive effect has been overestimated. There is a tendency for errors to occur in this particular direction when other variables that also have a positive effect on the dependent variable are omitted. Essentially, the variable that is in the equation is capturing both its own effect on the dependent variable, as well as some of the effect of the omitted variable.

Finally, if we use the data to estimate a regression equation with both independent variables, we find

$$\hat{y} = -1993 + 21.6x_1 + 9333x_2$$
$$(-0.13)\ (7.12)\ \ (7.24)$$

(9.7)

Both variables have significant effects on housing prices, and, more importantly, we have estimated their effects on housing prices quite accurately, since the coefficient of 21.6 is

Table 9.2 Regression results associated with the house price example

	Coefficient	Standard deviation	Confidence interval
Intercept	−1993	14,881	(−37,181, 33,195)
X_1	21.6	2.98	(14.6, 28.7)
X_2	9333	1310	(6234, 12,432)

close to its true value of 20, and the coefficient of 9333 is close to its true value of 10,000. The intercept is not too close to its true value of 4000, but we note from Table 9.2 that all of the true coefficients are within two standard deviations of their estimated values, and that all true values of the coefficients lie within their estimated confidence intervals.

9.3 DUMMY VARIABLES

It is sometimes necessary to include explanatory, independent variables that are categorical. For instance, income is often not reported exactly, but is instead classified into a category. Similarly, locations may be classified into, for example, central city, suburb, or rural categories.

To handle independent variables that have, say, k categories in regression analysis, we create $k - 1$ variables. One category is arbitrarily omitted; this is often the first (e.g., lowest income) or the last (e.g., highest income). Each of these new variables represents a category and is a binary, 0–1 *dummy variable*. An observation is assigned a value of one for a particular dummy variable if the observation *is* in the corresponding category, and it is assigned a value of zero if it *is not* in that category.

Consider the example in Table 9.3, where individuals either report, or are assigned one of three locations – central city, suburb, or rural. We will arbitrarily choose the rural region as the omitted category. We define $k - 1 = 2$ categories, the first associated with the central city and the second associated with the suburb.

Table 9.3 Coding scheme for dummy variables

Individual	Location	x_1	x_2
1	Central city	1	0
2	Central city	1	0
3	Suburb	0	1
4	Rural	0	0
5	Suburb	0	1

The first two individuals live in the city – they are each assigned $x_1 = 1$ since they live in the city, and $x_2 = 0$ since they do not live in the suburb. Individuals 3 and 5 live in the suburb, so they are assigned $x_1 = 0$ since they do not live in the city, and $x_2 = 1$ since they live in the suburb. Note that individual 4 lives in the rural region, and is assigned a value of 0 on both x_1 and x_2, since the person does not live in the city or the suburb.

The reason that a category is always omitted when dummy variables are employed has to do with multicollinearity. If all categories were included, there would be perfect multicollinearity, and this violates an assumption of multiple regression. Perfect multicollinearity would occur if we defined k dummy variables, since the sum of the k columns in any given row would always equal one. In our example above, we have included only two of the three categories; there is no reason to include a separate column for the third category, since it supplies us with redundant information (for example, we know that if individual 4 does not live in the central city or the suburb, he or she must live in the rural area).

Dummy variables are coded as 0 or 1 only, and not, for example, 1 or 2. The 0/1 coding is a result of the fact that the dummy variable is a nominal, categorical variable, and 0/1 coding corresponds to absence/presence. As we will see, it also facilitates interpretation of regression coefficients.

Once dummy variables are defined, regression analysis proceeds in the usual way. Suppose that individuals 1 through 5 are observed to make 3, 4, 7, 1, and 5 weekly shopping trips, respectively. The resulting, best-fitting regression equation is found from a computer program to be:

$$y = 1 + 2.5x_1 + 5x_2 \qquad (9.8)$$

Table 9.4 displays the observed and predicted values.

Table 9.4   Weekly shopping trip frequency

Individual	Observed	Predicted
1	3	3.5
2	4	3.5
3	7	6
4	1	1
5	5	6

The regression results in (9.8) may be interpreted as follows. Being located in the rural region implies that x_1 and x_2 are zero, and so the predicted value of y is simply the intercept (equal to 1 in this example). Thus, when there are no other variables in the equation, the intercept in dummy variable regression is the predicted value of the dependent variable for the omitted category. Estimated regression coefficients (i.e., slopes) are interpreted relative to the omitted category. For example, being located in the central city is 'worth' an extra 2.5 shopping trips, relative to the omitted category. Therefore, we predict that someone located in the central city will shop an average of 3.5 times per week. Being in the suburb is 'worth' an extra five shopping trips per week, again, relative to the omitted (rural) category. Therefore, individuals residing in the suburb are predicted to shop $1 + 5 = 6$ times per week.

You may be wondering at this point what would have happened if we had omitted a category other than the rural region. Suppose that the central city had been chosen as the omitted category. Then our data would look like those in Table 9.5.

Here we define $x_1 = 1$ if the individual lives in the suburb, and $x_2 = 1$ if the individual lives in the rural region. Using these data in a multiple regression analysis yields:

$$\hat{y} = 3.5 + 2.5x_1 - 2.5x_2 \qquad (9.9)$$

Table 9.5 An alternative coding scheme

Individual	Location	x_1	x_2	Weekly shopping trips
1	Central city	0	0	3
2	Central city	0	0	4
3	Suburb	1	0	7
4	Rural	0	1	1
5	Suburb	1	0	5

The coefficients are different, but when we interpret them in light of the new variable definitions, we come to the same conclusions as before (as we should!). For instance, the intercept of 3.5 is the predicted number of weekly shopping trips made by those in the central city (the omitted category). The coefficient of +2.5 is the extra number of shopping trips made by suburban residents, relative to the omitted category. Thus, suburban residents are predicted to make 3.5 + 2.5 = 6 shopping trips per week, the same as in the previous example. Similarly, the coefficient of −2.5 means that rural residents make 2.5 fewer shopping trips than central city residents do each week (3.5 − 2.5 = 1.0).

9.3.1 Dummy Variable Regression in a Recreation Planning Example

Part of the statewide recreation planning process is to generate estimates and forecasts of recreation activity. Annual participation in a specific recreation activity is taken to be a function of variables such as age, income, and population density. In New York state, a survey of approximately 7500 people was undertaken; individuals were asked about their participation frequencies in various activities, and their age, income, and location were recorded. The independent, explanatory variables were recorded as dummy variables. The dependent variable is the number of times the individual participated in the activity at organized facilities (public or private), over the course of a year. A multiple regression analysis was run for each recreation activity. Table 9.6 displays the results.

For each independent variable, the highest category is omitted (i.e., high income, elderly, and urban locations). Recall that the coefficients are to be interpreted relative to these omitted categories. Thus, a person in the low-income category swims, on average, 5.55 times less per year than a person in the omitted high-income category. By using these coefficients, one can estimate the participation frequency for any activity and any set of categories. For example, how often does a middle-income, young adult, living in the rural regions of the state, participate in court games at organized facilities each year? The answer is equal to 1.34 (the intercept, called the 'base participation rate' in the table), plus 0.43 (the coefficient associated with those in the middle-income category), plus 5.76 (for being a young adult), minus 3.65 (for those living in the rural area). Our estimate is therefore equal to 3.88 times per year. It should of course be kept in mind that this is an *average* participation rate, across all individuals in that category.

Table 9.6 Recreation participation coefficients

Activity	Base participation rate[a]	Income				Age				Population density		
		Low	Low-middle	Middle	High-middle	Youth	Young adult	Adult	Middle aged	Rural	Exurban	Suburban
Swimming	2.45	-5.55	-4.37	-0.73	3.22	22.83	9.83	8.94	2.91	8.51	5.98	4.78
Biking	-0.06	-0.94	-0.11	-0.92	0.07	21.63	6.21	3.82	1.17	1.64	2.95	1.31
Court games	1.34	-0.55	0.63	0.43	1.31	16.41	5.76	1.65	0.12	-3.65	-2.11	-1.54
Camping	0.27	-0.13	-0.01	0.34	0.39	1.93	0.44	-0.01	-0.19	1.25	0.41	0.80
Tennis	0.74	-2.30	-2.42	-2.45	-0.46	7.76	4.28	3.31	0.61	1.48	1.78	2.14
Picnicking	0.66	-0.15	0.31	0.48	0.67	1.87	2.23	1.84	0.67	2.30	0.97	1.10
Golf	0.93	-1.44	-1.38	-0.71	-0.58	-0.30	-0.30	0.28	0.70	1.02	1.34	0.42
Fishing	-0.21	1.35	0.17	0.57	0.83	3.36	1.26	2.01	0.72	1.77	1.41	1.17
Hiking	1.31	0.22	0.20	-0.04	0.77	2.49	0.65	0.47	0.18	0.53	0.54	0.30
Boating	0.24	-1.06	-0.77	-0.09	0.45	3.58	1.42	1.28	0.50	2.36	1.68	1.86
Field games	-0.17	-0.39	-0.19	1.53	0.86	8.22	2.73	0.94	-0.13	1.48	0.83	0.56
Skiing	0.70	-0.98	-1.10	-0.86	-0.36	0.85	0.37	0.63	0.03	0.19	0.78	0.31
Snowmobiling	-0.07	-0.50	-0.48	-0.08	-0.37	0.33	0.85	0.20	0.14	1.52	0.97	0.44
Local winter	0.66	-1.49	-1.18	-0.80	0.15	5.13	1.34	0.82	0.06	1.84	-0.32	1.05

[a] The control group selected was the highest income, age, and density group. This column gives the estimated base annual participation rate for this group. The other columns give the amounts to be added to this amount to obtain participation rates for any other income, age, and density group. Negative participation rates are possible and should be interpreted as zero.

Source: New York State Office of Parks and Recreation (1978).

If an individual happens to be in an omitted category, the implied coefficient is equal to zero. Thus, high-middle-income elderly residents of urban areas swim, on average, 2.45 (base participation rate, or intercept), plus 3.22 (income coefficient), plus 0 (since they are in the omitted, elderly category), plus 0 (since they are also in the omitted, urban category), which is equal to 5.67 times per year.

Recreation planners use these coefficients to plan for future use. Demographic projections provide forecasts of the number of people in each age/income/population density category. If we can assume that the coefficients in the table constitute reliable future estimates of individual recreation participation, we can use the coefficients together with the demographic forecasts to project how much demand there will be for each recreation activity. The state can then prioritize recreation projects in a way that will best meet the anticipated demand.

9.4 MULTIPLE REGRESSION ILLUSTRATION: SPECIES IN THE GALÁPAGOS ISLANDS

The data in Table 9.7 contain two possible dependent variables related to the number of species found on 30 islands (total species, and number of native species), as well as five

Table 9.7   Galápagos Islands: species and geography

Island	Observed species		Area km²	Elevation (m)	From nearest island	Distance (km) From Santa Cruz	Area of adjacent island (km²)
	Number	Native					
1 Baltra	58	23	25.09	–	0.6	0.6	1.84
2 Bartolomé	31	21	1.24	109	0.6	26.3	572.33
3 Caldwell	3	3	0.21	114	2.8	58.7	0.78
4 Champion	25	9	0.10	46	1.9	47.4	0.18
5 Coamaño	2	1	0.05	–	1.9	1.9	903.82
6 Daphne Major	18	11	0.34	119	8.0	8.0	1.84
7 Daphne Minor	24	–	0.08	93	6.0	12.0	0.34
8 Darwin	10	7	2.33	168	34.1	290.2	2.85
9 Eden	8	4	0.03	–	0.4	0.4	17.95
10 Enderby	2	2	0.18	112	2.6	50.2	0.10
11 Espanola	97	26	58.27	198	1.1	88.3	0.57
12 Fernandina	93	35	634.49	1494	4.3	95.3	4669.32
13 Gardner*	58	17	0.57	49	1.1	93.1	58.27

(Continued)

Table 9.7 (Continued)

Island	Observed species		Area km²	Elevation (m)	From nearest island	Distance (km) From Santa Cruz	Area of adjacent island (km²)
	Number	Native					
14 Gardner[†]	5	4	0.78	227	4.6	62.2	0.21
15 Genovesa	40	19	17.35	76	47.4	92.2	129.49
16 Isabela	347	89	4669.32	1707	0.7	28.1	634.49
17 Marchena	51	23	129.49	343	29.1	85.9	59.56
18 Onslow	2	2	0.01	25	3.3	45.9	0.10
19 Pinta	104	37	59.56	777	29.1	119.6	129.49
20 Pinzón	108	33	17.95	458	10.7	10.7	0.03
21 Las Plazas	12	9	0.23	–	0.5	0.6	25.09
22 Rabida	70	30	4.89	367	4.4	24.4	572.33
23 San Cristóbal	280	65	551.62	716	45.2	66.6	0.57
24 San Salvador	237	81	572.33	906	0.2	19.8	4.89
25 Santa Cruz	444	95	903.82	864	0.6	0.0	0.52
26 Santa Fé	62	28	24.08	259	16.5	16.5	0.52
27 Santa Maria	285	73	170.92	640	2.6	49.2	0.10
28 Seymour	44	16	1.84	–	0.6	9.6	25.09
29 Tortuga	16	8	1.24	186	6.8	50.9	17.95
30 Wolf	21	12	2.85	253	34.1	254.7	2.33

* Near Espanola. † Near Santa Maria

Note: Some island names are not accurate (e.g., Seymour is actually North Seymour) and others have alternative names.

Source: Andrews and Herzberg (1985). The values marked '-' are not known.

potential independent variables that may help us to understand why different islands have different numbers of species. A map of the region is shown in Figure 9.2.

In our example, we will use the total number of species as the dependent variable. We will now explore some of the choices and questions that one is faced with in arriving at a suitable regression model. All of the output shown in the tables is from *SPSS for Windows*.

9.4.1 Model 1: The Kitchen-sink Approach

One idea would be to simply put all five independent variables on the right-hand side, and see what happens – i.e., everything is put into the equation except the kitchen sink! This

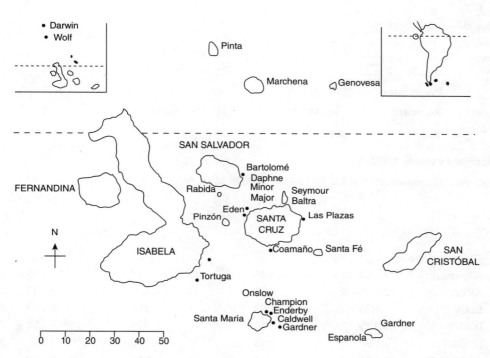

Figure 9.2 Galápagos Islands

Table 9.8 The kitchen-sink model

Variables entered/removed[a]

Model	Variables entered	Variables removed	Method
1	AREAADJ, DISSC, AREA, DISNISL, ELEV[b]		Enter

[a] Dependent variable: SPECIES.

[b] All requested variables entered.

Model summary

Model	R	R square	Adjusted R square	Std. error of the estimate
1	.877[a]	.768	.707	65.9482

[a] Predictors: (Constant), AREAADJ, DISSC, AREA, DISNISL, ELEV.

(Continued)

Table 9.8   (Continued)

ANOVA[a]

Model		Sum of squares	df	Mean square	F	Sig.
1	Regression	274097.4	5	54819.479	12.605	.000[b]
	Residual	82634.046	19	4349.160		
	Total	356731.4	24			

[a] Dependent variable: SPECIES.

[b] Predictors: (Constant), AREAADJ, DISSC, AREA, DISNISL, ELEV.

Coefficients[a]

Model		Unstandardized coefficients		Standardized coefficients	t	Sig.
		B	Std. error	Beta		
1	(Constant)	11.485	25.626		.448	.659
	AREA	−2.85E−02	.025	−.220	−1.141	.268
	ELEV	.330	.063	1.212	5.270	.000
	DISNISL	−.155	1.149	−.019	−.134	.894
	DISSC	−.260	.243	−.149	−1.068	.299
	AREAADJ	−7.96E−02	.020	−.611	−3.903	.001

[a] Dependent variable: SPECIES

approach is *not* recommended as a way to arrive at a final regression equation, and it is shown here for illustrative purposes only. The output in Table 9.8 shows the value of r^2 to be 0.768 (in the output, the correlation coefficient is depicted using an upper-case 'R'). The adjusted r^2 value is lower (0.707), in keeping with the observation made in section 9.3.1 that r^2 overestimates the true correlation and should be adjusted downward.

It is also of interest to look at the signs and significance of the individual variables. There are two significant variables – elevation and the area of the adjacent island. This can be seen by noting that these two variables have entries of less than 0.05 in the 'Sig.' column (in *SPSS*, p-values are given in a column labeled 'Sig.'). Elevation has a positive effect on species number, and the area of the adjacent island has a negative effect. Note that the sign of the area variable is negative, although the magnitude of the effect is insignificant, since the p–value is greater than 0.05. This is counter to one's intuition that *more* species would be found on larger islands. These low p-values are associated with high t-values – the t-statistic of 5.27 for elevation and the t-value of –3.903 for area of the adjacent island are clearly significant. Note that values in the 't' column may be derived by dividing the coefficient (in the 'B' column) by the associated standard error (where the latter quantity is identical to the standard deviation of the slope – see

Equation 8.25 and Section 8.6 as a reminder of how this was discussed for the case of one independent variable).

The standard error of the estimate is 66, which is approximately equivalent to the average absolute value of a residual (in this case, the actual average absolute value of a residual is 44). This is pretty high, since half of the islands have fewer than 44 species! Finally, note that the ANOVA table is similar to that in the univariate case, with the regression degrees of freedom equal to the number of independent variables.

It is often tempting to use the kitchen-sink approach because, when an independent variable is added to the regression equation, the r^2 value *always* increases. It is important to realize that a high r^2 value is *not* the primary goal of regression analysis; if it was, we could simply keep adding explanatory variables until we achieved our desired value of r^2! A more reasonable strategy often involves either (a) deleting variables that do not reduce the value of r^2 very much, and/or (b) adding variables only when they increase r^2 appreciably. Variable selection is discussed in more detail in Section 9.5.

9.4.2 Missing Values

Delving right into the analysis without a consideration of certain questions can lead to misinterpretation. Before we start, we should decide on how we are going to treat missing values. In the table, there are five missing values of elevation. All of the missing values, with the exception of the first observation, are for islands that have extremely small areas (and, with the exception of Seymour, small numbers of species). There are various ways we can proceed, including:

1. Delete all observations with missing values. This is the default option used by many software packages (and the option used in deriving the results shown in Table 9.8).

2. Replace missing values with the mean. Most statistical software packages have an option that allows missing values to be replaced with the mean of the remaining values. This is often not a good option; in many instances (including the present one), values are missing because they are *not* typical observations, and hence replacing them with the mean may be misleading.

3. Use the other independent variables, or some subset of them, to predict the missing value. We could perform an initial regression of elevation on the other four independent variables for the nonmissing cases. Then we can use the results to predict the value of elevation, based upon the values of the other independent variables, for the missing observations.

Which option, or combination of options, should we choose? Option two is not a reasonable one here. The mean elevation is 412 m, and it would be foolish to suppose that the unknown elevations are this great on the very small islands for which we have no data.

For the very small islands (observations 5, 9, and 21), one can justifiably exclude them from the analysis. Although we might also exclude Baltra and Seymour, here we will estimate their elevations from a regression of elevation on area. The resulting regression equation is:

$$\text{Elevation} = 300 + 0.358(\text{Area}) \tag{9.10}$$

We estimate Baltra's elevation as $300 + 25.09(0.358) = 309$ m, and Seymour's elevation as $300 + 1.84(0.358) = 301$ m. (In this particular case, one can find the missing elevation values from other sources; the actual values are considerably less than the estimates we have derived here.)

9.4.3 Outliers and Multicollinearity

A cursory examination of the data reveals that there are a small number of large islands. The presence of outliers is not an unusual feature of many studies, and it is important that we know whether these observations are exerting a significant effect on the results. *Leverage values* are designed to indicate how influential particular observations are in regression analysis. One guideline is that if the leverage value exceeds $2p/n$, where p is the number of independent variables, the observation may be considered as an outlier.

We also want to make sure that multicollinearity is not exerting an undue influence on the results. An examination of the correlations among the independent variables will reveal those where high correlations may exist. The *tolerance* is equal to the proportion of variance in an independent variable that is not explained by the other independent variables. It is equal to $1 - r^2$, where the r^2 is associated with the regression of the independent variable on all other independent variables. A low tolerance indicates problems with multicollinearity, since the variable in question has a high correlation with the other independent variables. The reciprocal of the tolerance is the *variance inflation factor* (VIF); a common rule of thumb is that if it is greater than about 5, this indicates potential multicollinearity problems. However, a word of caution is in order here. As we saw in Chapter 7, correlation depends upon sample size and hence such rules of thumb are not always indicative of the importance of correlation.

9.4.4 Model 2

In this second model, we have accounted for the missing elevation data, and have collected information on outliers and multicollinearity.

From the output (Table 9.9), we see that the inclusion of Baltra and Seymour has not changed the results very much. The r^2 value is 0.755, the standard error of the residuals is about 65, and elevation and the area of the adjacent island are still the only significant independent variables. We learn, however, that the variance inflation factor is slightly high (though not greater than the rule-of-thumb value of 5) for elevation and area. This is not surprising when we also inspect the correlation matrix (not shown here), which reveals a very high correlation between the two variables.

Table 9.9 Regression estimation with outliers removed

Variables entered/removed[a]

Model	Variables entered	Variables removed	Method
1	AREAADJ, DISSC, AREA, DISNISL, ELEV[b]		Enter

[a] Dependent variable: SPECIES.
[b] All requested variables entered.

Model summary[a]

Model	R	R square	Adjusted R square	Std. error of the estimate
1	.869[b]	.755	.697	64.8830

[a] Dependent variable: SPECIES.
[b] Predictors: (Constant), AREAADJ, DISSC, AREA, DISNISL, ELEV.

ANOVA[a]

Model		Sum of squares	df	Mean square	F	Sig.
1	Regression	272396.7	5	54479.338	12.941	.000[b]
	Residual	88405.975	21	4209.808		
	Total	360802.7	26			

[a] Dependent variable: SPECIES.
[b] Predictors: (Constant), AREAADJ, DISSC, AREA, DISNISL, ELEV.

Coefficients[a]

Model		Unstandardized coefficients		Standardized coefficients	t	Sig.	Collinearity statistics	
		B	Std. error	Beta			Tolerance	VIF
1	(Constant)	3.501	24.263		.144	.887		
	AREA	−2.50E−02	.024	−.192	−1.023	.318	.331	3.026
	ELEV	.325	.061	−1.191	5.291	.000	.230	4.339
	DISNISL	−7.93E−03	1.124	−.001	−.007	.994	.592	1.689

(Continued)

Table 9.9   (Continued)

Model		Unstandardized coefficients		Standardized coefficients			Collinearity statistics	
		B	Std. error	Beta	t	Sig.	Tolerance	VIF
	DISSC	−0.222	.237	−.130	−.936	.360	.604	1.654
	AREAADIJ	−7.76E−02	.020	−.594	−3.880	.001	.498	2.009

[a] Dependent variable: SPECIES.

Collinearity diagnostics[a]

Model	Dimension	Eigenvalue	Condition Index	Variance proportions				
				(Constant)	AREA	ELEVL	DISNISL	DISSC
1	1	3.170	1.000	.02	.01	.01	.02	.02
	2	1.416	1.496	.00	.06	.01	.06	.04
	3	.764	2.037	.00	.12	.00	.00	.00
	4	.347	3.021	.45	.09	.00	.23	.05
	5	.231	3.702	.00	.05	.02	.59	.77
	6	7.222E−02	6.625	.52	.67	.95	.10	.12

[a] Dependent variable: SPECIES.

One option for treating multicollinearity is to exclude from the analysis one of the highly correlated variables. Here area and elevation are correlated, and we might decide to drop one of the two, since they are close to redundant (and since the sign of one of them is not correct). Which one should we drop? The choice should come primarily from a consideration of the underlying process, and, secondarily, the magnitude of the variance inflation factors. Both area and elevation might reasonably affect species number, but we will choose to exclude area because it might be argued that elevation is relatively more important in terms of species diversity. A note of caution is again in order here. Dropping variables from the analysis should only occur after a well-reasoned consideration of the underlying process. It does little to advance understanding of the process when important variables are dropped from the regression equation solely because they don't perform well.

The leverage values (output from *SPSS*, and not shown in the table) also reveal that several outliers have an important impact upon the results. Leverage values exceed the rule-of-thumb value of $2p/n = 10/27 = 0.37$ for observations 8, 12, 15, and 16. Fernandina (observation 12) and Isabela (observation 16) have by far the two highest elevations among the 30 islands. There seems to be less justification for deleting the other two observations. Darwin (observation 8) and Genovesa (15) are geographic outliers, but there are also other geographic outliers with low leverage values.

9.4.5 Model 3

In this third model, we have deleted area as an independent variable, and have deleted the observations for Fernandina and Isabela. Note that we have chosen to delete only two of the outliers; we are able to justify their removal because of their very large elevations. One should avoid removing outliers from the analysis unless there is a reasonably compelling rationale. The output (Table 9.10) shows that only elevation remains significant. The value of r^2 remains high at 0.736, and the standard error of the estimate is slightly lower, at about 62. Multicollinearity is not an issue, since all of the VIFs are less than 5. Three observations (Bartolomé, Darwin, and Rabida) are still

Table 9.10   Regression with missing data removed or estimated, and outliers and area variable removed

Variables entered/removed[a]

Model	Variables entered	Variables removed	Method
1	AREAADJ, ELEV, DISNISL, DISSC[b]		Enter

[a] Dependent variable: SPECIES.
[b] All requested variables entered.

Model summary[a]

Model	R	R square	Adjusted R square	Std. error of the estimate
1	.858[b]	.736	.684	62.2581

[a] Dependent variable: SPECIES.
[b] Predictors: (Constant), AREAADJ, ELEV, DISNISL, DISSC.

ANOVA[a]

Model		Sum of squares	df	Mean square	F	Sig.
1	Regression	216670.7	4	54167.666	13.975	.000[b]
	Residual	77521.336	20	3876.067		
	Total	294192.0	24			

[a] Dependent variable: SPECIES.
[b] Predictors: (Constant), AREAADJ, ELEV, DISNISL, DISSC.

(Continued)

Table 9.10　(Continued)

Coefficients[a]

Model		Unstandardized coefficients		Standardized coefficients			Collinearity statistics	
		B	Std. error	Beta	t	Sig.	Tolerance	VIF
1	(Constant)	5.179	24.690		−.210	.836		
	ELEV	.344	.050	.831	6.936	.000	.919	1.088
	DISNISL	−.267	1.086	−.036	−.246	.808	.602	1.662
	DISSC	−.170	.232	−.109	−.730	.474	.591	1.691
	AREAADJ	−.5.15E–02	.081	−.073	−.632	.534	.981	1.020

[a] Dependent variable: SPECIES.

Collinearity diagnostics[a]

Model	Dimension	Eigenvalue	Condition index	Variance proportions				
				(Constant)	ELEV	DISNISL	DISSC	AREAADJ
1	1	3.030	1.000	.02	.03	.03	.02	.02
	2	.932	1.804	.00	.00	.03	.03	.72
	3	.618	2.214	.03	.32	.07	.10	.16
	4	.274	3.325	.22	.09	.64	.25	.03
	5	.146	4.556	.73	.56	.23	.59	.07

[a] Dependent variable: SPECIES.

outliers. This is likely due to their extreme values on some of the independent varia-bles. Bartolomé and Rabida are both adjacent to large islands, and Darwin is a long way from Santa Cruz.

9.4.6 Model 4

To conclude with a parsimonious model, we can remove the variables that are not signifi-cant. Thus, we regress species number on elevation. The result is the equation:

$$\text{Species} = -24.27 + 0.35(\text{Elevation}) \tag{9.11}$$

From Table 9.11, the value of r^2 is still high at 0.716 (it is necessarily lower than before since we have removed variables, but it has not declined very much). The standard error of the estimate is about 60. Furthermore, a check of the leverage values reveals no outliers.

Table 9.11 Final regression equation for species data

Variables entered/removed[a]

Model	Variables entered	Variables removed	Method
1	ELEV[b]		Enter

[a] Dependent variable: SPECIES.
[b] All requested variables entered.

Model summary[a]

Model	R	R square	Adjusted R square	Std. error of the estimate
1	.846[a]	.716	.703	60.3225

ANOVA[a]

Model		Sum of square	df	Mean square	F	Sig.
1	Regression	210499.4	1	210499.4	57.848	.000[a]
	Residual	83692.624	23	3638.810		
	Total	294192.0	24			

Coefficients[a]

Model		Unstandardized coefficients		Standardized coefficients		
		B	Std. error	Beta	t	Sig.
1	(Constant)	−24.270	18.640		−1.302	.206
	ELEV	.350	.046	.846	7.606	.000

[a] Dependent variable: SPECIES.

9.5 VARIABLE SELECTION

As seen in the example, a common issue in regression analysis is the selection of variables that will appear as explanatory variables on the right-hand side of the regression equation. A brute-force approach to this question would be to try all possible combinations. With p potential independent variables, this would mean that we would try p

separate regressions that have just one independent variable, all $\binom{p}{2} = p(p-1)/2$ equations that have two variables, all $\binom{p}{3} = p(p-1)(p-2)/6$ equations that have three variables, and so on. If p is large, this is a lot of equations. Even with $p = 5$, one would have to try 31 regression equations. But perhaps the real drawback is that this is even more of a 'kitchen-sink' approach than is including all of the variables on the right-hand side. It amounts to an admission that we don't know what we are doing, and that our strategy is to just go with what looks best. It is always desirable to start from hypotheses and underlying processes first, in keeping with the principles of the scientific method described in the first chapter. In a more exploratory spirit, however, there may be some cases where we really have little in the way of a priori hypotheses – in this case, the all-possible regressions approach might be viewed as a potential way to generate new hypotheses.

An alternative way to select variables for inclusion in a regression equation is the *forward selection* approach. The variable that is most highly correlated with the dependent variable is entered first. Then, given that that variable is already in the equation, a search is made to see whether there are other variables that would be significant if added. If so, the one with the greatest significance is added. In this way, a regression equation is built up. The procedure terminates when there are no variables in the set of potential variables that would be significant, if entered into the equation.

Backward selection starts with the kitchen-sink equation, where all of the possible independent variables are in the equation. Then the one that contributes least to the r^2 value is removed, if the reduction in r^2 is not significant. The process of removing variables continues until the removal of any variable in the equation would constitute a significant reduction in r^2.

Stepwise regression is a combination of the forward and backward procedures. Variables are first added in the manner of forward selection. However, as each variable is added, variables entered on earlier steps are re-checked to see if they are still significant. If they are not still significant, they are removed. The process terminates when the addition of any variable not already in the equation would result in it being insignificant.

9.6 CATEGORICAL DEPENDENT VARIABLE

There are many situations where the dependent variable will be a categorical variable. For example, we may wish to model whether individuals patronize a park as a function of the distance to the park, or whether individuals commute by train as a function of automobile travel time. We may want to estimate the probability that a customer patronizes any of, say, four supermarkets as a function of characteristics of the stores and characteristics of the customers.

In each of these examples, the dependent variable is categorical, in the sense that possible outcomes may be placed into categories. An individual either goes to the park, or does not go. An individual either commutes by train, or does not. If there are only four choices of supermarkets in an area, the consumers may be classified according to which one they patronize.

When the dependent variable is categorical, special consideration must be given to how regression analyses are carried out. In this section, we will examine why this is the case, and we will learn how *logistic regression* may be used in such situations.

9.6.1 Binary Response

In the simplest case, there are two possible responses. For example, we may assign the dependent variable a value of $y = 1$ if the individual takes the train to work, and $y = 0$ otherwise. Suppose, for instance, we had the data in Table 9.12 for $n = 12$ respondents.

A cursory examination of the table reveals that there seems to be a tendency to take the train when the travel time by automobile is high. Where auto travel time is not as high, there is more of a tendency for the y variable to be zero, indicating that the individual drives to work.

We could begin by running an ordinary least squares analysis; we would find:

$$\hat{y} = -0.396 + 0.0153x \tag{9.12}$$

Table 9.12 Data on travel mode and automobile travel time

y	x: Auto travel time (min)
0	32
1	89
0	50
1	49
0	80
1	56
0	40
1	70
1	72
1	76
0	32
0	58

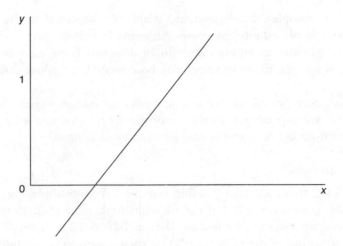

Figure 9.3 **Predicted probabilities outside the (0, 1) interval**

The value of $\hat{y}$, which is a continuous variable, may be interpreted as the predicted probability of taking the train, given an automobile travel time equal to x.

There are several problems with this approach. One is that the assumption of homoscedasticity is not met; the estimated variance about the regression line is equal to $y(1 - y)$, and therefore is not constant. Perhaps more troubling is that the predicted probabilities ($\hat{y}$) do not have to stay on the (0, 1) interval; the reader might confirm that values of x less than around 25 will yield negative probabilities, and values of x greater than about 100 will yield probabilities greater than one! The problem is as shown in Figure 9.3.

How then should we proceed? One idea is to make the probabilities related to x in a nonlinear way. The logistic curve in Figure 9.4 has the following equation:

$$\hat{y}_i = \frac{e^{\alpha + \beta x_i}}{1 + e^{\alpha + \beta x_i}} \tag{9.13}$$

Note that when $\alpha + \beta x$ is a large negative number, the predicted probability is near zero, while if $\alpha + \beta x$ is a large positive number, the predicted probability is near one. In these cases, the predicted probabilities approach their asymptotes of 0 or 1, but never actually reach them. Thus, it is no longer possible to predict probabilities that are either negative or greater than one.

While we've solved one problem by keeping the dependent variable on the (0,1) interval, we've created another. How can we estimate the parameters (i.e., α and β)? We can't use linear regression, since the equation and the curve in Figure 9.4 are clearly not linear. One approach is to use nonlinear least squares. Specifically, we want to find α and β to minimize the sum of squared deviations between observed and predicted values:

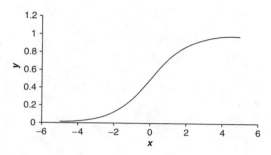

Figure 9.4 The logistic curve

$$\min \sum_{i=1}^{n} (y_i - \hat{y}_i)^2 \qquad (9.14)$$

where the predicted value, $\hat{y}_i$, is given by Equation 9.13 above. One way to achieve this is through statistical software. Another would be to systematically test many various combinations of α and β to see which combination results in the lowest sum of squared residuals. The answer, when minimizing the sum of squared residuals for the data in Table 9.12, is $\alpha = -4.501$ and $\beta = 0.080$.

Although the predicted probabilities are not linearly related to x, we can transform the predicted probabilities into a new variable, z, which *is* linearly related to x. The transformation is called the *logistic* transform, and it is carried out by first finding $\hat{y} / (1 - \hat{y})$, and then taking the natural logarithm of the result. The quantity $\hat{y} / (1 - \hat{y})$ is known as the odds (in favor of the event), and so the new variable is known as the 'log-odds'. Thus, using z to define our new variable, we have:

$$z = \ln\left(\frac{\hat{y}}{1 - \hat{y}}\right) = \alpha + \beta x \qquad (9.15)$$

The logistic regression model is therefore one that assumes that the log-odds increases (or decreases) linearly, as x increases (see Figure 9.5).

The logarithm of a number is the power to which the 'base' is raised to equal that number. For example, suppose that we are working with logarithms to the base 10 (usually the notation in this case is 'log'). Then $\log 100 = 2$, since $10^2 = 100$. Similarly $\log 1000 = 3$ (since $10^3 = 1000$), $\log 10 = 1$, and so on. Natural logarithms use a base of $e = 2.71 \ldots$ (and use the notation 'ln'). So, e.g., $\ln 5 = 1.609$, since $e^{1.609} = (2.71\ldots)^{1.609} = 5$.

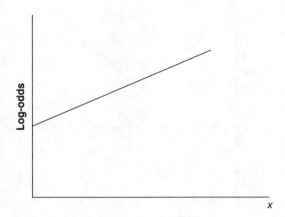

Figure 9.5 Linear relationship between log-odds and x

Odds are often stated in place of probabilities for events such as horse racing. If the probability that a horse wins a race is $y = 0.2$, the probability that it loses is $1 - y = 0.8$. While the odds in favor of an event are equal to $y/(1 - y)$, the odds against an event are given as $(1 - y)/y$. Therefore, the odds against the horse winning are stated as 4 to 1 ($= 0.8/0.2$). Suppose that five people each bet \$1; one bets that the horse will win, and the other four that the horse will lose. If the horse wins, the person betting on the horse will collect \$5, equal to the total amount bet (of course, in reality, the winner will have to give a share of his winnings to the race track, and to the government in payment of taxes!). If the probability that the horse wins rises to 0.333, the probability that it loses declines to $y = 0.667$, and the odds against it decline to 2 to 1 ($= 0.667/0.333$). Though it is less common to do so in horse racing, we could also state the odds in the other direction. When the horse has a winning probability of 0.2, the odds that the horse wins are $y/(1 - y) = 0.2/0.8 = 0.25$ to 1. When the probability of winning rises to 0.33, the odds in favor of the horse rise to $0.33/0.67 = 0.5$ to 1.

Returning to our example, the log-odds that an individual takes the train is given by:

$$z = \ln\left(\frac{\hat{y}}{1 - \hat{y}}\right) = -4.501 + 0.080x \tag{9.16}$$

We see that the slope coefficient β tells us how much the log-odds will change when x changes by one unit. In the present example, when $x = 32$ minutes, the predicted probability of taking the train is 0.1262, using $x = 32$ and the estimated values of α and β. The odds of taking the train are then given as 0.1444 ($= \hat{y} / (1 - \hat{y}) = 0.1262/(1 - 0.1262)$) to 1. Stated another way, the odds against taking the train are 6.92 ($= 1/0.1444$) to 1.

Should the individual experience an increase in auto travel time of one minute, the log of the odds of choosing the train would go up by 0.080. What does it mean to say that the log of the odds has gone up by 0.080? We can 'undo' the log by exponentiating; if $a = \ln(b)$, then $b = e^{a}$. This means that if the log of the odds has increased by 0.080, the odds of choosing the train have increased by a multiplicative factor of $e^{.080} = 1.084$. The new odds in favor of taking the train are now equal to 0.1565 = 0.1444 × 1.084 to 1. Equivalently, the odds against the train have declined by a factor of 1.084, and are now 6.39 ($= 6.92/1.084$) to 1.

When the automobile travel time increases by one minute, from $x = 32$ to $x = 33$, the probability of taking the train increases from 0.1262 to 0.1353. This may be verified by using Equation 9.13 with the estimated values of α and β, and $x = 33$:

$$\hat{y}_i = \frac{e^{\alpha + \beta x_i}}{1 + e^{\alpha + \beta x_i}} = \frac{0.1565}{1 + 0.1565} = .1353. \tag{9.17}$$

Also note that $0.1353/(1 - 0.1353) = 0.1444$, and $(1 - 0.1353)/0.1353 = 6.39$. Finally, note that in logistic regression, when $\alpha + \beta x = 0$, the predicted probability is equal to 1/2 (since $e^{0} = 1$). This is equivalent to $x = -\alpha/\beta$. In our example, $-\alpha/\beta = 4.501/0.080$, which is approximately equal to 56. When automobile travel time is about 56 minutes, the probability that an individual takes the train is about 0.5 (or '50–50').

9.7 A SUMMARY OF SOME PROBLEMS THAT CAN ARISE IN REGRESSION ANALYSIS

Regression analysis carried out on spatial data raises special issues. One particularly difficult problem is that associated with the modifiable area unit problem, discussed in Chapter 7. A regression of a dependent variable on a set of independent variables may yield substantially different conclusions when carried out on spatial units of differing sizes. Fotheringham and Wong (1991) note that with multiple and logistic regression, the magnitude and significance of regression coefficients can be very sensitive to the size and configuration of areal units. If feasible, the sensitivity of one's results to changes in the size and/or shape of the spatial units should be explored.

Table 9.13, adapted from Haining (1990a), summarizes some of the problems that can plague regression analyses. The table describes the consequences of the problems, and, in addition, describes how they may be diagnosed and corrected. Section numbers refer to other sections of this text that provide relevant discussion of the problem.

Table 9.13 Some problems that can arise in regression analysis

Problem	Consequences	Diagnostic	Corrective action
Residuals:			
Nonnormal	Inferential tests may be invalid	Shapiro–Wilk test	Transform y values
Heteroscedastic	Biased estimation of error variance, leading to invalid inference	Plot residuals against y and x's	Transform y values
Not independent	Underestimate variance of regression coefficients. Inflated R^2	Moran's I (10.3.3)	Spatial regression (11.3)
Nonlinear relationship	Poor fit and nonindependent residuals	Scatterplots of y against x's Added variable plots (11.2)	Transform y and/or x variables
Multicollinearity (9.1.1)	Variance of regression estimates is inflated	Variance inflation factor (9.4)	Delete variable(s)
Incorrect set of explanatory variables (9.2)	Difficulties in performing efficient analysis, and poor regression estimates	Added variable plots (11.2)	Stepwise regression (9.5)
Outliers (9.4)	May severely affect model estimates and fit	Plots. Leverage values (9.4)	Data deletion
Categorical response variable	Linear regression model inappropriate		Logistic regression (9.6)
Spatially varying parameters	Invalid estimation and inference if deleted	Moran's I (10.3.3)	Expansion method; Geographically weighted regression (11.4)
Missing data (nonrandom)	Possibly invalid estimation and inference Could waste other case information		Missing data at random Estimate missing values (9.4) Delete observation (9.4)

Note: Relevant sections of text are given in parentheses.
Source: Adapted from Haining (1990a), pp. 332–3.

9.8 MULTIPLE AND LOGISTIC REGRESSION IN *SPSS 21 FOR WINDOWS*

9.8.1 Multiple Regression

Data input is similar to that for simple, linear regression (Chapter 8). Each observation is represented by a row in the data table, and each variable is represented by a column.

Click on `Analyze/Regression/Linear`. Then move the dependent variable into the box labeled `Dependent` on the right, and move the desired independent variables into the box labeled `Independents`. Then click on `OK`.

There are a number of common options that you may wish to choose before clicking on `OK`. Under `Save`, it is common to check the boxes to save predicted values, residuals, leverage values to detect outliers, and confidence intervals for either the mean (i.e., the regression line) or individual predictions. All of these saved quantities will be attached as new columns in the dataset. Under `Statistics`, it is desirable to check `Collinearity` diagnostics, to check for multicollinearity. Under the box where variables that are in the regression are indicated, one may choose the method by which independent variables are entered onto the right-hand side. The default is `enter`, which means that all independent variables will be entered. A common alternative is to choose `stepwise`, which enters and removes variables one at a time, depending upon their significance. Note that it is not necessary to make a choice here if the desired independent variables have already been selected by moving them into the `Independent` box.

9.8.2 Logistic Regression

There are two ways that logistic regression may be carried out using *SPSS for Windows*. The first approach is to use nonlinear least squares. This is easiest to understand, since, as the previous section indicates, we are simply looking for the values of α and β that will make the sum of squared residuals as small as possible.

9.8.2.1 Data input

In both cases, the approach to data input is the same. As in linear regression, the dependent variables and independent variables are arranged in columns. Each row represents an observation. It is common, but not necessary, to have the dependent variable in the first column. Make sure that the column containing the dependent variable consists of a column of '0's and '1's, consistent with its binary response nature.

9.8.2.2 Using SPSS for Windows 21 *and Nonlinear Least Squares*

1. Choose `Analyze, Regression, Nonlinear`.

2. Click on `Parameters`, define α and β, give estimated values, and choose `Continue`. Since entering Greek letters is not an option, you can, for example, enter 'a' (without the quotes), and then give it a starting value of zero, and then click on `Add`. Do the same for the other parameter, perhaps calling it 'b'. Choosing good estimated values is sometimes important, and not always easy to do. It may require a bit of trial and error. Using $\alpha = 0$ and $\beta = 0$ is often not a bad way to start.

3. Select the dependent variable.

4. Set up the model; this refers to the equation for the predicted values of the dependent variable. For logistic regression, you should define the model as in Equation 9.13. More specifically, when there is one independent variable, you would enter the following: `Exp(a+b*VAR)/(1+Exp(a+b*VAR))`, replacing 'VAR' with the name of your independent variable.

5. Choose `OK` to run the nonlinear least squares analysis.

For the data in Table 9.12, you should obtain the results described in Section 9.6.1.

The nonlinear least squares approach, however, is a bit more awkward to implement in *SPSS* than its alternative, known as the *maximum likelihood* approach to finding α and β. In addition, maximum likelihood is, to a statistician, generally a preferable alternative since it produces estimates that are unbiased and that have relatively smaller sampling variances, at least when the sample sizes are large.

Many statistical packages, including *SPSS 21 for Windows*, make use of maximum likelihood estimation. The likelihood of observing $y_i = 1$ when $x = x_i$ is:

$$pr(y_i = 1) = \frac{e^{\alpha + \beta x_i}}{1 + e^{\alpha + \beta x_i}} \tag{9.18}$$

Similarly, the likelihood of observing $y = 0$ is:

$$pr(y_i = 0) = \frac{1}{1 + e^{\alpha + \beta x_i}} \tag{9.19}$$

The likelihood of the sample is therefore:

$$L = \prod_{i=1}^{n} pr(y_i) = \prod_{i=1}^{n} \left(\frac{e^{\alpha + \beta x_i}}{1 + e^{\alpha + \beta x_i}} \right)^{y_i} \left(\frac{1}{1 + e^{\alpha + \beta x_i}} \right)^{1 - y_i} \tag{9.20}$$

Many programs, such as *SPSS for Windows*, choose α and β to maximize this likelihood of obtaining the observed sample.

9.8.2.3 Using logistic regression in SPSS for Windows

1. Choose `Analyze, Regression, Binary Logistic`.
2. Choose the dependent variable and the covariates (independent variables).
3. Choose `OK`.

Using the *SPSS 21 for Windows* logistic regression routine and the data in Table 9.12, we find that the maximum likelihood estimates yield:

$$z = -4.5362 + 0.077x \qquad (9.21)$$

Note that these values of $\alpha = -4.5362$ and $\beta = 0.077$ found via the maximum likelihood method are similar to those found via nonlinear least squares.

9.8.2.4 Interpreting output from logistic regression

Tables 9.14 and 9.15 display the output from the logistic regression analysis of the commuting behavior data in Table 9.12. We are of course interested in the slope and intercept, and these are displayed in the same part of the output where we found them in linear regression. They are given along with their standard deviations (also known as standard errors; see the column headed 'S.E.'). If the coefficients are more than twice their corresponding standard errors (approximately), they may be regarded as significantly different from zero. In this example, the coefficients are not significantly different from zero; this is also reflected in the column headed 'Sig.', where we find that the p-value associated with each coefficient is greater than 0.05.

Note that the output also contains a column headed Exp(B); this is the exponentiated slope referred to in the text, and it tells us by how much the odds will change when the x variable is increased by one unit. In this example, an increase of one minute in the commuting time leads to the odds of taking the train increasing by a multiplicative factor of 1.0804.

Another interesting part of the output is the two-by-two classification table. It shows us that there were six observations where $y = 0$ (the individual did not take the train). Of these, five were predicted correctly by the logistic regression equation, and one was predicted incorrectly. (A prediction is classified as 'correct' if the model predicts that the actual outcome has a likelihood of greater than 0.5.) Note that the fifth individual has an observed auto travel time of $x = 80$ minutes. The model predicted that there would be a 0.839 probability that the individual would take the train (Table 9.15), yet we observed that he/she did not ($y = 0$).

Table 9.14 **Logistic regression output**

```
Dependent Variable..  TRAIN

Beginning Block Number      0.  Initial Log Likelihood Function

-2 Log Likelihood    16.635532

* Constant is included in the model.

Beginning Block Number      1.  Method: Enter

Variable(s) Entered on Step Number
1..        AUTOTT

Estimation terminated at iteration number 4 because Log Likelihood
decreased by less than .01 percent.

 -2 Log Likelihood        12.543
 Goodness of Fit          11.629
 Cox & Snell - R^2          .289
 Nagelkerke - R^2           .385

                    Chi-Square       df      Significance

 Model                  4.093         1          .0431
 Block                  4.093         1          .0431
 Step                   4.093         1          .0431

Classification Table for TRAIN
The Cut Value is .50
```

```
                        Predicted
                        .00   1.00  Percent Correct
                         0     1
            observed
              .00    0 | 5  |  1  |  83.33%
                       |----|-----|
              1.00   1 | 2  |  4  |  66.67%
                     Overall 75.00%
```

```
-------------------------Variables in the Equation--------------------

Variable      B        S.E.     Wald      df    Sig.      R      Exp(B)

AUTOTT      .0773     .0456    2.8744·     1    .0900    .2293   1.0804
Constant  -4.5362    2.7641    2.6933      1    .1008
```

Of the six individuals who *did* take the train, the model predicted four correctly, and there were two cases where the model predicted that the individual would not take the train when in fact they did. Individuals 4 and 6 both took the train, yet the model predicted probabilities of less than 0.5 that they would do so. This table summarizes how successful the model is in predicting actual outcomes.

Table 9.15 Summary of results

Y	X: Auto travel time (min)	Predicted probabilities	
		Linear OLS	Logistic
0	32	.093	.113
1	89	.963	.913
0	50	.368	.339
1	49	.352	.321
0	80	.826	.839
1	56	.459	.449
0	40	.215	.191
1	70	.673	.706
1	72	.704	.737
1	76	.765	.793
0	32	.093	.113
0	58	.490	.487

EXERCISES

1. The following data are collected in a study of park attendance.

Park visit? 1 = Yes; 0 = No	Distance from park (km)
0	8
0	6
1	1
0	4
1	3
0	2
0	6
1	5
1	7
1	2
1	1
1	3
1	5

Park visit? 1 = Yes; 0 = No	Distance from park (km)
1	7
0	8
0	9
0	8
0	6
1	4
1	4
0	7
0	9

Use logistic regression to determine how the likelihood of visiting the park varies with the distance that an individual resides from the park.

2. In American football, the likelihood of a successful field goal declines with increasing distance. The following data were collected for one week from games played by teams in the National Football League.

Made? 1 = Yes; 0 = No	Yards
0	34
1	20
0	51
1	32
0	51
0	29
1	19
0	37
0	43
1	47
1	24
1	31
1	41
1	22
1	26

(Continued)

(Continued)

Made? 1 = Yes; 0 = No	Yards
1	34
1	41
1	24
1	39
1	43

(a) Use logistic regression to determine how the odds of making a field goal change as distance increases.

(b) Use the results to draw a graph depicting how the predicted probability of making a field goal changes with distance.

(c) In the waning seconds of SuperBowl XXV, Scott Norwood missed a 47-yard field goal that would have carried the Buffalo Bills to victory over the New York Giants. Use your model to predict the likelihood that a kicker is successful at a 47-yard field goal attempt. Did Norwood really deserve the criticism he received for missing the attempt?

3. What is multicollinearity? How can it be detected? Why is it a potential problem in regression analysis? How might its effects be ameliorated?

4. The number of times per year a person uses rapid transit is a linear function of income:

$$Y = 1.2 + 2.4 \, X_1 + 8.4 \, X_2 + 15.6 \, X_3$$

where X_1, X_2, and X_3 are dummy variables for medium, high, and very high incomes, respectively (the low-income category has been omitted). What is the predicted number of annual transit trips per year for each of the four income categories?

5. Given the following data:

Y	X
0	8
1	6
0	9
1	4
1	3

(Continued)

(Continued)

is β (the 'slope' of the logistic curve) positive or negative? Answer without actually finding the coefficients, and justify your answer.

6. Suppose, for a given set of data, we find that a logistic regression yields $\beta = -0.43$. What is the change in odds, for a unit change in x?

7. The following results were obtained from a regression of $n = 14$ housing prices (in dollars) on median family income, size of house, and size of lot:

	Sum of squares	df	Mean square	F
Regression SS:	4234	3	___	___
Residual SS:	3487	___	___	
Total SS:	___	___		

	Coefficient (b)	Standard error (sb)	VIF
Median family income	1.57	0.34	1.3
Size of house (sq. ft)	23.4	11.2	2.9
Size of lot (sq. ft)	−9.5	7.1	11.3
Constant	40,000	1000	

(a) Fill in the blanks.

(b) What is the value of r^2?

(c) What is the standard error of the estimate?

(d) Test the null hypothesis that $R^2 = 0$ by comparing the F-statistic from the table with its critical value.

(e) Are the coefficients in the direction you would hypothesize?

If not, which coefficients are opposite in sign from what you would expect?

(f) Find the t-statistics associated with each coefficient, and test the null hypotheses that the coefficients are equal to zero. Use $\alpha = 0.05$, and be sure to give the critical value of t.

(g) What do you conclude from the variance inflation factors (VIFs)? What (if any) modifications would you recommend in light of the VIFs?

(h) What is the predicted sales price of a house that is 1500 square feet, on a lot 60´×100´, and in a neighborhood where the median family income is $40,000?

8. Find any dataset and choose a dependent variable and two or three independent variables. The variables chosen should be defined spatially. There should be at least 15–20, and, preferably, at least 30 observations.

   (a) State any null hypotheses you may have, as well as the alternative hypotheses.

   (b) Graph the dependent variable (y) vs each independent (x) variable. Describe any obvious outliers.

   (c) Graph the dependent variables against each other, and comment on any obvious multicollinearity.

   (d) Regress y on each of the independent variables separately. Also regress y on the combined set of all independent variables. If you have three independent variables, you may also wish to regress y on pairs of independent variables. Comment on the results.

9. Use the data in Table 9.7 to study how the number of native species on islands varies with the size of the island, the maximal elevation of the island, and the distance to nearby islands. There are many choices you will need to make; there is no single 'correct' answer to this question. Some considerations you should think about include the following: (a) Is the total number of species, or the number of native species, the more appropriate dependent variable? (b) What should be done about the missing elevation values that occur in some cases? (c) Are there outliers? If so, how can they be identified? (d) What about multicollinearity? Should any variables be eliminated from the analysis? One goal you should have is to come up with a 'best' equation, in the sense that variables in the equation are both significant and meaningful.

10. (a) Using *SPSS* or *Excel* and the Hypothetical UK Housing Prices dataset, construct a regression equation using housing price as the dependent variable, and bedrooms, bathrooms, date built, garage, fireplace, floor area, and whether the home is detached as the independent variables. Investigate the importance of multicollinearity and outliers. Comment on the weaknesses of this specification, and on the results.

    (b) Attempt to improve the regression equation found in (a). Justify your decisions in constructing and carrying out the analysis.

11. (a) Using *SPSS* or *Excel* and the Milwaukee dataset described in Section 1.9.2, construct a regression equation using housing sales price as the dependent variable, and number of bedrooms, lot size, finished square footage in the house, age of house, and number of bathrooms, as the independent variables. Investigate the importance of multicollinearity and outliers. Comment on the weaknesses of this specification, and on the results.

    (b) Attempt to improve the regression equation found in (a). Justify your decisions in constructing and carrying out the analysis.

On the Companion Website

To support Chapter 9, both the **Home Sales in Milwaukee, Wisconsin** data-set and the **Hypothetical UK Housing Prices** dataset are available within the 'Datasets' section of the website. The website also includes information about their respective fields and formats, and provides links to other resources in the 'Further Resources' section.

10

SPATIAL PATTERNS

10.1 INTRODUCTION

One assumption of regression analysis as applied to spatial data is that the residuals are independent and thus are not spatially autocorrelated – that is, there is no spatial pattern to the errors. Residuals that are not independent can affect estimates of the variances of the coefficients, and hence make it difficult to judge their significance.

We have also seen in previous chapters that lack of independence among observations can affect the outcome of t-tests, ANOVA, correlation, and regression, often leading one to find significant results where none in fact exists. One reason for learning more about spatial patterns and their detection, then, is an indirect one – we seek to assess spatial dependence so that we may ultimately correct our statistical analyses based upon dependent spatial data.

Understanding the complicating effects of spatially dependent observations in statistical analysis provides an important motivation for learning more about spatial patterns. The analysis of spatial patterns is also important when there is a direct interest in the geographic phenomenon and/or process itself. For example, crime analysts wish to know if clusters of criminal activity exist. Health officials seek to learn about disease clusters and their determinants. In these instances, it is useful to discern whether clusters of activity could have arisen by chance alone or, alternatively, whether they might be attributed to statistically significant manifestations of a process of clustering (of, e.g., crime or disease).

In this chapter, we will investigate statistical methods aimed at detecting spatial patterns and assessing their significance. We will in particular focus upon statistical tests of the null hypothesis that a spatial pattern is random. The structure of the chapter follows from the fact that data are typically in the form of either point locations (where exact locations of, e.g., disease or crime are available), or in the form of aggregated areal information (where, e.g., information is available only on regional rates).

10.2 THE ANALYSIS OF POINT PATTERNS

Carry out the following experiment.

Draw a rectangle that is 6 inches by 5 inches on a sheet of paper. Locate 30 dots at random within the rectangle. This means that each dot should be located independently of the other dots. Also, for each point you locate, every subregion of a given size should have an equal likelihood of receiving the dot.

Then draw a six-by-five grid of 30 square cells on top of your rectangle. You can do this by making little tick marks at one-inch intervals along the sides of your rectangle. Connecting the tick marks will divide your original rectangle into 30 squares, each having a side of length one inch.

Give your results a score, as follows. Each cell containing no dots receives one point. Each cell containing one dot receives 0 points. Each cell containing two dots receives 1 point. Cells containing three dots receive 4 points, cells containing four dots receive 9 points, cells containing 5 dots receive 16 points, cells containing 6 dots receive 25 points, and cells containing 7 dots receive 36 points. Find your total score by adding up the points you have received in all 30 cells.

DO NOT READ ON UNTIL YOU HAVE COMPLETED THE INSTRUCTIONS ABOVE!

Classify your pattern as follows.

If your score is 16 or less, your pattern is significantly more uniform or regular than random (i.e., spatially dispersed).

If your score is between 17 and 45, your pattern is characterized as random.

If your score is greater than 45, your pattern exhibits significant clustering.

On average, a set of 30 randomly placed points will receive a score of 29. With a random pattern, 95% of the time the score will be between 17 and 45. The majority of people who try this experiment produce patterns that are more uniform or regular (i.e., dispersed) than random, and hence their scores are less than 29. Their point patterns are more spread out than a truly random pattern. When individuals see an empty space on their diagram, there is an almost overwhelming urge to fill it in by placing a dot there! Consequently, the locations of dots placed on a map by individuals are not independent of the locations of previous dots, and hence an assumption of spatial randomness is violated.

Consider next Figures 10.1 and 10.2, and suppose you are a crime analyst looking at the spatial distribution of recent crimes. Make a photocopy of the page, and indicate in pencil where you think the clusters of crime are. Do this by simply encircling the clusters (you may define more than one cluster on each diagram).

DO NOT READ THE NEXT PARAGRAPH UNTIL YOU HAVE COMPLETED THIS EXERCISE!

How many clusters did you find? It turns out that both diagrams were generated by locating points at random within the rectangle! In addition to having trouble drawing random patterns, individuals also have a tendency to 'see' clusters where no statistically significant cluster exists. This results from the mind's strong desire to organize spatial information.

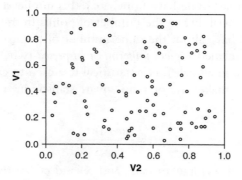

Figure 10.1   Hypothetical spatial pattern of crime

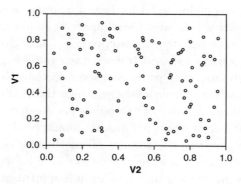

Figure 10.2   Another hypothetical spatial pattern of crime

Both of these exercises point to the need for objective, quantitative measures of spatial pattern – it is simply not sufficient to rely on one's visual interpretation of a map. Crime analysts cannot necessarily pick out true clusters of crime just by looking at a map, nor can health officials always pick out significant clusters of disease from map inspection alone.

10.2.1 Quadrat Analysis

The experiment involving the scoring of points falling within the '6 × 5' rectangle is an example of *quadrat analysis*, developed primarily by ecologists in the first half of the twentieth century. In quadrat analysis, a grid of square cells (quadrats) of equal size is used as an overlay, on top of a map of incidents (see, for example, Figure 10.3. One then counts the number of incidents in each cell. In a random pattern, the mean number of points per cell will be roughly equal to the variance of the number of points per cell.

If there is a large amount of variability in the number of points from cell to cell (i.e., some cells have many points; some have none, etc.), this implies a tendency toward *clustering*. If there is very little variability in the number of points from cell to cell (i.e., when all or almost all of the cells have about the same number of points), this implies a tendency toward a pattern that is termed *regular, uniform,* or *dispersed* (where the number of points per cell is about the same in all cells). The statistical test used to evaluate the null hypothesis of spatial randomness makes use of a chi-square statistic involving the variance-mean ratio:

$$\chi^2 = \frac{(m-1)\sigma^2}{\bar{x}} \qquad (10.1)$$

where m is the number of quadrats or cells, and $\bar{x}$ and σ^2 are the mean and variance of the number of points per quadrat, respectively. This value is then compared with a critical value from a chi-square table, with $m - 1$ degrees of freedom.

Quadrat analysis is easy to employ, and it has been a mainstay in the spatial analyst's toolkit of pattern detectors over several decades. One important issue is the size of the quadrat; if the cell size is too small, there will be many empty cells, and if clustering exists on any but the smallest spatial scales, it may be missed. If the cell size is too large, one may miss patterns that occur *within* cells. One may find patterns at some spatial scales and not at others, and thus the choice of quadrat size can seriously influence the results. Curtiss and McIntosh (1950) suggest an 'optimal' quadrat size of two points per quadrat. Bailey and Gatrell (1995) suggest that the mean number of points per quadrat should be about 1.6. These suggestions reflect a concern with maximizing the amount of information contained in the cells (i.e., as indicated above, cells that are too small convey too little information since so many of them are empty, and cells that are too large convey little information about the actual location of the points).

10.2.1.1 Summary of the Quadrat Method

1. Divide a study region into m cells of equal size.

2. Find the mean number of points per cell ($\bar{x}$). This is equal to the total number of points divided by the number of cells (m).

3. Find the variance of the number of points per cell, s^2, as follows:

$$s^2 = \frac{\sum_{i=1}^{i=m}(x_i - \bar{x})^2}{m-1} \qquad (10.2)$$

where x_i is the number of points in cell i.

4. Calculate the variance–mean ratio (VMR):

$$\mathrm{VMR} = \frac{s^2}{\bar{x}} \qquad (10.3)$$

5. Interpret the results as follows:

   (a) If $s^2/\bar{x} < 1$, the variance of the number of points is less than the mean. In the extreme case where the ratio approaches zero, there is very little variation in the number of points from cell to cell. This characterizes situations where the distribution of points is spread out, or uniform, across the study area.

   (b) If $s^2/\bar{x} > 1$, there is a good deal of variation in the number of points per cell – some cells have substantially more points than expected (i.e., $xi > \bar{x}$ for some cells i), and some cells have substantially fewer than expected (i.e., $x_i < \bar{x}$). This characterizes situations where the point pattern is more clustered than random.

   (c) A value of $s^2/\bar{x}$ near one indicates that the points are close to randomly distributed across the study area.

6. Hypothesis testing.

   (a) Multiply the VMR by $m - 1$; the quantity $\chi^2 = (m - 1)\mathrm{VMR}$ has a chi-square distribution, with $m - 1$ degrees of freedom, when H_0 is true. This fact allows us to obtain critical values, χ^2_L and χ^2_H, from a chi-square table. In particular, for a two-sided test we will reject H_0 if either $\chi^2 < \chi^2_L$ or $\chi^2 > \chi^2_H$. If the number of cells (m) is greater than about 30, $(m - 1)\mathrm{VMR}$ will, when H_0 is true, have a distribution that is approximately normal, with mean $m - 1$ and variance equal to $2(m - 1)$. This means that we can treat the quantity:

$$z = \frac{(m-1)\mathrm{VMR} - (m-1)}{\sqrt{2(m-1)}} = \sqrt{(m-1)/2}(\mathrm{VMR} - 1) \qquad (10.4)$$

as an approximation of a normal random variable with mean 0 and variance 1. With $\alpha = 0.05$, the critical values are $z_L = -1.96$ and $z_H = 1.96$. The null hypothesis of no pattern is rejected if $z < z_L$ (implying uniformity – that is, spatial dispersion) or if $z > z_H$ (implying clustering).

Example

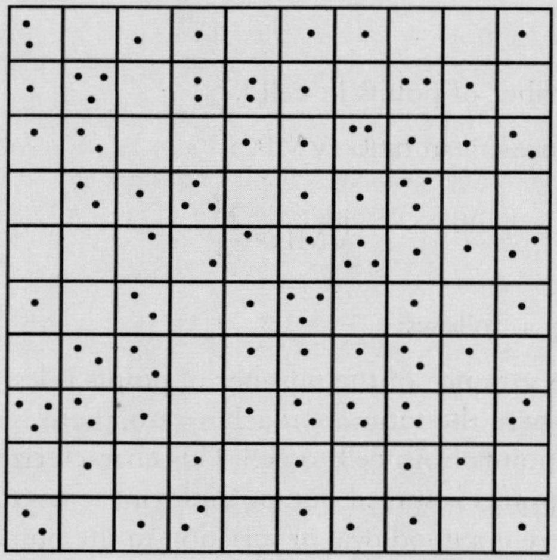

Figure 10.3 A spatial point pattern

We wish to know whether the pattern observed in Figure 10.3 is consistent with the null hypothesis that the points are located randomly. We first calculate the VMR. There are 100 points on the 10 × 10 grid, implying a mean of one point per cell. There are six cells with three points, 20 cells with two points, 42 cells with one point, and 32 cells with no points. The variance is:

$$\frac{\left\{6(3-1)^2 + 20(2-1)^2 + 42(1-1)^2 + 32(0-1)^2\right\}}{99} = \frac{76}{99}$$
$$= 0.77 \qquad (10.5)$$

and, since the mean is equal to one, this is also our observed VMR. Since VMR < 1, there is a tendency toward a uniform pattern. How unlikely is a value 0.77 if the null hypothesis is true – is it unlikely enough that we should reject the null hypothesis?

Since the number of degrees of freedom (df) is large, the sampling distribution of $\chi^2 = (m-1)$ VMR begins to approach the shape of a normal distribution. In particular, we can use Equation 10.4, so that in our example, we have:

$$z = \frac{99(0.77) - 99}{\sqrt{2(99)}} = \sqrt{99/2}(0.77 - 1) = -1.618 \qquad (10.6)$$

This falls within the critical values of z and hence we do not have strong enough evidence to reject the null hypothesis.

If cells of a different size had been used, the results, and possibly the conclusions, would have been different. By aggregating the cells in Figure 10.3 to a 5×5 grid of 25 cells, the VMR declines to 0.687 (based on a variance of 1.658^2 and a mean of four points per cell). The χ^2 value is $24(.687) = 16.5$. Since the degrees of freedom are less than 30, we will use the chi-square table (Table A.6) to assess significance. With 24 degrees of freedom, using interpolation to find the critical values at $p = 0.025$ and $p = 0.975$ yields $\chi^2_L = 12.73$ and $\chi^2_U = 40.5$ (these values were actually found using a more detailed table). Since our observed value of 16.5 falls between these limits, we again fail to reject the hypothesis of randomness.

To summarize, after finding VMR in steps 1–4 above, calculate $\chi^2 = (m - 1)$ VMR, and compare it with the critical values found in a chi-square table, using df $= m - 1$.

If $m - 1$ is greater than about 30, you can use the fact that $z = \sqrt{(m-1)/2}(\text{VMR} - 1)$ has, approximately, a normal distribution with mean 0 and variance 1, implying that, for a two-sided test with $\alpha = 0.05$, one may compare z with the critical values $z_L = -1.96$ and $z_H = 1.96$.

It is interesting to note that the quantity $\chi^2 = (m - 1)$VMR may be written as:

$$\chi^2 = (m-1)\text{VMR} = \frac{(m-1)s^2}{\overline{x}} \qquad (10.7)$$

$$= \frac{(m-1)\Sigma(x_i - \overline{x})^2}{\overline{x}(m-1)} = \frac{\Sigma(x_i - \overline{x})^2}{\overline{x}}$$

The final quantity $\Sigma(x_i - \overline{x})^2 / \overline{x}$ is the sum across cells of the squared deviations of the observed numbers from the expected numbers of points in a cell, divided by the expected number of points in a cell. This is commonly known as the chi-square goodness-of-fit test.

10.2.2 Nearest Neighbor Analysis

Clark and Evans (1954) developed nearest neighbor analysis to analyze the spatial distribution of plant species. They developed a method for comparing the observed average distance between points and their nearest neighbors with the distance that would be expected between nearest neighbors in a random pattern.

We begin by defining R_0 to be the observed average distance between points and their nearest neighbors. Let R_e be the expected distance between points and their nearest neighbors when points are distributed randomly. Intuitively, if R_0 is small relative to R_e, the pattern will be clustered; if R_0 is large relative to R_e, the pattern will be more dispersed than random.

R_0 may be calculated as $\sum_{i=1}^{n} d_i / n$ where n is the number of points in the study area, and where d_i is the distance from point i to its nearest neighbor. Note that nearest neighbors may be reflexive – that is, they may be nearest neighbors of each other.

R_e is calculated as one over twice the square root of the density of points:

$$R_e = \frac{1}{2\sqrt{\rho}} = \frac{1}{2\sqrt{n / A}} \tag{10.8}$$

where ρ is the density of points, and A is the size of the study area.

The nearest neighbor statistic, R, is defined as the ratio between the observed and expected values:

$$R = \frac{R_0}{R_e} = \frac{\overline{d}}{1 / (2\sqrt{\rho})} = 2\overline{d}\sqrt{\rho} \tag{10.9}$$

R varies from 0 (a value obtained when all points are in one location, and the distance from each point to its nearest neighbor is zero), and a theoretical maximum of about 2.14, for a perfectly uniform or systematic pattern of points maximally spread out on an infinitely large two-dimensional plane. A value of $R = 1$ indicates a random pattern, since the observed mean distance between neighbors is equal to that expected in a random pattern. It is also known that if we examined many random patterns, we would find that the variance of the nearest neighbor statistic, R, is:

$$V[R] = \frac{4 - \pi}{\pi n} \tag{10.10}$$

where n is the number of points. Thus, we can form a z-test, to test the null hypothesis that the pattern is random with the now familiar process of starting with the statistic (R), subtracting its expected value (1), and dividing by its standard deviation to obtain a z-score:

$$z = \frac{(R - 1)}{\sqrt{V[R]}} = \frac{\sqrt{\pi n}(R_0 - 1)}{\sqrt{4 - \pi}} \approx 1.913(R - 1)\sqrt{n} \tag{10.11}$$

The quantity z has, approximately, a normal distribution with mean 0 and variance 1, and hence tables of the standard normal distribution may be used to assess significance. A value of $z > 1.96$ implies that the pattern has significant uniformity, and a value of $z < -1.96$ implies that there is a significant tendency toward clustering.

The strength of this approach lies in its ease of both calculation and comprehension. Several cautions should be noted in the interpretation of the nearest neighbor statistic. The statistic, and its associated test of significance, may be affected by the shape of the region. Long, narrow, rectangular shapes may have relatively low values of R simply because of the constraints imposed by the region's shape. Points in long, narrow rectangles are *necessarily* close to one another.

The location of points relative to the boundary of the study region can also make a difference in the analysis. One solution to the boundary problem is to place a buffer area around the study area. The nearest neighbors are found for all points within the study area (but not for the points in the buffer area). Points inside of the study area (such as point A in Figure 10.4) may have nearest neighbors that fall into the buffer area, and these distances should also be used in the analysis.

If information on the location of points just outside of the study area cannot be obtained, then one can create a similar buffer region, but with it now lying *inside* of the study area. This has the effect of creating a new study area that is effectively smaller than the original one. While having the benefit of accounting for edge effects, it has the necessary disadvantage of using less information due to the smaller size of the modified study area.

Yet another approach to addressing edge effects is to use Monte Carlo simulation. With the Monte Carlo approach, the null hypothesis (in this case, that the point pattern is random) is simulated by choosing the same number of points as is observed, and randomly placing them within the study area. This random placement is carried out with the aid of software, a calculator, or a table of random numbers such as that found in Appendix A.1. Then the statistic (R) is found for this simulated pattern. This step is then repeated a large number of times to generate many values of R, preferably using a computer. The value of R observed for the actual map is compared with these simulated values to see how unusual it is, if the null hypothesis is true. If the observed R is unusual relative to the simulated values (e.g., if a value more extreme than the one observed would occur less than 5% of the time), the null hypothesis is rejected. This Monte Carlo method of simulating the null hypothesis many times, and then comparing the observed result to the simulations is a very general approach, and can be used in a wide range of situations to test hypotheses.

Illustration of the Monte Carlo method

Dominik Hašek, the former goalie for the gold-medal, Czech ice hockey team in the 1998 Olympics, saved 92.4% of all shots he faced when he played professionally for the Buffalo Sabres of the National Hockey League (NHL). The average save percentage of other goalies in the NHL is about 90%. Hašek faced about 31 shots per game, while the Sabres managed just 25 shots per game on the opposing

(Continued)

(Continued)

goalie. To evaluate how much Hašek meant to the Sabres, compare the outcomes of 1000 games using Hašek's statistics with the outcomes of 1000 games assuming the Sabres had an 'average' goalie, who stopped 90% of the shots against him.

Solution

Take 31 random numbers between 0 and 1. Count those greater than 0.924 as goals against the Sabres with Hašek in goal. Take 25 numbers from a uniform distribution between 0 and 1, and count those greater than 0.9 as goals for the Sabres. Record the outcome (win, loss, or tie). Repeat this 1000 times (preferably using a computer!), and tally the outcomes. Finally, repeat the entire experiment using random numbers greater than 0.9 (instead of 0.924) to generate goals against the Sabres without Hašek. Each time the experiment is performed, a different outcome will be obtained. In one comparison, the results were as follows:

	Wins	Losses	Ties
Scenario 1 (with Hašek)	434	378	188
Scenario 2 (without Hašek)	318	515	167

To evaluate Hašek's value to the team over the course of an 82-game season, the outcomes above may first be converted to percentages, multiplied by 82, and then rounded to integers yielding:

	Wins	Losses	Ties
Scenario 1	36	31	15
Scenario 2	26	42	14

Thus, Hašek is 'worth' about ten wins; that is, they win about ten games a year that they would have lost if they had an 'average' goalie.

Another potential difficulty with the nearest neighbor statistic is that, since only nearest neighbor distances are used, clustering is only detected on a relatively small spatial scale. To overcome this, it is possible to extend the approach to second- and higher-order nearest neighbors; discussion of that extension is beyond the scope of this text.

Most importantly, it is often of interest to ask not only whether clustering exists, but whether clustering exists over and above some background factor (such as population). For

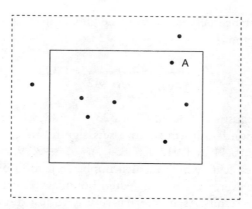

Figure 10.4 Boundary effects in nearest neighbor analysis

instance, using the nearest neighbor statistic to determine whether crime is clustered in an urban area is often not too enlightening – the nearest neighbor analysis likely reveals clustering because the population itself tends to be clustered. Nearest neighbor methods are not particularly useful in these situations because they only relate to the spatial location of the points, and do not account for other factors that are already known to influence the spatial distribution of points. The approaches to the study of pattern that are described in Section 10.3 do not have this limitation.

10.2.2.1 Illustration

For the point pattern in Figure 10.5, there are six point locations (A through F).

Distances between points are given along the lines connecting the points. The mean distance between nearest neighbors is $R_0 = (1 + 2 + 3 + 1 + 3 + 3)/6 = 13/6 = 2.167$.

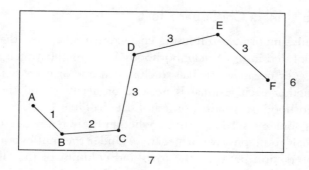

Figure 10.5 Nearest neighbor distances

The expected mean distance between nearest neighbors in a pattern of six points placed randomly in a study region with area $7 \times 6 = 42$ is:

$$R_e = \frac{1}{2\sqrt{\rho}} = \frac{1}{2\sqrt{6/42}} = 1.323 \qquad\qquad (10.12)$$

The nearest neighbor statistic is $R = 2.167/1.323 = 1.638$, which means that the pattern displays a tendency toward uniformity. To assess significance, we can calculate the z-statistic from Equation 10.11 as $1.913(1.638 - 1) \sqrt{6} = 2.99$, and this is much greater than the critical value of 1.96, which in turn implies rejection of the null hypothesis of a random pattern. However, we have neglected boundary effects, and these can have a significant effect on the results. As an alternative way to test the null hypothesis, we can randomly choose six points by choosing random x-coordinates in the range $(0, 7)$ and random y-coordinates in the range $(0, 6)$. Then we compute the mean distance from each of the six points to their nearest neighbors, and repeat the whole process many times. Simulating the random placement of six points in the 7×6 study region 10,000 times led to a mean distance between nearest neighbors of 1.62. This is greater than the expected distance of $R_e = 1.323$ noted above. This greater-than-expected distance can be attributed directly to the fact that points near the border of the study region are relatively farther from other points in the study region than they presumably would have been to points just outside of the study region. Ordering the 10,000 mean distances to nearest neighbors from lowest to highest reveals that the 500th highest one is 2.29. This implies that only 5% of the time would we expect a mean nearest neighbor distance greater than 2.29. Our observed distance of 2.167 is less than 2.29, and so we, having accounted for boundary effects through our Monte Carlo simulation, fail to reject the null hypothesis.

10.3 GEOGRAPHIC PATTERNS IN AREAL DATA

10.3.1 An Example Using a Chi-Square Test

In a regression of housing prices on housing characteristics, suppose that we have data on 51 houses that are located in three neighborhoods. How might we tell whether there is a tendency for positive or negative residuals to cluster in one or more neighborhoods? One idea is to note whether each residual is positive or negative, and then to tabulate the residuals by neighborhood (see the hypothetical data in Table 10.1).

We can use a chi-square test to determine whether there is any neighborhood-specific tendency for residuals to be positive or negative. Under the null hypothesis of no spatial pattern (i.e., no interaction between the rows and columns of the table), the expected values of the table entries are equal to the product of the row and column totals, divided by the overall total. For example, we expect 23(16)/51 = 7.22 positive residuals in neighborhood 1. These expected values are given in parentheses in Table 10.2.

Table 10.1 Hypothetical residuals

Sign of residual	Neighborhood			Total
	1	2	3	
+	10	6	7	23
–	6	15	7	28
Total	16	21	14	51

Table 10.2 Observed and expected frequencies of residuals

Sign of residual	Neighborhood			Total
	1	2	3	
+	10	6	7	23
	(7.22)	(9.47)	(6.31)	
–	6	15	7	28
	(8.78)	(11.53)	(7.69)	
Total	16	21	14	51

Note: Expected values are given in parentheses.

The chi-square statistic is:

$$\chi^2 = \sum_{i=1}^{n} \frac{(O_i - E_i)^2}{E_i} \tag{10.13}$$

where O_i and E_i are the observed and expected frequencies in cell i, and there are n cells in the table. When the null hypothesis is true, this statistic has a χ^2 distribution with degrees of freedom equal to $(r-1)(c-1)$, where r and c are, respectively, the number of rows and columns in the table.

In this example, the observed value of chi-square is

$$\chi^2 = \frac{(10-7.22)^2}{7.22} + \frac{(6-9.47)^2}{9.47} + \frac{(7-6.31)^2}{6.31} + \frac{(6-8.78)^2}{8.78} + \tag{10.14}$$

$$\frac{(15-11.53)^2}{11.53} + \frac{(7-7.69)^2}{7.69} = 4.40$$

This is less than the critical value of 5.99, found by using the chi-square table with $\alpha =$ 0.05, and 2 degrees of freedom. Therefore, the null hypothesis of no pattern is not rejected. That is, there is no evidence here for spatial autocorrelation in the residuals.

If spatial autocorrelation in the residuals is detected, what can be done about it? One possibility is to include a new, location-specific dummy variable. This will serve to capture the importance of an observation's location in a particular neighborhood. In our present housing price example, we could add two variables, one each for two of the three neighborhoods (following the usual practice of omitting one category). You should also note that if there are k neighborhoods, it is *not* necessary to have $k-1$ dummy variables; rather, you might choose to have only one or two dummy variables for those neighborhoods having large deviations between the observed and predicted values. Adoption of this approach will yield better estimates of the regression coefficients that represent the effects of housing characteristics on housing prices.

10.3.2 Moran's *I*

Moran's I statistic (1948, 1950) is one of the classic (as well as one of the most common) ways of measuring the degree of spatial autocorrelation in areal data. Moran's I is calculated as follows:

$$I = \frac{n\Sigma_i^n\Sigma_j^n w_{ij}(y_i - \overline{y})(y_j - \overline{y})}{(\Sigma_i^n\Sigma_j^n w_{ij})\Sigma_i^n(y_i - \overline{y})^2} , \tag{10.15}$$

where there are n regions and w_{ij} is a measure of the spatial proximity between regions i and j. It is interpreted much like a correlation coefficient. Values near +1 indicate a strong spatial pattern (high values tend to be located near one another, and low values tend to be located near one another). Values near −1 indicate strong negative spatial autocorrelation; high values tend to be located near low values. (Examples of spatial patterns with negative autocorrelation are extremely rare.) Finally, values near 0 indicate an absence of spatial pattern.

Although Equation 10.15 is perhaps daunting at first glance, it is helpful to realize that if the variable of interest is first transformed into a z-score $\{z = (x - \overline{x})/s\}$, a much simpler expression for I results:

$$I = \frac{n\Sigma_i\Sigma_j w_{ij}z_i z_j}{(n-1)\Sigma_i\Sigma_j w_{ij}} \tag{10.16}$$

The conceptually important part of the formula is the numerator, which sums the products of z-scores in nearby regions. Pairs of regions where *both* regions exhibit above-average scores (or below-average scores) will contribute positive terms to the numerator, and these pairs will therefore contribute toward positive spatial autocorrelation. Pairs where one region is above average and the other is below average will contribute negatively to the numerator, and hence to negative spatial autocorrelation.

The weights $\{w_{ij}\}$ may be defined in a number of ways. Perhaps the most common definition is one of *binary connectivity*; $w_{ij} = 1$ if regions i and j are contiguous, and $w_{ij} = 0$ otherwise. 'Contiguous' can in turn be defined as requiring regions to share at least a common point (termed '*queen's case* contiguity'), or, more restrictively, a common boundary of nonzero length (termed '*rook's case* contiguity'). Sometimes the w_{ij} defined in this way are then standardized to define new w_{ij}^* by dividing the weight by the number of regions that i is connected to; i.e., $w_{ij}^* = w_{ij} / \Sigma_j w_{ij}$. In this case, all regions i are characterized by a set of weights linking i to other regions that sum to one; i.e., $\Sigma_j w_{ij} * = 1$.

Alternatively, $\{w_{ij}\}$ may be defined as a function of the distance between i and j (e.g., $w_{ij} = d_{ij}^{-\beta}$ or $w_{ij} = \exp[-\beta d_{ij}]$), where the distance between i and j could, for example, be measured along the line connecting the centroids of the two regions. It is conventional to use $w_{ii} = 0$. It is also common, though not necessary, to use symmetric weights, so that $w_{ij} = w_{ji}$.

It is important to recognize that the value of I is very dependent upon the definition of the $\{w_{ij}\}$. Using a simple binary connectivity definition for the map in Figure 10.6 gives us:

$$W = \{w_{ij}\} = \begin{matrix} 0 & 1 & 1 & 0 & 0 \\ 1 & 0 & 1 & 1 & 0 \\ 1 & 1 & 0 & 1 & 1 \\ 0 & 1 & 1 & 0 & 1 \\ 0 & 0 & 1 & 1 & 0 \end{matrix} \qquad (10.17)$$

In this instance, the definition of $\{w_{ij}\}$ causes the neighborhood around region 5 to be much smaller than the neighborhood around region 2. This is not necessarily 'wrong', but suppose that we were interested in the spatial autocorrelation of a disease that was

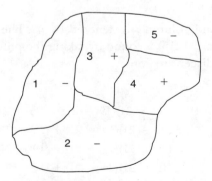

Figure 10.6   Positive and negative residuals in a five-region system

characterized by rates that were strongly associated over small distances, but not correlated over large distances. Our observed value of I would be a combined measure of strong association between close adjacent pairs (such as regions 4 and 5) and weak association between distant adjacent pairs (such as regions 2 and 3). In such instances, it might be more appropriate to use a distance–based definition of $\{w_{ij}\}$.

10.3.2.1 Illustration

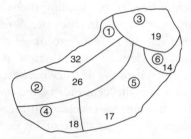

Figure 10.7　Hypothetical six-region system

Consider the six-region system in Figure 10.7. Using a binary connectivity definition of the weights leads to:

$$\mathbf{W} = \begin{bmatrix} 0 & 1 & 1 & 0 & 0 & 0 \\ 1 & 0 & 1 & 1 & 1 & 0 \\ 1 & 1 & 0 & 0 & 1 & 1 \\ 0 & 1 & 0 & 0 & 1 & 0 \\ 0 & 1 & 1 & 1 & 0 & 1 \\ 0 & 0 & 1 & 0 & 1 & 0 \end{bmatrix} \tag{10.18}$$

where an entry in row i and column j is denoted by w_{ij}. The double summation in the numerator of I (see Equation 10.15) is found by taking the product of the deviations from the mean, for all pairs of adjacent regions:

$$
\begin{aligned}
(32 - 21)(26 - 21) &+ (32 - 21)(19 - 21) + (26 - 21)(32 - 21) \\
&+ (26 - 21)(19 - 21) + (26 - 21)(18 - 21) \\
&+ (26 - 21)(17 - 21) + (19 - 21)(32 - 21) \\
&+ (19 - 21)(26 - 21) + (19 - 21)(17 - 21) \\
&+ (19 - 21)(14 - 21) + (18 - 21)(26 - 21)
\end{aligned}
\tag{10.19}
$$

$$+ (18 - 21)(17 - 21) + (17 - 21)(19 - 21)$$
$$+ (17 - 21)(26 - 21) + (17 - 21)(18 - 21)$$
$$+ (17 - 21)(14 - 21) + (14 - 21)(19 - 21)$$
$$+ (14 - 21)(17 - 21) = 100$$

Since the sum of the weights in (10.18) is 18, and since the variance of the six regional values is 224/5, based on (10.15), Moran's I is equal to:

$$I = \frac{6(100)}{18(224)} = 0.1488 \tag{10.20}$$

In addition to this descriptive interpretation, there is a statistical framework that allows one to decide whether any given pattern deviates significantly from a random pattern. If the number of regions is large, the sampling distribution of I, under the hypothesis of no spatial pattern, approaches a normal distribution, and the mean and variance of I can be used to create a z-statistic in the usual way:

$$z = \frac{I - E[I]}{\sqrt{V[I]}} \tag{10.21}$$

where $E[I]$ and $V[I]$ denote, respectively, the expected value (i.e., the theoretical mean) and variance of I, when the null hypothesis of no spatial pattern is true. The value is then compared with the critical value of z found in the normal table (e.g., $\alpha = 0.05$ would imply critical values of -1.96 and $+1.96$).

The mean and variance are equal to:

$$E[I] = \frac{-1}{n-1} \tag{10.22}$$

$$V[I] = \frac{n^2(n-1)S_1 - n(n-1)S_2 + 2(n-2)S_0^2}{(n+1)(n-1)^2 S_0^2}$$

where:

$$S_0 = \sum_{i}^{n} \sum_{j \neq i}^{n} w_{ij}$$

$$S_1 = 0.5 \sum_{i}^{n} \sum_{j \neq i}^{n} (w_{ij} + w_{ij})^2 \tag{10.23}$$

$$S_2 = \sum_{k}^{n} (\sum_{j}^{n} w_{kj} + \sum_{i}^{n} w_{ik})$$

Computation is not complicated, but it is tedious enough to not want to do it by hand! Unfortunately, few software packages that calculate the coefficient and its significance are available. An exception is Anselin's *GeoDa* (Anselin et al. 2006).

There are however simplifications and approximations that facilitate the use of Moran's *I*. An alternative way of finding Moran's *I* is to simply take the ratio of two regression slope coefficients (see Griffith 1996). The numerator of *I* is equal to the regression slope obtained when the quantity $a_i = \sum_{j=1}^{n} w_{ij} z_j$ is regressed on z_i, and the denominator of *I* is equal to the regression slope obtained when the quantity $b_i = \sum_{j=1}^{n} w_{ij}$ is regressed on $c_i = 1$. Here the z's represent the z-scores of the original variables, and the slope coefficients are found using no-intercept regression (i.e., constraining the result of the regression so that the intercept is equal to zero).

In addition, Griffith gives $2 / \Sigma\Sigma w_{ij}$ as an approximation for the variance of the Moran coefficient. This expression, though it works best only when the number of regions is sufficiently large (about 20 or more), is clearly easier to compute than the one given in Equations 10.22 and 10.23! Alternatively, when observational units are on a square grid, and connectivity is indicated by the four adjacent cells, the variance may be approximated by $1/(2n)$, where n is the number of cells. Based on either a grid of hexagonal cells or a map displaying 'average' connectivity with other regions, the variance may be approximated by $1/(3n)$. An example is given in Section 10.5.

The use of the normal distribution to test the null hypothesis of spatial randomness requires that one of two assumptions holds:

1. Normality. It can be assumed that regional values are generated from identically distributed normal random variables (i.e., the variables in each region arise from normal distributions that all have the same mean and same variance).

2. Randomization. It can be assumed that all possible permutations (i.e., regional rearrangements) of the regional values are equally likely.

The formulae given above (Equations 10.22 and 10.23) for the variance assume that the normality assumption holds. The variance formula for the randomization assumption is algebraically more complex, and gives values that are only slightly different than those given above (see, e.g., Griffith 1987).

If either of the two assumptions above holds, the sampling distribution of *I* has a normal distribution if the null hypothesis of no pattern is true. One of these two assumptions must hold to generate the sampling distribution of *I* so that critical values of the test statistic may be established. For example, if the first assumption was used to generate regional values, *I* could be computed; this could then be repeated many times, and a histogram of the results could be produced. The histogram would have the shape of a normal distribution, a mean of E[*I*], and a variance of V[*I*]. Similarly, the observed regional values on a map could be randomly rearranged many times, and the value of I computed each time. Again, a histogram could be produced; it would again have the shape of a normal distribution with mean $E[I]$ and a variance slightly different than $V[I]$. If we can rely on one of these two assumptions, we do not need to perform these experiments to generate histograms, since we know beforehand that they will produce normal distributions with known mean and variance.

Unfortunately, there are many circumstances in geographical applications that lead the analyst to question the validity of both assumptions. For example, maps of counties by township are often characterized by high population densities in the townships corresponding to or adjacent to the central city, and by low population densities in outlying townships. Rates of crime or disease, though they may have equal means across townships, are unlikely to have equal variances. This is because the outlying towns are characterized by greater uncertainty – they are more likely to experience atypically high or low rates simply because of the chance fluctuations associated with a relatively smaller population base. Thus, the first assumption is not satisfied, since all regional values do not come from identical distributions – some regional values, namely the outlying regions, are characterized by higher variances. Likewise, not all permutations of regional values are equally likely – permutations with atypically high or low values out in the periphery are more likely than permutations with atypically high or low values near the center.

How can we test the null hypothesis of no spatial pattern in this instance? One approach is to use Monte Carlo simulation. Since the z-test described above by Equation (10.21) is no longer valid, we need an alternative way to come up with critical values. The idea is to first assume that the null hypothesis of no spatial pattern is true. Suppose we have data on the number of diseased individuals (n_i) and the population (p_i) in each region. For each individual, assign disease to that individual with probability $\sum_i n_i / \sum_i p_i$, which is the overall disease rate in the population. Then calculate Moran's I. This is repeated many times, and the resulting values of Moran's I may be used to create a histogram depicting the relative frequencies of I when the null hypothesis is true. Furthermore, the values can be arranged from lowest to highest, and this list can be used to find critical values of I. For example, if the simulations are carried out 1000 times, and critical values are desired for a test using $\alpha = 0.05$, they can be found from the ordered list of I values. The lower critical value would be the 25th item on the list, and the upper critical value would be the 975th item on the list.

10.4 LOCAL STATISTICS

10.4.1 Introduction

Besag and Newell (1991) classify the search for geographic clusters into three primary areas. First are 'general' tests, designed to provide a single measure of overall pattern for a map consisting of point locations. These general tests are intended to provide a test of the null hypothesis that there is no underlying pattern, or deviation from randomness, among the set of points. Examples include the nearest neighbor test, the quadrat method, and the Moran statistic, all outlined above. In other situations, the researcher wishes to know whether there is a cluster of events around a single or small number of prespecified foci. For example, we may wish to know whether disease clusters around a toxic waste site, or we may wish to know whether crime clusters around a set of liquor establishments. These are "local" or "focused" tests. Finally, Besag and Newell describe 'tests for the detection of clustering'. Here there is no a priori idea of where the clusters may be; the methods are aimed at looking at many local statistics, scanning the map and uncovering the size and location of any possible clusters.

General tests are carried out with what are called 'global' statistics; again, a single summary value characterizes any deviation from a random pattern. 'Local' statistics are used to evaluate whether clustering occurs around a particular point or points, and hence are employed for both focused tests and tests for the detection of clustering. Local statistics have been used in both a confirmatory manner, to test hypotheses, and in an exploratory manner, where the intent is more to suggest, rather than confirm, hypotheses.

Local statistics may be used to detect clusters, either when the location is prespecified (focused tests) or when there is no a priori idea of cluster location. In the latter case, all local statistics on the map are tested, and an adjustment to the critical value is made for the multiple testing that occurs. Without such an adjustment, if a large number of local statistics are being tested, some would exceed their usual critical values by chance alone. One form of adjustment is the so-called 'Bonferroni adjustment' – instead of using a Type I error of α for each local statistic, the critical value is determined using α/n, where n is the number of local tests to be carried out.

When a global test finds no significant deviation from randomness, local tests may be useful in uncovering isolated hotspots of increased incidence. When a global test does indicate a significant degree of clustering, local statistics can be useful in deciding whether (a) the study area is relatively homogeneous in the sense that local statistics are quite similar throughout the area, or (b) there are local outliers that contribute to a significant global statistic. Anselin (1995) discusses local tests in more detail.

10.4.2 Local Moran Statistic

The local Moran statistic for location i is:

$$I_i = n(\gamma_i - \overline{\gamma})\Sigma_j w_{ij}(\gamma_j - \overline{\gamma}_j) \qquad (10.24)$$

where the sum is over all other locations (j). The sum of these local Moran's is equal, up to a constant of proportionality, to the global Moran; i.e, $\Sigma I_i = I$. For example, the local Moran statistic for region 1 in Figure 10.7 is:

$$I_1 = (32 - 21)[(26 - 21)] + (19 - 21)] = 33 . \qquad (10.25)$$

The expected value of the local Moran statistic is:

$$E[I_i] = \frac{-\Sigma_j w_{ij}(y_j - \overline{y})}{n - 1} \qquad (10.26)$$

and the expression for its variance is more complicated. Anselin gives the variance of I_i, and assesses the adequacy of the assumption that the test statistic has a normal distribution (which would allow one to carry out a z-test) under the null hypothesis.

10.4.3 Getis' G_i Statistic

To test whether a particular location i and its surrounding regions have higher than average values on a variable (x) of interest, Ord and Getis (1995) have used the statistic:

$$G_i^* = \frac{\Sigma_j w_{ij}(d)x_j - W_i^* \overline{x}}{s\{[nS_{1i}^* - W_i^{*2}] / (n - 1)\}^{1/2}} , \qquad (10.27)$$

where s is the sample standard deviation of the x values, n is the number of regions, and $w_{ij}(d)$ is equal to 1 if region j is within a distance of d from region i, and 0 otherwise. Also:

$$W_i^* = \sum_j w_{ij}(d) \qquad (10.28)$$

$$S_{1i}^* = \sum_j w_{ij}^2$$

One can see that the numerator of Equation (10.27) represents, for region i, the difference between the weighted value of x in the neighborhood of i and the value that would be expected if the neighborhood was 'average' in its x characteristics. Ord and Getis note that, when the underlying variable has a normal distribution, so does the test statistic. Furthermore, the distribution is asymptotically normal even when the underlying distribution of the x-variables is not normal, if the distance d is sufficiently large. Since the statistic (10.27) is written in standardized form, it can be taken as a standard normal random variable, with mean 0 and variance 1.

For region 1 in Figure 10.7, we will use weights equal to 1 for regions 1, 2, and 3, and weights equal to 0 for other regions. The G_i statistic is:

$$G_1^* = \frac{77 - 3(21)}{6.69\sqrt{\dfrac{6(3) - 9}{5}}} = 1.56 \tag{10.29}$$

Since this variable has a normal distribution with mean 0 and variance 1 under the null hypothesis that region 1 is not located in a region of particularly high values, we can use a one-sided test with $\alpha = 0.05$ and $z = 1.645$. We therefore fail to reject the null hypothesis.

10.5 FINDING MORAN'S *I* USING *SPSS 21 FOR WINDOWS*

Consider the six-region system in Figure 10.7. With connectivity defined by a binary 0–1 weight for adjacent regions, we have the weight matrix given by Equation 10.18. To compute the value of Moran's *I* in *SPSS*, we first convert the six regional values to z-scores. These may be found by using `Analyze`, `Descriptives`, and `Descriptives`, and then clicking on `Save standardized scores as variables`. For the six regions, the z-scores are 1.64, 0.747, −0.299, −0.448, −0.598, and −1.046, respectively. Then the quantities $a_i = \sum_j w_{ij} z_j$ are found. These are simply weighted sums of the z-scores, and with binary weights, this means that the regions that *i* is connected to are those z-scores that are summed. For example, region 1 is connected to region 2 and 3. For region 1, a_1 = 0.747 − 0.299 = 0.448. The six a_i scores are 0.448, 0.299, 0.747, 0.149, −1.046, and −0.896, respectively. Now perform a regression, using the *a*'s as the dependent variable and the *z*'s as the independent variable. In *SPSS*, click on `Analyze`, `Regression`, `Linear`, and then define the dependent and independent variables. Then, under `Options`, make sure the box labeled `Include constant in equation` is NOT checked. This yields a regression coefficient of 0.446 for the numerator.

For the denominator, we again use no-intercept regression to regress six *y*-values on six *x*-values. The six '*y*-values' are the sum of the weights in each row (2, 4, 4, 2, 4, and 2 for rows 1–6 of the matrix in 10.18, respectively). The six *x*-values are 1, 1, 1, 1, 1, and 1 (this will always be a set of *n* ones, where *n* is the number of regions). After again making sure that a constant is NOT included in the regression equation, one finds the regression coefficient is 3.0. Moran's *I* is simply the ratio of these two coefficients: 0.446/3 = 0.1487.

The variance of *I* in this example may be found from Equation 10.22:

$$V[I] = \frac{(36)(5)(36) - (6)(5)(240) + 2(4)(18)^2}{7(5)^2(18)^2} = 0.033 \tag{10.30}$$

The z-value associated with a test of the null hypothesis of no spatial autocorrelation is $(0.1487 - (-0.2)/ \sqrt{0.033} = 1.92$. This would exceed the critical value of 1.645 under a one-sided test (which we would use, for example, if our initial alternative hypothesis was that positive autocorrelation existed), and would be slightly less than the critical value of 1.96 in a two-sided test. We note, however, that we are on shaky ground in assuming that this test statistic has a normal distribution, since the number of regions is small. We also note that, in this case, the approximation of $1/(3n)$ described in Section 10.3.3 for the variance of I would have yielded a variance of $1/18 = 0.0555$, which is not too far from that found above using Equation 10.22. The approximation of two divided by the sum of the weights, also described in Section 10.3.2, would have yielded $2/18 = 0.1111$. This latter approximation works better for systems with a greater number of regions.

10.6 FINDING MORAN'S *I* USING *GEODA*

GeoDa is a freely downloadable software package that facilitates spatial analysis. At the time of this writing (September, 2013), it (Version 1.4.6) runs on different versions of Windows (including XP, Vista, 7 and 8), Mac OS, and Linux. In this section, we will describe the use of *GeoDa* for calculating and testing Moran's *I*.

The *GeoDa* website has several sample datasets; here we will use a dataset (sids.zip) on the incidence of deaths from Sudden Infant Death Syndrome (SIDS) in North Carolina. The dataset consists primarily of the number of SIDS cases and the number of births in each of the 100 counties in North Carolina for the periods 1974–78 and 1979–84; more detail is given by Cressie (1993). To begin, choose File, then Open Shapefile, and choose the file sids.shp – this is a map of the counties of North Carolina. Then click on Open. The dataset comes with information on the number of SIDS cases, and the number of births; if we are interested in a map of SIDS rates, we will first have to create the rates from the given information (the file sids2. zip contains information on rates, but here we will demonstrate how to calculate them within *Geoda*). To do this, click on the spreadsheet icon, then on Table, and then choose Variable Calculation from the submenu. Click on the Rates tab. Click on the Add Variable box, and create a name for the rate variable (e.g., sidrate74). In the first Event Variable box, use the drop-down menu to choose sid74. In the Base Variable box, choose bir74. (Leave the Weight box empty, and leave the default method of *Raw rate*. Then click on Apply. A new column will be created on the spreadsheet, and this will contain the newly defined sidrate74 variable, defined here as the fraction of all births that result in SIDS cases (or, alternatively, the probability that a birth results in a SIDS case, assuming a homogeneous population).

Next, we will display a map, and then calculate and interpret Moran's *I*. Begin by clicking on the blue bar at the top of the map window to make that window active. Then choose Map and Quantile Map; here you can choose the number of classes for the choropleth map (the default is 4 classes, or colors, for the map). Choose the newly defined variable (sidrate74) from the drop-down menu listing the variables in the table. To find Moran's *I*, we need to specify a weight matrix. If you want to use a previously created weight matrix, choose Tools, then Weights, and then Select. To create a new weight matrix, choose Tools, then Weights, and then Create. The new window asks you for a 'Weights File ID Variable'; choose CNTY_ID from the drop-down menu, since we are using county IDs for our geographic units. (Then we click on rook contiguity, and we will leave the order of contiguity set at its default value of 1. An order of contiguity of 2 would imply that our neighborhood around any county would be defined not only by adjacent counties that share a common border with it, but also by counties adjacent (i.e., sharing a border) to those adjacent counties. Other options exist here for alternative definitions for weights – for example, definitions that are based upon distance between county centroids. Finally, click on Create to create the file. You are asked for a file name – let's call the file of weights sidswt – the software will by default assign this file an extension of 'gal', so that the result is a weight file named sidswt.gal. Now select Space, and then Univariate Moran's I. In the new window, select the variable of interest (i.e., the created SIDS rate, sidrate 74), and the previously created weight file (e.g., sidswt.gal) and click OK. A graph (termed a Moran scatterplot) will appear, with the value of Moran's *I* across the top – in this case, the value is $I = 0.2477$. The graph has 100 points on it – one for each county. The horizontal axis gives the value of the (standardized) SIDS rate for the county (if it is to the left of the vertical axis, the SIDS rate is below average, and if it is to the right, it is above average). The vertical axis gives the weighted sum of the SIDS rates in surrounding counties, where the weights are those defined previously. Counties in the 'northeast' quadrant contribute to positive spatial autocorrelation, since these are counties with above average SIDS rates, and they are in a 'neighborhood' that has above-average rates. Similarly, counties in the southwest quadrant contribute toward positive spatial autocorrelation, since these represent counties with below average rates that have surrounding neighborhoods of below average rates. Counties lying in the other two quadrants on the graph contribute toward negative spatial autocorrelations. The slope of the best-fitting line through these points is equal to the value of Moran's *I*. *GeoDa* features linked windows, so that by clicking on a county on the map, the corresponding county will be highlighted on the graph and the table. Similarly, using the mouse to highlight a rectangle containing points on the Moran scatterplot will cause the corresponding counties to be highlighted on the map and in the table.

EXERCISES

1. The following table represents the number of residuals observed in a regression of wheat yields on precipitation and temperature over a six-county area:

County:	1	2	3	4	5	6	
+	7	10	12	9	14	15	Number of positive residuals
−	12	8	19	10	10	10	Number of negative residuals

Use the chi-square test to determine whether there is any interaction between location and the tendency of residuals to be positive or negative. If you reject the null hypothesis of no pattern, then describe how you might proceed in the regression analysis.

2. A regression of sales on income and education leaves the following residuals:

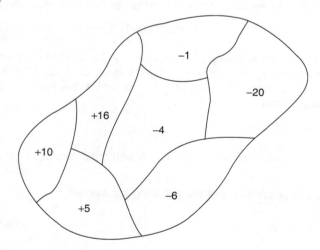

   Use Moran's *I* to determine whether there is a spatial pattern to the residuals. Assume binary connectivity to determine the weights. If you reject the null hypothesis, describe how you would proceed with the regression analysis.

3. (a) Find the nearest neighbor statistic for the following pattern assuming a study area of 40 km²:

(Continued)

(Continued)

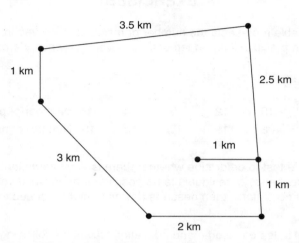

(b) Test the null hypothesis that the pattern is random by finding the z-statistic:

$$z = 1.913(R-1)\ \sqrt{n}\ .$$

(c) Find the chi-square statistic, $\chi^2 = (m-1)\sigma^2 / \bar{x}$ for a set of 81 quadrats, where 1/3 of the quadrats have 0 points, 1/3 of the quadrats have 1 point, and 1/3 of the quadrats have 2 points. Then find the z-value to test the hypothesis of randomness, where

$$z = \frac{\chi^2 - (m-1)}{\sqrt{2(m-1)}},$$

and where m is the number of cells. Compare it with critical values of $z = -1.96$ and $z = +1.96$.

4. Vacant land parcels are found at the following locations:

Find the variance and mean of the number of vacant parcels per cell, and use the variance–mean ratio to test the hypothesis that parcels are distributed randomly (against the two-tailed hypothesis that they are not).

5. Find the nearest neighbor statistic when n points are equidistant from one another on the circumference of a circle with radius r, and there is one additional point located at the center of the circle. Assume that travel between neighboring points on the circumference can only occur along the circumference. Note that you can break the solution into two parts – one where the distance between neighboring points along the circumference is less than r, and the other where that distance is greater than or equal to r. Hints: The area of a circle is πr^2 and the circumference of a circle is $2\pi r$

6. For the nearest neighbor test, prove that the following two z-scores are equivalent:

$$\frac{R-1}{\sigma_R} = \frac{r_o - r_e}{\sigma_r},$$

where:

$$\sigma_R = .52 / \sqrt{n}; \sigma_r = \frac{.26}{\sqrt{np}}; R = r_o / r_e$$

Thus there are two equivalent ways of carrying out the nearest-neighbor test.

7. Find the nearest neighbor statistic for four points located at the vertices of a rectangle of length 5 and width 4.

8. Using *GeoDa*, find Moran's *I* for the Sudden Infant Death Syndrome (SIDS) data, for the period 1979–84.

9. Find the nearest neighbor statistic for a study region that has an area of 100 km². There are ten points, and the mean distance between nearest neighbors is observed to be 0.8 km. Using $\alpha = 0.05$, test the null hypothesis that the point pattern is random, versus the alternative that it is clustered.

On the Companion Website

The website's 'Further Resources' section contains links to other material which supplements and expands upon information presented in Chapter 10, including links to the journal papers and chapters referenced in this chapter.

11

SOME SPATIAL ASPECTS OF REGRESSION ANALYSIS

11.1 INTRODUCTION

We have already noted that spatial autocorrelation presents difficulties in estimating regression relationships. In some cases, we may be interested in the pattern of spatially correlated residuals for its own sake. Figure 11.1 is a map I produced for an undergraduate project, showing the residuals from a regression of snowfall on temperature, elevation, and latitude. In this case, the primary purpose was to obtain a visual impression of the effect of the North American Great Lakes on snowfall patterns in New York State. One can clearly see two bands of excess snowfall, one downwind from Lake Erie, and the other downwind from Lake Ontario. The effects downwind of Lake Erie are particularly strong, ranging up to 50–60 inches a year greater than that predicted by temperature, elevation, and latitude alone. The remainder of the map has relatively small residuals. One might also speculate that the negative residuals along the northeast border of the state constitute a precipitation shadow effect, since this area is directly east of the Adirondack Mountains, and much of the moisture would have precipitated out before reaching the eastern border.

In the snowfall example, it was not necessary to have precise estimates of the effects of temperature, elevation, and latitude on snowfall, since primary interest was in the map

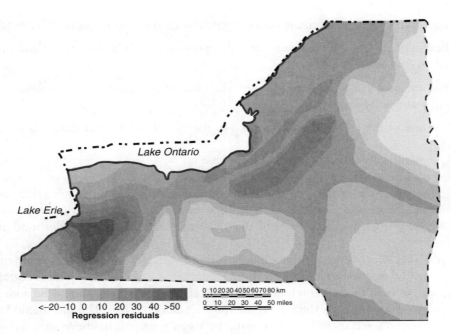

Figure 11.1 Regression residuals from snowfall analysis

pattern of the residuals. However, spatial autocorrelation in the residuals violates an underlying assumption of ordinary least squares regression, and so alternatives must be considered when reliable regression equations are desired. One approach to spatial regression models is to remedy the situation by adding to the list of explanatory variables the values of x and/or y in surrounding regions as well. This is treated in Section 11.2. Section 11.3 takes the approach of modeling spatial autocorrelation in the residuals directly.

Until this point, we have assumed that values of the regression coefficients were global, in the sense that they were thought to apply to the region as a whole. However, it is possible that the coefficients vary over space – that is, it is possible that different locations have different regression coefficients. Section 11.4 examines two approaches to spatially varying regression parameters. The final section provides an illustration of the various methods.

11.2 SPATIAL LAG MODEL, NEIGHBORHOOD-BASED EXPLANATORY VARIABLES, AND ADDED-VARIABLE PLOTS

When regression residuals exhibit spatial autocorrelation, this suggests that the regression results may benefit from additional explanatory variables. Haining (1990b) identifies four situations where spatial effects may be entered into the right-hand side of a regression equation:

1. The value of y depends upon values of y nearby (also known as a spatial lag model).

2. The value of y at a site depends not only upon values of x at the site but also upon values of x at nearby sites.

3. The value of y at a site depends upon the value of x at the site and on values of x and y at nearby sites.

4. The size of the error at a site is related to the size of the error at nearby sites.

Case (4) is statistically indistinguishable from case (3).

Thus, it may be worth trying new variables, including $x_i^* = \Sigma_j w_{ij} x_j$ and/or $y_i^* = \Sigma_j w_{ij} y_j$ on the right-hand side of the regression equation, where the w_{ij} are weights that serve to define the neighborhood around point i. Both x_i^* and y_i^* are defined for each spatial location, i, and consist of weighted sums of the value of y or x in the neighborhood around location i.

The potential effectiveness of these neighborhood variables (or any other omitted variable) can also be assessed graphically. Haining notes that *added variable plots* are 'graphical devices that are used to decide whether a new explanatory variable should be added to a regression' (see also Weisberg 1985; Johnson and McCulloch 1987). The idea behind added variable plots is to see if there is a relationship between y, once it has been adjusted for the variables already in the equation, and some omitted variable. Let x_p denote the omitted variable. The procedure is as follows:

1. Obtain the residuals of the regression of y on the x-variables, where the latter is the set of all variables already in the equation.

2. Obtain the residuals of the regression of x_p on the x-variables.

3. Plot the residuals obtained in (1) on the vertical axis and those from (2) on the horizontal axis.

The result is the relationship between x_p and y, adjusted for the other x's. If the points in the plot lie along or near a straight line, this suggests that the variable should be added to the regression equation.

11.3 SPATIAL REGRESSION: AUTOCORRELATED ERRORS

It is possible to specify a spatial regression model in the same way as the usual linear regression model, with the exception that the residuals are modeled as functions of the surrounding residuals (see, e.g., Bailey and Gatrell 1995). If we use ε to denote the usual residual or error term, the residual for a particular observation is written as a linear function of the other residuals:

$$\varepsilon_i = \rho \sum_{j=1}^{n} w_{ij} \varepsilon_j + u_i, \tag{11.1}$$

where w_{ij} is a measure of the connection between region i and region j (often taken as a binary connectivity measure), ρ is a measure of the strength of the correlation of the residuals, and u_i is the remaining error term, after the correlation among residuals has been accounted for. Note if $\rho = 0$, the model reduces to the ordinary linear regression model.

To estimate the model, one can define the quantities:

$$y^{\star} = y - \rho \sum_{j=1}^{n} w_{ij} y_j$$
$$x^{\star} = x - \rho \sum_{j=1}^{n} w_{ij} x_j \tag{11.2}$$

Then regressions of y^* versus x^* are tried for a variety of ρ values, beginning at zero. The residuals of each regression are inspected, and the value of ρ associated with the most suitable set of residuals is adopted (usually this means that the value of ρ chosen is the one that is associated with the regression that has the highest value of r^2). Bailey and Gatrell note that this estimation procedure is, strictly, not one that is the best from a statistical viewpoint, and more sophisticated approaches exist. However, this approach should give the analyst a good idea of the spatial effects that may be present in a model.

11.4 SPATIALLY VARYING PARAMETERS

11.4.1 The Expansion Method

With linear regression, the slope and intercept parameters are 'global', in the sense that they apply to all observations. The expansion method (Casetti 1972; Jones III and Casetti 1992) suggests that these parameters may themselves be functions of other variables. Thus, in a linear regression equation of house prices (y) on lot size (x_1) and number of bedrooms (x_2):

$$y = b_0 + b_1 x_1 + b_2 x_2 + \varepsilon \tag{11.3}$$

the effect of lot size on house prices (b_1) may itself depend upon whether there is a park nearby (for example, large lot sizes may be more valuable in a suburb if there is no other green space nearby). So, we add an expansion equation:

$$b_1 = c_0 + c_1 d \tag{11.4}$$

where d is the distance to the nearest park. We would expect c_1 to be positive; large distances to the nearest park would mean that b_1 is high, which in turn means that small increments in lot sizes have a large influence on house prices.

If we substitute this expansion equation into the original equation, we have:

$$y = b_0 + (c_0 + c_1 d)x_1 + b_2 x_2 + \varepsilon \qquad (11.5)$$
$$= b_0 + c_0 x_1 + c_1 d x_1 + b_2 x_2 + \varepsilon$$

To estimate the coefficients, we perform a linear regression of y on the variables x_1, x_2, and dx_1. In Equation 11.5, the new quantity, dx_1, may be thought of as a new variable, created by multiplying together distance to park (d) and lot size (x_1). When the coefficient c_1 is significant, this is known as an *interaction* effect; the effect of lot size on housing prices interacts with, or depends upon, the distance to the park (or alternatively, the effect of distance to the park depends upon the size of the lot).

To take an explicitly spatial example, the regression coefficient might be taken to be a function of the x-coordinate and/or y-coordinate; such a specification would be appropriate if it was felt that the size of the effect of an explanatory variable on the dependent variable varied in an east–west and/or north–south direction.

The edited collection of Jones III and Casetti (1992) contains a wide variety of applications of the expansion method. These include applications to models of welfare, population growth and development, migrant destination choice, urban development, metropolitan decentralization, and the spatial structure of agriculture. The collection also includes methodological contributions that focus upon statistical aspects of the model, including its relationship to spatial dependence in the data.

11.4.2 Geographically Weighted Regression

In a series of articles, Fotheringham and his colleagues have outlined an alternative approach to the expansion method that accounts for spatially varying parameters (see, e.g., Brunsdon et al. 1996, 1999; Fotheringham et al. 1998). Their geographically weighted regression (GWR) technique is based upon 'local' views of regression, as observed from each data location. For each particular location, one can estimate a regression equation, where weights are attached to observations surrounding the location. Relatively large weights are given to points near the location, and smaller weights are assigned to observations far from the location. As Fotheringham et al. (2000) note:

> There is a continuous surface of parameter values … In the calibration of the GWR model it is assumed that observed data near to point i have more of an influence in the estimation of the [regression coefficients] than do data located farther from i. (p. 108)

More formally, the dependent variable at location i is modeled as follows:

$$y_i = b_{i0} + \sum_{j=1}^{p} b_{ij}x_{ij} + \varepsilon_i \tag{11.6}$$

where, as is the case with simple linear regression, there are p independent variables, and x_{ij} represents the observation on variable j at location i. The important point to note is that the b coefficients have i subscripts, indicating that they are specific to the location of observation i.

One reasonable and common choice for the weights is a negative exponential function of squared distance:

$$w_{ij} = e^{-\beta d_{ij}^2} = \exp(-\beta d_{ij}^2) \tag{11.7}$$

so that points that are farther away will be assigned lower weights.

To estimate the regression coefficients at location i, one first defines the weights (w_{ij}), using an initial 'guess' for the value of β (one possibility would be to use $\beta = 0$, which corresponds to the ordinary least squares case). Then define the quantities:

$$\begin{aligned} y_j^\star &= \sqrt{w_{ij}}\,y_j \\ x_j^\star &= \sqrt{w_{ij}}\,x_j \qquad j = 1...,n \end{aligned} \tag{11.8}$$

These are the weighted observations. At location i, run a linear regression of the $y^\star$ on the $x^\star$, omitting observation i from the analysis. Use the resulting regression coefficients to predict the value of y at location i. Then find the squared difference between the observed value of y (denoted y_i) and this predicted value:

$$\left\{ y_i - \hat{y}_{\neq i}(\beta) \right\}^2 \tag{11.9}$$

where $\hat{y}_{\neq i}(\beta)$ is the predicted value of the dependent variable at location i when observation i has not been used in the estimation, and the β reminds us that this prediction was made using a specific value of β. After this has been repeated for each location i, one may compute the total sum of squared deviations between observed and predicted values as:

$$s(\beta) = \sum_{i=1}^{n} \left\{ y_i - \hat{y}_{\neq i}(\beta) \right\}^2 \tag{11.10}$$

The next step is to repeat this procedure for many values of β, choosing as 'best' the value that minimizes the score $s(\beta)$. This final value of β yields the best set of weights. The final regression coefficients at each location are given as follows: first use the final, optimal value of β to define the weights, and then regress $y*$ and $x*$ using *all* of the observations.

For further details, the reader is referred to Fotheringham et al. (2002); they focus exclusively on the subject of geographically weighted regression, and include software for implementing the technique.

11.5 ILLUSTRATION

Figure 11.2 displays the location of 30 hypothetical houses in a square study area that features a park at its center. The dataset in Table 11.1 was generated by assuming that housing prices were related to lot size, number of bedrooms, and the presence of a fireplace.

Furthermore, spatial effects were added in the generation of the data. The lot sizes were generated in such a way as to be spatially autocorrelated, and the effect of lot size on housing prices was made to be a function of how distant the house was from the centrally located park. More specifically, houses were assigned fireplaces with probability 0.3, and were assigned a number of bedrooms by allowing integers in the range 2–6 with equal likelihoods. Lot sizes were normally distributed with mean 6 and standard deviation 0.8. Housing prices were generated using the equations:

$$p = 20,000 + b_1 x_1 + 20,000 x_2 + 20,000 x_3 + \varepsilon$$
$$b_1 = 10,000 + 20,000 d \tag{11.11}$$
$$\varepsilon \sim N(0, 20,000^2)$$

where p is price, x_1 is lot size (in thousands of square feet), x_2 is number of bedrooms, and x_3 is a dummy variable indicating the presence or absence of a fireplace (1 = presence; 0 = absence). The quantity d is the distance from the centrally located park, and ε is a normally distributed error term with mean 0 and standard deviation equal to 20,000. All digits have been retained in the generated prices, though in practice one would expect them to be rounded to, say, the nearest hundred.

Thus, the 'true' data follow quite closely an expansion-equation model, and we will expect that such a model will perform quite well. But for now, let us assume that we are

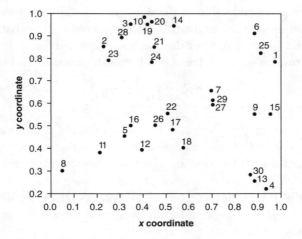

Figure 11.2 Location of 30 hypothetical houses

simply faced with the data in Table 11.1, and we want to model housing prices as a function of the independent variables.

11.5.1 Ordinary Least Squares

Table 11.2 shows the results from the ordinary least squares regression of housing price on lot size, number of bedrooms, and presence of fireplace. The coefficients are all significant.

Table 11.1   Hypothetical data on 30 houses

CASE	XCOORD	YCOORD	PRICE	LOTSIZE	BEDRMS	FIREPLC
1	.9619	.7817	224323	5.987	3	1
2	.2378	.8520	143510	4.241	2	1
3	.3481	.9440	233533	5.039	6	1
4	,9329	.2235	192328	3.100	4	1
5	.3258	.4532	158553	5.133	2	0
6	.8847	.9136	297893	6.397	5	1
7	.7063	.6176	150054	7.590	2	1
8	.0473	.2902	193785	4.848	6	0
9	.8927	.5538	206744	4.272	6	0
10	.4131	.9766	159585	5.126	3	0
11	.2189	.3649	212046	4.583	5	0
12	.3957	.3827	171795	4.589	3	1
13	.8909	.2550	125737	2.343	3	0
14	.5363	.9402	253078	7.138	4	1
15	.9574	.5488	189896	5.016	4	0
16	.3571	.5017	228830	6.783	5	1
17	.5396	.4733	163033	8.169	2	1
18	.5687	.3996	202935	5.199	2	1
19	.4256	.9444	205478	6.426	3	0
20	.4431	.9568	207324	6.199	2	1
21	.4555	.8451	249965	7.895	3	0
22	.5191	.5430	193800	8.689	4	0
23	,2518	.7851	203844	5.000	5	0
24	.4458	.7717	153122	5.654	2	0
25	.9242	.8261	252367	6.959	4	0
26	.4457	.4998	101089	5.376	2	0
27	.7138	.5867	196954	6.806	4	0
28	.3222	.8879	158972	4.352	2	1
29	.7107	.6137	191339	8.170	4	0
30	.8657	.2810	184990	3.377	4	0

The r^2 value is 0.562, and the standard error of the estimate is 29,080. The residuals display positive spatial autocorrelation (determined by mapping the residuals), indicating potential problems with the estimation. The coefficient on lot size (10,324) is a bit low, since we know from the way the data were generated that it ranges from a low of 10,000 near the park to a high of about 20,000 (= 10,000 + 20,000(0.5)) near the periphery.

11.5.2 Added-variable Plots

We begin by making the rather arbitrary decision that neighbors are defined in this example as the three closest observations. Thus, $w_{ij} = 1$ if observation j is one of the three nearest neighbors of i, and 0 otherwise.

Following Haining's example, we will consider the addition of new explanatory variables. Those we will consider are:

Table 11.2   Results from OLS regression

Variables entered/removed[a]

Model	Variables entered	Variables removed	Method
1	FIREPLC, LOTSIZE, BEDRMS[b]		Enter

[a] Dependent variable: PRICE.
[b] All requested variables entered.

Model summary

Model	R	R square	Adjusted R square	Std. error of the estimate
1	.749[a]	.562	.511	29080.13

[a] Predictors: (Constant), FIREPLC, LOTSIZE, BEDRMS.

ANOVA[a]

Model		Sum of squares	df	Mean square	F	Sig.
1	Regression	2.8E +10	3	9.4E + 09	11.108	.000[b]
	Residual	2.2E +10	26	8.5E + 08		
	Total	5.0E +10	29			

[a] Dependent variable: PRICE.
[b] Predictors: (Constant), FIREPLC, LOTSIZE, BEDRMS.

Coefficients[a]

Model		Unstandardized coefficients		Standardized coefficients		
		B	Std. error	Beta	t	Sig.
1	(Constant)	51249.856	26866.791		1.908	.068
	LOTSIZE	10323.904	3444.409	.391	2.997	.006
	BEDRMS	20628.873	4153.976	.661	4.966	.000
	FIREPLC	24834.295	10942.393	.301	2.270	.032

[a] Dependent variable: PRICE.

$$y_i^* = \sum_{j=1}^{n} w_{ij} y_j$$

$$x_i^* = \sum_{j=1}^{n} w_{ij} x_j$$

(11.12)

The first suggests that y at a location is a function not only of x at that location, but also of the y values in surrounding locations. The second equation suggests that the y value at a location may also be a function of the x-values in surrounding locations. To construct added variable plots for each of these potential additions to the regression equation, we need (a) the residuals of the ordinary least squares regression (from Section 11.5.1), and (b) the residuals from regressions of y_i^* and x_i^* on x. These residual plots are shown in panels (a) and (b) of Figure 11.3. Neither plot shows a significant correlation, and so we conclude that these variables would not significantly improve the specification of the regression equation.

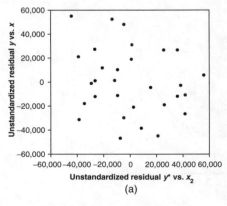

(a)

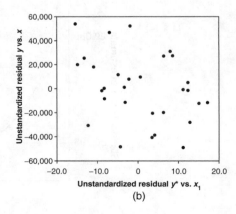

(b)

Figure 11.3   Added variable plots

11.5.3 Spatial Regression: Autocorrelated Errors

Following the autocorrelated errors model of Section 11.3, and using the same definition of the weights (w) used in Section 11.5.2, we define the quantities:

$$y^{\star} = y - \rho \sum_{j=1}^{n} w_{ij} y_j$$

$$x^{\star} = x - \rho \sum_{j=1}^{n} w_{ij} x_j.$$

(11.13)

We would like to choose a value of ρ that is associated with a 'good' set of residuals. Although there are different ways that this could be done, after trying different ρ values, we find that $\rho = 0.18$ minimizes the standard error of the estimate, which is also defined as the standard deviation of the residuals (see Figure 11.4). When $y^{\star}$ is regressed on $x^{\star}$ using this value of $\rho = 0.18$, we obtain the results in Table 11.3. The standard error of the estimate has been reduced to 26,795, and the value of r^2 is now 0.883. We also note here that minimizing the standard error of the estimate is the same as maximizing the value of r^2. In the example here, all variables are significant as before, and the t-values for all coefficients are higher than those found using ordinary least squares (Section 11.5.1). In addition, the coefficient on lot size in this equation is equal to 17,910, which is closer to its average value of about 15,000 (recall that we generated the data so that the true lot size coefficient varied from 10,000 to about 20,000).

11.5.4 Expansion Method

Next, we estimate the expansion model:

$$p = b_0 + b_1 x_1 + b_2 x_2 + b_3 x_3 + \varepsilon$$

$$b_1 = \gamma_0 + \gamma_1 d$$

(11.14)

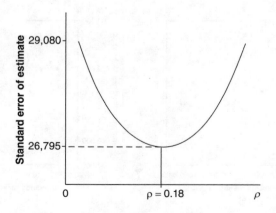

Figure 11.4   Minimizing the standard error of the estimate

Table 11.3 Spatial regression results with $\rho = 0.18$

	Coefficient	Standard error	t
Intercept	2926.9	4914	0.60
Lot size	17910	3341	5.30
No. of bedrooms	22921	3139	7.30
Fireplace	27233	9003	3.02

where the variables are as defined above, and where γ_0 and γ_1 are the regression coefficients that tell us how the influence of lot size on housing prices varies with distance from the park. This may be rewritten as:

$$p = b_0 + (\gamma_0 + \gamma_1 d)x_1 + b_2 x_2 + \varepsilon \tag{11.15}$$

which is identical to:

$$p = b_0 + \gamma_0 x_1 + \gamma_1 d x_1 + b_2 x_2 + b_3 x_3 + \varepsilon \tag{11.16}$$

The results obtained when using ordinary least squares on this equation are shown in Table 11.4.

Table 11.4 Results from expansion method

Variables entered/removed[a]

Model	Variables entered	Variables removed	Method
1	DPLOT FIREPLC, LOTSIZE, BEDRMS[b]		Enter

[a] Dependent variable: PRICE.
[b] All requested variables entered.

Model summary[a]

Model	R	R square	Adjusted R square	Std. error of the estimate
1	.864[b]	.747	.706	22552.50

[a] Dependent variable: PRICE.
[b] Predictors: (Constant), DPLOT, FIREPLC, LOTSIZE, BEDRMS.

(Continued)

ANOVA[a]

Model		Sum of squares	df	Mean squares	F	Sig.
1	Regression	3.7E +10	4	9.4E + 09	18.409	.000[b]
	Residual	1.3E +10	25	5.1E + 08		
	Total	5.0E +10	29			

[a] Dependent variable: PRICE.
[b] Predictors: (Constant), DPLOT, FIREPLC, LOTSIZE, BEDRMS.

Coefficients[a]

Model		Unstandardized coefficients		Standardized coefficients		
		B	Std. error	Beta	t	Sig.
1	(Constant)	42361.839	20939.718		2.023	.054
	FIREPLC	20632.446	8543.021	.250	2.415	.023
	BEDRMS	16482.800	3364.706	.528	4.899	.000
	LOTSIZE	8313.839	2712.410	.315	3.065	.005
	DPLOT	19743.607	4624.268	.454	4.270	.000

[a] Dependent variable: PRICE.

The r^2 value is equal to 0.747, and all parameters, including those associated with the expansion equation, are significant. Furthermore, all parameter values are near their 'true' values, and the standard error of the estimate is 22,552.

Of course, it should be kept in mind that one reason this particular approach has worked relatively well here is that the estimated model is consistent with the way in which the data were generated. We helped ourselves out by choosing to expand the model using the relation between lot size effects and distance from the park – this was a good choice because that is how the data were created!

11.5.5 Geographically Weighted Regression

Using the weights defined in Equation 11.7 and the method outlined in Section 11.4.2, the optimal value of β was found to be 0.13. This defines a set of weights that are associated with the variables in Equation 11.8. The regressions are then run once for each data point, using these weights. Figure 11.5 displays a map of the coefficient on lot size. From the figure, one can see that the parameter is higher away from the park, and this is in keeping with our expectations, since the effect of lot size on house prices was made to be greater in peripheral locations. A check of the correlation between a location's lot size coefficient and its distance to the park reveals that it is equal to $r = 0.361$, which is significant at the 0.05 level.

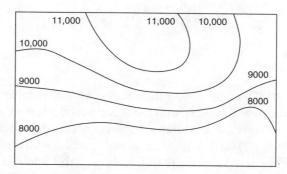

Figure 11.5   Spatial variation in lot size coefficient

11.6 SPATIAL REGRESSION WITH *GEODA* 1.4.6

In this section, we will illustrate how some of the methods for spatial regression that are described in this chapter are implemented in *GeoDa*. A full understanding of both spatial regression and all of the facilities available in *GeoDa* is beyond the scope of this text, but the software is sufficiently user-friendly that it is relatively straightforward to carry out regression using spatial data.

The *GeoDa* website has a number of sample datasets, and here we will use the file 'baltimore. zip' to look at a regression of house prices on the age of the house, in Baltimore, Maryland (USA). We begin the illustration by opening *GeoDa* and clicking on File, and Open Shapefile. Then select 'baltim' and click on Open. The first step is to define a weight matrix. Since this particular file contains the centroids of the geographic regions, but not the boundaries of those subregions, we cannot define the weights using, for example, rook or queen contiguity. Instead, we will define the 'neighborhood' around each centroid as consisting of the closest four centroids (this choice of four is arbitrary here; ultimately, the choice should be made based upon considerations of the appropriate spatial scale of the process. Often this is unknown, and it is not a bad idea to explore the results across various spatial scales by carrying out the analysis several times). To create the weights, click on Tools, then Weights, and then Create. Then click on Add ID Variable, use the default of Poly_ID, and click on Add Variable which appears at the bottom of the box. Click the Rook Contiguity dialog button. Click on Create. Name the output file; the default here is baltim, and a file will be created when you click on Save.

Next, click on the Methods and Regression tab. From the window that opens, choose PRICE and click on the arrow that moves it to the Dependent Variable box. Choose AGE and click on the arrow to move it to the Covariates box. Click on the Weights File box, choose the previously created weight matrix (baltim.gal), and then choose Run. This will run the Classic ordinary least squares regression model (as described in Chapter 8), and will yield output that includes the Moran's *I* value associated with the residuals. The regression equation is:

$$\text{Price} = 55.08 - 0.358(\text{Age})$$
$$(t = -4.56)$$
$$p < 0.001 \tag{11.17}$$

The r^2 value is 0.0904. The negative coefficient on the AGE variable implies that housing prices tend to go up for newer houses (which of course have younger ages). The coefficient is highly significant. However, the Moran's I value associated with the residuals is high (0.427) and significant ($p < 0.001$). This suggests that spatial considerations are important; if it was not significant, we could terminate our analysis at this point. In addition, Moran's I was found for both the PRICE variable ($I = 0.5105$) and the AGE variable ($I = 0.4348$), using the steps described at the end of Chapter 10. These are both highly significant, and, as was suggested in section 7.6.1, it is possible that the apparent relationship between these two variables is at least in part due to the spatial autocorrelation exhibited by each of them.

There are two alternative spatial regression models that offer ways to proceed within *GeoDa*. These are the spatial lag, and spatial error models. The former corresponds to the idea of adding to the regression an explanatory variable that consists of a weighted combination of the dependent variable (i.e., the first equation in Equation 11.12). The spatial error model is the same as what we have termed the autocorrelated errors model in Section 11.3. To carry out the spatial lag approach (i.e., to include an 'added variable' consisting of the weighted dependent variable), return to the Regress tab, click OK, choose the variables and weights as before, and then choose the Spatial Lag dialog button before choosing Run. The resulting regression equation is:

$$\text{Price} = 25.92 - 0.188(\text{Age}) + 0.547(\text{w\_price})$$
$$(t = -2.95) \qquad (t = 8.82)$$
$$p = 0.003 \qquad (p < 0.001) \tag{11.18}$$

The r^2 value has increased to 0.469. Note also that the AGE variable has a magnitude that is lower in absolute value (0.132, vs 0.358 in the first model), and it is less significant than it was in the first model (it is still highly significant, but the t value is much lower, and the p-value is higher). In addition, the new variable constructed for each region (namely, price in the neighborhood of that region) is also very significant – prices depend not only on the age of the house, but prices in the vicinity of each subregion.

The spatial error (or autocorrelated error) model is run in the same way, with the exception that the Spatial Errors dialog is chosen before choosing Run and OK. The result is:

$$\text{Price} = 48.82 - 0.137(\text{Age})$$
$$(t = -1.78)$$
$$p = 0.075 \tag{11.19}$$

The r^2 value here is equal to 0.464. The age variable is no longer significant. The measure of autocorrelation associated with the error terms (as used in Equation 11.1 and 11.13) is equal to 0.717. (Note that this quantity is given using the notation ρ in Equations 11.1 and 11.3, while it is called lambda, or λ within *GeoDa*.) It is highly significant ($t = 11.98$, and $p < 0.001$).

This illustration serves to demonstrate that ignoring spatial autocorrelation can result in both the misinterpretation of regression coefficients, and incorrect assessment of the importance of independent variables.

To determine which of the two spatial models is more appropriate, the following procedure may be adopted:

After the Classic model is run, examine the p-values (given under the 'PROB') column for Lagrange Multiplier (lag), and Lagrange Multiplier (error). If they are both insignificant (i.e., greater than 0.05), no further analysis is needed. If one is significant ($p < 0.05$), and the other is not, run the spatial model corresponding to the significant p-value (spatial lag, or spatial error). If they are both significant, examine the p-values associated with the Robust LM (lag) and Robust LM (error). These will usually be such that one is significant and the other is not; run the spatial regression model associated with the significant one. Further detail is provided in the workbook that can be found on the *GeoDa* website (Anselin 2005). For this example, this procedure points to the spatial lag model as the best choice.

EXERCISES

1. Use an added variable plot to determine whether distance to the park should be added to a regression of housing price on lot size, number of bedrooms, and presence or absence of a fireplace. Use the data in Table 11.1.

2. Using the data in Table 11.1 and geographically weighted regression, produce a map showing the spatial variation in the coefficient on number of bedrooms. Alternatively, you may provide a table showing the regression coefficient for the number of bedrooms at each of the 30 sample locations.

3. With the data in Table 11.1, first perform an ordinary least squares regression, with housing price as the dependent variable and lot size as the independent variable. Then use the expansion method, with the lot size coefficient depending upon the number of bedrooms. Interpret the results.

4. Use the spatial regression method outlined in Sections 11.3 and 11.5.3 with the data in Table 11.1 for a regression of housing prices on lot size and number of bedrooms.

(Continued)

On the Companion Website

The website's 'Further Resources' section contains links to other material which supplements and expands upon information presented in Chapter 11, including links to the journal papers and chapters referenced in this chapter.

12

DATA REDUCTION: FACTOR ANALYSIS AND CLUSTER ANALYSIS

12.1 INTRODUCTION

Many studies of complex geographic phenomena begin with a set of data and notions of hypotheses and theories that are vague at best. Often large datasets are organized in tables, where the rows consist of observations and the columns represent variables. For example, each census tract in a county may be represented by a row of data; the entries in the row consist of, e.g., data on a number of socioeconomic and demographic variables. A column of data would represent the set of observations on one of these variables, across all census tracts.

A first step is to reduce the dataset in some manner, so that it is more interpretable. In this chapter, we will learn about two common approaches to data reduction – namely, *factor analysis* and *cluster analysis*. Factor analysis collapses the columns of the dataset to construct a smaller number of new factors or indices that are linear combinations of the original variables. Cluster analysis collapses the data row-wise by finding rows of data that are similar to one another. In this way, clusters of similar observations are created.

12.2 FACTOR ANALYSIS AND PRINCIPAL COMPONENTS ANALYSIS

Factor analysis may be used as a data reduction method, to reduce a dataset containing a large number of variables down to one of more manageable size. When many of the original variables are highly correlated, it is possible to reduce a large number of original variables to a small number of underlying factors.

A geometric interpretation helps one to understand the purpose of factor analysis. A dataset consisting of n observations on p variables may be represented as n points plotted in a p-dimensional space. This is easiest to imagine when $p = 1, 2$, or 3, and the latter case is illustrated in Figure 12.1. The figure attempts to depict a three dimensional ellipsoidal figure that contains the majority of the data points. The idea behind factor analysis is to construct factors that represent a large proportion of the variability of the dataset. The original axes correspond to the original variables; the longest axis of the ellipsoid is a *new* variable, which is a linear combination of the original variables. This new variable, or factor, captures as much of the variability in the dataset as possible.

Thus, the first factor corresponds, geometrically, to the longest axis of the ellipsoid. A second factor is derived by finding the second-longest axis of the ellipsoid, such that this second axis is perpendicular to the first axis. The fact that the axes of the ellipsoid are perpendicular implies that the newly defined factors will be uncorrelated with one another – they represent separate and independent aspects of the underlying data.

A dataset characterized by an extremely elongated ellipsoid would be well represented by a single factor – that combination of variables would explain almost all of the variability in the original data. In the extreme case, the plotted data would fall along a single line, which would constitute the axis or single factor that would capture *all* of the variability in the data. At the other extreme, the data ellipsoid could be spherical; in this case, all factors explain an equal amount of the variability in the original data (since all axes would have equal lengths), and there are no dominant factors.

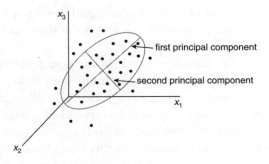

Figure 12.1   Data ellipsoid in $p = 3$ dimensions

In this discussion, we will focus more upon the interpretation of the outputs of factor analysis, and less on its mathematical aspects. The next subsection addresses the interpretation of factor analysis results through an example using 1990 census data from Erie County, New York, which contains the city of Buffalo.

12.2.1 Illustration: 1990 Census Data for Erie County, New York

Geographers often use many census variables in their analyses, and the set of variables can often contain subsets that measure essentially the same phenomenon. The following example illustrates, for a small set of census data, how the number of original variables can be collapsed into a smaller number of uncorrelated factors.

A 235 × 5 data table was constructed by collecting and deriving the following information for the 235 census tracts in Erie County, New York (variable labels are in parentheses):

(a) Median household income (medhsinc).

(b) Percentage of households headed by females (femaleh).

(c) Percentage of high-school graduates who have a professional degree (educ).

(d) Percentage of housing occupied by owner (tenure).

(e) Percentage of residents who moved into their present dwelling before 1959 (lres).

These five variables capture different aspects of the socioeconomic and demographic character of census tracts. Do they represent separate dimensions of socioeconomic and demographic structure, or is there significant redundancy in what they measure, indicating that the variables might be reduced to a smaller number of underlying indices or factors? A natural place to start is with the correlation matrix. Table 12.1 reveals that the highest correlations are with the median household income variable; areas of high income have low percentages of households headed by females, high percentages of homeowners, high percentages of graduates with professional degrees, and a relatively

Table 12.1   Correlation among variables

		MEDHSINC	FEMALEH	EDUC	TENURE	LRES
				Correlation matrix		
Correlation	**MEDHSINC**	1.000	−.595	.415	.569	−.455
	FEMALEH	−.595	1.000	−.348	−.531	.221
	EDUC	.415	−.348	1.000	.117	−.161
	TENURE	.569	−.531	.117	1.000	−.438
	LRES	−.455	.221	−.161	−.438	1.000

low proportion of long-term residents. Using the test of significance described in Chapter 5, all correlations with absolute value greater than $2 / \sqrt{235} = 0.130$ are significant.

The second step is to examine the outcome of describing the data as an ellipsoid, as described above. The method of *principal components* is used to describe the p axes of the ellipse (which in turn is constructed in a p–dimensional space, where p is the number of variables). The relative lengths of the axes are called *eigenvalues*. They are referred to in the *SPSS* output of Table 12.2 as 'extraction sums of squared loadings'. A 'loading' is the correlation between a component or factor and the original variable. If one were to sum the squared correlations between a factor and all of the original variables, this would be equal to the eigenvalue, or the length of the ellipse's corresponding axis. From the table, we see that the highest eigenvalue is 2.6, and the second highest is 0.96. Note that the column displaying these values sums to five – the eigenvalues (i.e., 'extraction sums of squared loadings') will always sum to the number of variables. In an extreme case, there would be a single component with perfect correlations with all of the original variables. The eigenvalue for this component would be equal to $1^2 + 1^2 + \ldots + 1^2 = p$. All of the other eigenvalues would be equal to zero, and the ellipse would collapse to a single line.

Table 12.2   **Variance explained by each component**

	Total variance explained					
	Extraction sums of squared loadings			Rotation sums of squared loadings		
Component	Total	% of variance	Cumulative %	Total	% of variance	Cumulative %
1	2.602	52.032	52.032	1.035	20.707	20.707
2	.957	19.149	71.181	1.032	20.637	41.344
3	.741	14.826	86.007	1.018	20.358	61.702
4	.362	7.244	93.251	1.005	20.110	81.812
5	.337	6.749	100.000	.909	18.188	100.000

Extraction Method: Principal Component Analysis.

This table also provides us with valuable information concerning how many factors are necessary to adequately describe the data. There are two 'rules of thumb' that are used to decide on the number of factors. One such rule is to retain components with eigenvalues greater than one. This would be an unfortunate rule to apply in this instance, since the second eigenvalue is just slightly less than one (0.96). An alternative is to plot the eigenvalues on the vertical axis and the factor number (ranging from 1 to p) on the horizontal axis of a graph. Then inspect the graph to locate a point that occurs just before the graph (termed a *scree plot*) flattens out; such a feature implies that the additional factors do not contribute much to the explanation of variability in the dataset. Figure 12.2 displays a scree

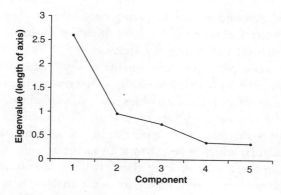

Figure 12.2 Scree plot for Erie County example

plot for our present example. Some judgment is called for, and we could in this instance justify the extraction of either two or three factors.

Suppose that we decide to extract two factors. The next step is to inspect the loadings, or correlations between the factors and the original variables. This is a key step in the analysis, since it is where the 'meaning' and interpretation of each factor occurs.

To aid in this interpretation, the extracted component solution is rotated in the p-dimensional space, so that the loadings tend to be either high (near plus or minus one) or low (near zero).

Table 12.3 shows that the two-factor solution may be described as follows. The first factor is one where income, tenure, and length of residence all 'load highly'. We can think of these variables as being combined to form a single index (the factor), which describes with a single number what the three variables represent. The second factor is associated with the other two variables – education and family structure. It is common practice to attempt to

Table 12.3 Factor loadings

	Rotated component matrix[a]	
	Component	
	1	2
MEDHSINC	.668	.562
FEMALEH	−.515	−.608
EDUC	−3.79E−02	.912
TENURE	.848	.154
LRES	−.766	−2.78E−03

Extraction Method: Principal Component Analysis.
Rotation Method: Varimax with Kaiser Normalization.
[a] Rotation converged in three iterations.

give the factors snazzy, descriptive names. Having noted this, it is often difficult to come up with something creative! The first factor here might be thought of as a housing/economic factor and the second a sociological factor.

The difference between principal components analysis and factor analysis may be summarized as follows. Principal components is a descriptive method of decomposing the variation among a set of p original variables into p components. The components are linear combinations of the original variables. It is often used as a prelude to factor analysis, which attempts to *model* the variability in the original set of variables via a reduced number of factors, which is less than p. In factor analysis, values of the original variables may be reconstructed by writing them as linear combinations of the factors, plus a 'uniqueness' term. Alternatively stated, in factor analysis, part of the variability in an original variable is captured by the factors (this portion of the variability is termed the communality), and part is *not* captured by the factors (this portion is termed the uniqueness). Table 12.4 shows the communalities for the two-factor solution. The highest communality is for education (0.833) and the lowest is for length of residence (0.587). The communality for a variable is equal to the sum of the squared correlations of the variable with the factors. For example, the communality for education is equal to its squared correlation with factor one (0.0379^2), plus its squared correlation with factor two (0.912^2). Length of residence has the highest uniqueness, since it is not highly correlated with the two factors.

Table 12.4   Communalities

	Extraction
MEDHSINC	.762
FEMALEH	.635
EDUC	.833
TENURE	.742
LRES	.587

Extraction Method: Principal Component Analysis.

It is important to realize that the output of a factor analysis is a strong function of the input. The fact that length of residence is not strongly related to either factor does not mean that it is not an important feature of urban structure. The factors that emerge from a factor analysis are not necessarily the 'most important' ones, but rather the ones that capture the nature of the dataset. If we had a dataset with 15 variables, and 11 of the 15 variables were alternative measures of income, we can be certain that an income factor would emerge as the principal factor, simply because so many variables were highly intercorrelated.

Finally, one of the outputs from factor analysis is the *factor scores*. Instead of making *p* separate maps describing the spatial pattern of each variable, one is now interested in making a number of maps equal to the number of underlying factors. For each factor, and for each observation, a score may be computed as a linear combination of the original variables. The result is a new data table; instead of the original *n* by *p* table, we now have an *n* by *k* table, where *k* is the number of factors. Figures 12.3 and 12.4 display the factor scores on each of our two factors for the Erie County census tracts.

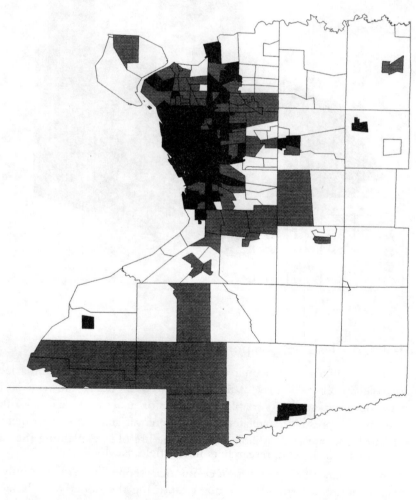

Figure 12.3   Factor 1 scores

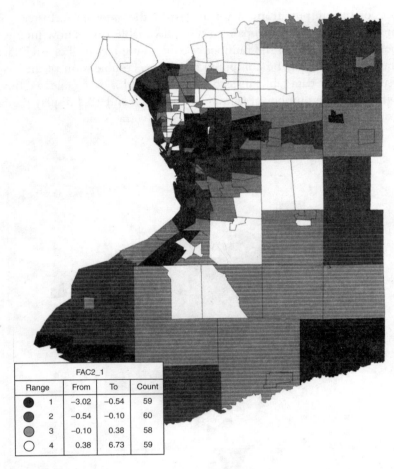

FAC2_1			
Range	From	To	Count
● 1	−3.02	−0.54	59
◕ 2	−0.54	−0.10	60
◔ 3	−0.10	0.38	58
○ 4	0.38	6.73	59

Figure 12.4   Factor 2 scores

12.2.2 Regression Analysis on Component Scores

As we have seen, the chief use of principal components analysis is to summarize a large number of variables in terms of a set of uncorrelated components. This can be ideal for choosing independent, explanatory variables for regression analysis, since one is often faced with a large number of possibly correlated independent variables, and the objective is to use a small subset of uncorrelated variables that will be useful in explaining the variability in the dependent variable.

Regression analysis may be carried out on component scores, ensuring not only that the independent variables are a parsimonious subset capturing the underlying dimensions of the full set of potential independent variables, but that they are uncorrelated as well. This idea for eliminating multicollinearity is one that is quite commonly employed (for example, see Ormrod and Cole 1996; Ackerman 1998; O'Reilly and Webster 1998). One disadvantage is that it is somewhat more difficult to interpret the regression coefficients. They now indicate how much the dependent variable changes when the component score changes by one unit, and it is more difficult to conceptualize just what a one unit increase in the component score really implies. Hadi and Ling (1998) also note some pitfalls in the use of principal components regression.

12.3 CLUSTER ANALYSIS

Whereas factor analysis works by searching for similar *variables*, cluster analysis has as its objective the grouping together of similar *observations*. Since it is conventional to represent each observation as a row in a data table, and each variable as a column, cluster analysis has at its core the search for similar rows of data. Factor analysis is based upon similarities among columns of data.

Like factor analysis, cluster analysis may be thought of as a data reduction technique. We seek to reduce the n original observations into g groups, where $1 \leq g \leq n$. In achieving this reduction of n observations into a smaller number of groups, a general goal is to minimize the within-group variation, and maximize the between-group variation. In Figure 12.5, there is relatively little variability within groups, as measured by the variation in the location of points around their group centroids. Relative to this within-group variability, there is much more variation in the locations of the group centroids in relation to the centroid for the entire dataset.

One of the more widespread applications of cluster analysis in geography has been in the area of geodemographics, where analysts seek to reduce a large number of subregions

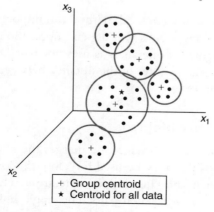

Figure 12.5   **Clustering in $p = 3$ dimensions**

(e.g., census tracts) by classifying them into a small number of types (see, e.g., Chapter 10 of Plane and Rogerson 1994). Cluster analysis has also been used as a method of regionalization, where the objective is to divide a region into a smaller number of contiguous subregions. In this case, it is necessary to modify traditional approaches to cluster analysis slightly to ensure that the created groups are comprised of contiguous subregions (see, e.g., Murtagh 1985; Duque et al. 2011; Duque et al. 2012).

Approaches to cluster analysis may be categorized into two broad types. *Agglomerative* or *hierarchical* methods start with n clusters (where n is the number of observations); each observation is therefore in its own cluster. Then two clusters are merged, so that $n - 1$ clusters remain. This process continues until only one cluster remains (this cluster contains all n observations). The process is hierarchical because the merger of two clusters at any stage of the analysis cannot be undone at later stages. Once two observations have been placed together in the same cluster, they stay together for the remainder of the grouping process.

In contrast, *nonhierarchical* or *nonagglomerative* methods begin with an a priori decision to form g groups. Then one begins with either an initial set of g seed points or an initial partition of the data into g groups. If one starts with a set of seed points, a partition of the data into g groups is achieved by assigning each observation to the nearest seed point. If one begins with a partition of the data into g groups, g seed point locations are calculated as the centroids of these g partitioned groups. In either case, an iterative process then takes place, where new seed points are calculated from partitions, and then new partitions are created from the seed points. This process continues until no reassignments of observations from one group to another occur. The convergence of this iterative process is usually very rapid.

The nonhierarchical methods have the advantage of requiring less computational resources, and for this reason they are the preferred method when the number of observations is very large. They have the disadvantage that the number of groups must be specified prior to the analysis, though in practice it is not uncommon to find solutions for a range of g values.

For both hierarchical and nonhierarchical methods, an important first step is to standardize the data – if this is not done, results will depend upon the unit of measurement (e.g., different results will be obtained if one analysis uses dollars and the other uses pounds to represent income). In addition, measures of the distance between observations will depend more on some variables than others.

12.3.1 More on Agglomerative Methods

With agglomerative methods, at each stage one merges the closest pair of clusters. There are many possible definitions that may be used for 'closest'. Consider all pairs of distances between elements of cluster A and cluster B. If there are n_A elements in cluster A and n_B elements in cluster B, there are $n_A n_B$ such pairs. The single linkage (or nearest neighbor) method defines the distance between clusters as the minimum distance among all of these pairs. The complete linkage (or furthest neighbor) method defines the distance between clusters as the maximum distance among all of these pairs.

One of the more commonly used methods is Ward's method. At each stage, all potential mergers will reduce the number of current clusters by one. Each of these potential mergers will result in an increase in the overall within sum of squares. (The within sum of squares may be thought of as the amount of scatter about the group centroids. With n clusters, the within sum of squares is equal to zero, since there is no scatter of other members about the group centroids. With one cluster, the within sum of squares is maximal.) Ward's method chooses that merger that results in the smallest increase in the within sum of squares. This is conceptually appealing, since we would like the within–group variability to remain as small as possible.

12.3.2 Illustration: 1990 Census Data for Erie County, New York

Here we will illustrate some of the features of cluster analysis using the dataset described above in the illustration of factor analysis.

Table 12.5 displays the results of a nonhierarchical k-means cluster analysis, where solutions range from $k = 2$ to $k = 4$. Three variables were used as clustering variables: the education variable, median household income, and the percentage of households headed by females.

Table 12.5   (a) Two-cluster solution; (b) three-cluster solution; (c) four-cluster solution

(a) Final cluster centers

	Cluster	
	1	2
Zscore (EDUC)	−.49757	.58646
Zscore (FEMALEH)	.51941	−.62329
Zscore (MEDHSINC)	−.55645	.76299

ANOVA

	Cluster		Error			
	Mean square	df	Mean square	df	F	Sig.
Zscore (EDUC)	67.302	1	.639	229	105.276	.000
Zscore (FEMALEH)	74.784	1	.678	229	110.335	.000
Zscore (MEDHSINC)	99.708	1	.496	229	201.103	.000

The F tests should be used only for descriptive purposes because the clusters have been chosen to maximize the differences among cases in different clusters. The observed significance levels are not corrected for this and thus cannot be interpreted as tests of the hypothesis that the cluster means are equal.

(Continued)

Table 12.5   (Continued)

Number of cases in each cluster

Cluster		
Cluster	1	126.000
	2	105.000
Valid		231.000
Missing		5.000

(b) Final cluster centers

	Cluster		
	1	2	3
Zscore(EDUC)	−.58897	−.27850	1.39280
Zscore(FEMALEH)	1.60968	−.27974	−.68224
Zscore(MEDHSINC)	−1.05910	.03641	1.11887

ANOVA

	Cluster		Error			
	Mean square	df	Mean square	df	F	Sig.
Zscore(EDUC)	57.715	2	.431	228	133.904	.000
Zscore(FEMALEH)	73.226	2	.366	228	199.830	.000
Zscore(MEDHSINC)	53.347	2	.467	228	114.152	.000

The F tests should be used only for descriptive purposes because the clusters have been chosen to maximize the differences among cases in different clusters. The observed significance levels are not corrected for this and thus cannot be interpreted as tests of the hypothesis that the cluster means are equal.

Number of cases in each cluster

Cluster		
Cluster	1	44.000
	2	141.000
	3	46.000
Valid		231.000
Missing		5.000

(c) Final cluster centers

	Cluster			
	1	2	3	4
Zscore(EDUC)	4.76906	−.30937	.99199	−.62440
Zscore(FEMALEH)	−1.02424	−.17113	−.68459	1.98402
Zscore(MEDHSINC)	−.87318	−.11596	1.19929	−1.13847

ANOVA

	Cluster		Error			
	Mean square	df	Mean square	df	F	Sig.
Zscore(EDUC)	41.594	3	.392	227	106.188	.000
Zscore(FEMALEH)	52.519	3	.319	227	164.566	.000
Zscore(MEDHSINC)	40.720	3	.401	227	101.478	.000

The F tests should be used only for descriptive purposes because the clusters have been chosen to maximize the differences among cases in different clusters. The observed significance levels are not corrected for this and thus cannot be interpreted as tests of the hypothesis that the cluster means are equal.

Number of cases in each cluster

Cluster	1	2.000
	2	143.000
	3	54.000
	4	32.000
Valid		231.000
Missing		5.000

After standardization, the resulting z-scores were used in the cluster analysis. For the two-cluster solution, the final cluster centers reveal that the first cluster is one where there are low z-scores on the education and median household income variables, and high z-scores on the percentage of households headed by females. The second cluster has the opposite characteristics, since the final cluster centroid is characterized by education and income values that are above average, and a percentage of female-headed households that is below average. There are 126 observations in the first cluster, and 105 in the second (and there are five observations with missing data).

The ANOVA table reveals that all of the variables are contributing strongly to the success of the clustering, since all of the F-values are extremely high and significant. It is important to note that, since the cluster analysis is *designed* to make the F-statistic large by minimizing within-group variation, these F-statistics should not be interpreted in the usual way. In particular, we would *expect* the F-statistics to be large, since we are creating clusters to make F large. Still, they can be used as rough guidelines to indicate the success of the clustering, and the relative success that individual variables have in achieving the cluster solution.

The three-cluster solution is similar to the two-cluster solution, with the addition of a 'middle' group that has values on all three variables that are close to the countywide averages. There are 44 observations in the first cluster (characterized by low levels of education and income, and a high percentage of female-headed households), 141 observations in the middle group, and 46 tracts in the third group. Again, all of the F-statistics are high, implying that all three variables help to place the observations into clusters.

One of the groups in the four-cluster solution has only two observations. These two observations are characterized by an extremely high percentage of individuals with professional degrees.

It appears that there are two distinct clusters, with a third, fairly large group characterized by rather average values on the variables. In addition, the cluster analysis has been helpful in locating two census tracts that could be characterized as outliers, due to their high values on the education variable. Figure 12.6 depicts the location of tracts in the four-cluster solution. The light grey regions represent locations for observations falling in the low-income/education cluster. The white areas correspond to observations in the large second cluster, and contain relatively average values on the variables. The darker grey corresponds to the cluster of regions with high values on income and education, and a low percentage of female-headed households. The darkest shading is used to represent the two outlying observations.

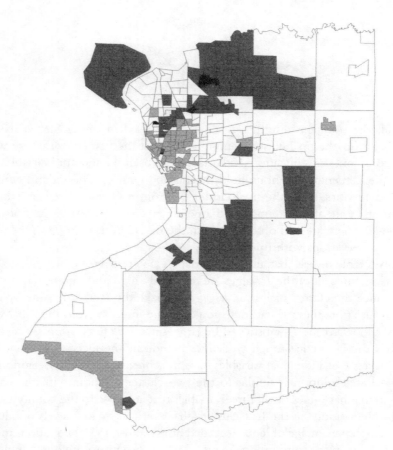

Figure 12.6   Three-cluster solution

An important piece of output from a hierarchical cluster analysis is the *dendogram*. As its name implies, a dendogram is a tree-like diagram. The dendogram captures the history of the hierarchical clustering process, as one proceeds from left to right along it. For illustrative purposes, it is rather difficult to show the dendogram that accompanies a hierarchical cluster analysis that has taken place for a very large number of observations. Instead, Figure 12.7 shows a dendogram for a subset of 30 tracts that have been selected at random from the dataset.

At the left of the dendogram, the branches that meet indicate observations that have clustered together. For example, tracts 70 and 146 were very close together in the three-variable space, and clustered together early in the process. In fact, the agglomeration schedule (shown in Table 12.6) indicates that these were the first two observations that were clustered together. The horizontal scale of the dendogram indicates the distance between the observations or groups that are clustered together. On the left of the dendogram, observations are close together when they cluster. On the right of the dendogram, there are only a small number of groups, and the distance between those groups is larger.

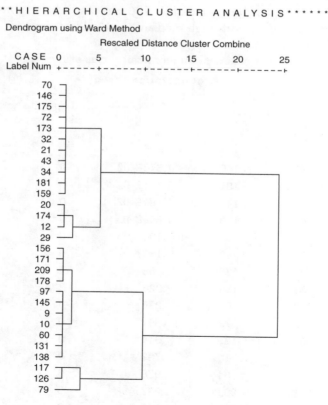

Figure 12.7   Dendogram

To decide on the number of clusters, one can imagine taking a vertical line, and proceeding from left to right along the dendogram. As one proceeds, the number of lines from the dendogram that intersect this vertical line decreases, from *n* to 1. A good choice for the number of groups is one where there is a fairly large horizontal range in the dendogram where the number of groups does not change. In Figure 12.7, it would make little sense to choose five groups, since these five groups could easily be simplified into four by proceeding just a little further to the right on the dendogram. The figure shows that there are two clear groups of tracts. The tracts that are in each of these groups may be found by proceeding to the left, from each of the two parallel, horizontal lines on the dendogram. Following these horizontal lines all the way to the left, through all of the branches, reveals all of the tracts in each cluster. For example, one of the two clusters consists of observations 156, 171, 209, 178, 97, 145, 9, 10, 60, 131, 138, 117, 126, and 79. Note that a three-cluster solution would subdivide this particular cluster into two subclusters, and one of those subclusters would be quite small (consisting only of observations 117, 126, and 79). The next step in this analysis would be to examine the characteristics of the observations in each cluster. For example, observations 117, 126, and 79 all have quite high values on the education variable, coupled with high median household incomes.

Table 12.6   Agglomeration schedule for hierarchical clustering

Agglomeration schedule

Stage	Cluster combined		Coefficients	Stage cluster first appears		Next stage
	Cluster 1	Cluster 2		Cluster 1	Cluster 2	
1	70	146	1.578E–02	0	0	7
2	34	181	3.158E–02	0	0	8
3	97	145	5.184E–02	0	0	9
4	72	173	7.689E–02	0	0	10
5	156	171	.104	0	0	14
6	21	43	.135	0	0	13
7	70	175	.168	1	0	17
8	34	159	.226	2	0	13
9	9	97	.295	0	3	12
10	32	72	.391	0	4	17
11	60	131	.514	0	0	19
12	9	10	.732	9	0	22
13	21	34	.978	6	8	21
14	156	209	1.237	5	0	20
15	20	174	1.545	0	0	16

Stage	Cluster combined		Coefficients	Stage cluster first appears		Next stage
	Cluster 1	Cluster 2		Cluster 1	Cluster 2	
16	12	20	1.858	0	15	23
17	32	70	2.259	10	7	21
18	117	126	2.721	0	0	25
19	60	138	3.193	11	0	22
20	156	178	3.782	14	0	24
21	21	32	4.642	13	17	26
22	9	60	5.718	12	19	24
23	12	29	7.126	16	0	26
24	9	156	8.771	22	20	27
25	79	117	11.396	0	18	27
26	12	21	17.329	23	21	28
27	9	79	28.934	24	25	28
28	9	12	58.941	27	26	0

12.4 DATA REDUCTION METHODS IN *SPSS 21 FOR WINDOWS*

12.4.1 Factor Analysis

Click on Analyze, then Dimension Reduction, and then Factor. Choose the variables that will enter into the factor analysis. Under Rotation, choose Varimax (it is *not* the default). This is the most commonly used rotation method, and you should use it unless you have a good reason to choose an alternative! Under Extraction, it is most common to choose Principal Components, and to choose Eigenvalues over 1 as significant. These are the defaults, and so, unless you wish to change them, you do not have to do anything. Under Scores, choose Save as Variables. This will save the factor scores as new variables by attaching a number of columns to your dataset that is equal to the number of significant factors. Under Descriptive, choose Univariate Descriptives if desired. It is also useful to check the box labeled coefficients under Correlation Matrix, to print out a table of the correlation coefficients among variables.

12.4.2 Cluster Analysis

12.4.2.1 Hierarchical methods

Choose Analyze, then Classify, then Hierarchical Cluster. Next choose the variables that are to be clustered. Then, under Method, choose the clustering method to be used. Note: Ward's method, though perhaps the most commonly used, is *not* the default choice. Next, choose the measure of distance that will be used to determine how far observations are from one another; squared Euclidean distance is the default, and is a reasonable (and readily understandable!) choice. Still in this section, you will likely want to choose *z*-scores under the box labeled `standardize`; again, it is *not* the default option. Under `Plots`, one will often want to turn off the default 'icicle plot', and check the box labeled `dendogram`. Under `Save`, you may save cluster membership, which adds a column of data to the data table indicating the cluster to which each observation belongs. This can be done for either a single predefined choice of the number of clusters, or for a predefined range of cluster numbers.

12.4.2.2 Nonhierarchical clustering

First click on `Analyze`, then on `Classify`, and then on `K-Means Cluster`. After choosing the variables to cluster (and recalling that it is usually a good idea to standardize the data by computing *z*-scores before doing this; this can be achieved by clicking on `Analyze`, `Descriptives`, `Descriptives`, and then checking the box to save the standardized scores as variables), choose the number of clusters desired. It is also usually a good idea to click on `Save`, and save cluster membership, which adds a column to the data table indicating the cluster membership of each observation. This is an iterative procedure, and the default number of iterations is equal to ten. One should check the output to see if the solution has converged (i.e., there are no changes in the locations of cluster centers from iteration to iteration). If it has not converged, the number of iterations should be increased (e.g., to 50), and then checked again for convergence. For example, the two cluster solution described in Section 12.3.2 required 15 iterations for convergence; the three- and four-cluster solutions each converged within the default number (10) of iterations.

EXERCISES

1. Explain and interpret the rotated factor loading table below.

	Factor 1	Factor 2
% < 15 years old	.88	.21
% Blue collar workers	.13	.86
% > 65 years old	−.92	−.11
% White collar workers	−.17	−.81
Median income	.24	−.71

---

2. Perform a hierarchical cluster analysis using the following data, and comment on the results.

Region	Mean Age	% Nonfamily	Median Income (× 000)
1	34	50	34
2	45	44	44
3	32	58	38
4	50	50	59
5	55	70	44
6	26	62	29
7	37	38	33
8	42	36	43
9	47	39	56
10	46	49	58
11	51	68	61
12	38	36	39
13	33	44	41
14	29	66	38

3. How many significant factors would be extracted in the following factor analysis? How many variables were in the original analysis? Explain your answer.

Factor	Eigenvalue	Cumulative Percent of Variance Explained
1	3.0	45
2	2.5	70
3	1.5	78
4	0.9	89
5	0.3	92
6	0.2	94
7	0.2	96
8	0.2	98
9	0.1	99
10	0.1	100

(Continued)

(Continued)

4. A researcher collects the following information for a set of census tracts:

Tract	Median Age	Income (× 000)	% Nonfamily	No. Autos	% New Residents	% Blue Collar
1	26	29	32	1	23	33
2	35	38	24	2	21	21
3	48	49	29	3	16	44
4	47	55	55	3	18	44
5	36	39	66	2	23	41
6	29	32	42	2	33	40
7	55	58	38	3	10	31
8	56	66	36	3	11	24
9	29	32	33	1	23	28
10	33	44	29	2	21	29
11	44	49	31	2	18	31
12	47	46	38	2	15	30
13	51	52	55	3	12	20
14	44	49	52	2	18	19
15	37	40	38	1	19	43
16	38	41	34	2	21	31

Use factor analysis to summarize the data.

Assume that the data come from a 4 x 4 grid laid over a map of the city as follows:

1	2	3	4
5	6	7	8
9	10	11	12
13	14	15	16

(a) Run a factor analysis to summarize the data above.

(b) How many factors are sufficient to describe the data (i.e., have eigenvalues greater than 1)?

(c) Describe the rotated factor loadings, describing each factor in terms of the most important variables that comprise it. Attempt to give names to the factors.

(Note: in this part of the question, discuss only those factors with eigenvalues greater than 1.)

(d) Save the factor scores, and make a map of the scores on factor 1.

5. Use the Hypothetical UK Housing Prices dataset to perform a cluster analysis. Use the nonhierarchical *k*-means method, using the factor scores found in question 5. Try *k* = 2, 3, 4, and 5 solutions, and comment on the relative merits of each.

6. Use the Milwaukee dataset to perform a factor analysis using the following variables: age of house, number of bedrooms, number of bathrooms, lot size, and number of finished square feet. Determine the appropriate number of factors, save the factor scores, and interpret the results.

7. Use the Milwaukee dataset to perform a cluster analysis. Use the nonhierarchical *k*-means method, using the factor scores found in question 7. Try *k* = 2, 3, 4, and 5 solutions, and comment on the relative merits of each.

8. Use the Singapore census dataset to carry out a factor analysis using the following variables: percent renters, percent with a long commute, percent with a university degree, percent with no schooling, percent of those in the labor force with a monthly income of over 8000 Singapore dollars, illiteracy rate, and percent of population over age 65. Determine the appropriate number of factors, save the factor scores, and interpret the results.

9. Use the Singapore census dataset to perform a cluster analysis. Use the nonhierarchical *k*-means method, using the factor scores found in question 9. Try *k* = 2, 3, 4, and 5 solutions, and comment on the relative merits of each.

On the Companion Website

To supplement Chapter 12, the website includes the **1990 Census Data for Erie County, New York** dataset used in the text as well as the **Home Sales in Milwaukee, Wisconsin**, **Hypothetical UK Housing Prices**, and **Singapore Census Data for 2010** datasets required for the Exercises under the 'Datasets' section of the website. The 'Further Resources' section includes links to other websites and videos that further explore and explain the concepts described here.

EPILOGUE

The primary purpose of this book has been to provide a foundation in some of the basic statistical tools that are used by geographers. The focus has been on inferential methods. Inferential statistical methods are attractive because they fit well into the time-honoured framework of the scientific method. There are of course limitations to the use of these methods. Many of the concerns are related to the nature of hypothesis testing. Why do we test whether two populations have exactly the same mean? An enumeration of two communities would almost certainly show that the true 'population' means of, say, commuting distances were in fact different. Why do we test whether a true regression coefficient is zero? Independent variables will commonly have *some* effect on the dependent variable, even if it is small. The principal point here is that, in many situations, the null hypothesis is often not going to be true, so why are we testing it? A response to this concern is that the inferential framework provides not only a way of testing hypotheses – it also provides a way of establishing confidence intervals around estimated parameters. Thus we can state with a given level of confidence the magnitude of the difference in commuting times, and we can specify with a given level of precision the magnitude of a regression coefficient.

Accompanying the increasing availability of large datasets has been an appropriate development of exploratory methods. Such exploratory methods are extremely useful in 'data mining' and 'data trawling' to suggest new hypotheses. Ultimately, confirmation of these new hypotheses is called for, and inferential methods become more appropriate.

Where does the student of quantitative methods in geography go from here? The books by Longley et al. (1998) and Fotheringham et al. (2000) represent two good examples of attempts to summarize developments in the field. They include accounts of new developments in the areas of exploratory spatial data analysis and geocomputation. The *Handbook of Spatial Analysis* (Fotheringham and Rogerson 2008) and the *Handbook of Applied Spatial Analysis* (Fischer and Getis 2009) bring together recent contributions to spatial analysis by leading researchers in the field. Keylock and Dorling (2004) have called for a broadening of the curriculum provided to students of quantitative methods, in terms of both relevance to specific geographic applications, and in terms of alternative statistical and mathematical methods, such as Bayesian approaches (for an overview in a geographic context, see Davies

Withers 2002). deSmith et al. (2011) provide a more general treatment of geospatial analysis, and they highlight the connections between spatial analysis and geographic information systems.

The books by Bailey and Gatrell (1995), Haining (2003), Cressie (2011), and Griffith (2013) are all worth exploring. These require more background than this book did, but that should not deter the interested student from exploring them. Even if one does not absorb all of the mathematical detail, it is possible to get a good sense of the types of questions and the range of problems that spatial analysis can address. Waller and Gotway (2004) provide a more focused book in the sense that the primary application is to health data, but the methods of spatial analysis and spatial statistics are covered at a reasonable level for a student with a year's training in the field. The following is a short list of some recent examples of the use of statistical methods in geography; these examples were chosen with the sole purpose of providing for the reader applications of the methods discussed in this book:

Two-sample tests
> Ligmann–Zielinska and Jankowski 2012
> Nelson 1997

Factor analysis in regression
> Ormrod and Cole 1996
> O'Reilly and Webster 1998
> Ackerman 1998

Logistic regression
> Bai et al. 2010
> Myers et al. 1997

Correlation
> Allen and Turner 1996
> Williams and Parker 1997

Spearman's correlation
> Dunhill 2011
> Cringoli et al. 2004

Correlation and regression
> Hendryx et al. 2010
> Takhteyev et al. 2012
> Wyllie and Smith 1996 (stepwise regression)
> Fan 1996

Spatial regression
> Shearmur et al. 2007
> Tu 2011 (Geographically Weighted Regression)
> Rey and Montouri 1999

Lloyd and Shuttleworth 2005
Gao et al. 2006
Principal components and factor analysis
IJmker et al. 2012
Clarke and Holly 1996
Hemmasi and Prorok 2002
Liu et al. 2003
Cluster analysis
Sasson et al. 2012
Dagel 1997
Staeheli and clarke 2003
Poon et al. 2004

ANSWERS FOR SELECTED EXERCISES

CHAPTER 2

2. mean = 27.2; std. dev. = 36.456

z-scores: 0.433 −0.582 −0.554 −0.444 2.6 0.379 −0.170 −0.280 −0.719 −0.664

7. (a)   30 (b) 220 (c) 174 (d) 210 (e) 20 (f) 23 (g) 992 (h) 94

$$\sum_{j=1}^{4} y_{ij}$$

9. $8 \times 7 \times 6 \times 5 \times 4 = 6720$

11. Entering each zone: 71 97 92 79

Leaving each zone: 81 75 101 82

Total: 339

13. grouped mean = 35.116; grouped variance = 195.819

15. 5 units

19. Location (4,4) is better: the total Manhattan distance is 32,500. For (4,5) the total distance is 34,000.

21a. (3.34, 3.48)

21b. (2.94, 2.48)

21c. 0.818, 2.24, 3.15, 1.14, and 3.27 for cities A through E, respectively.

21d. 1.689

21e. $1.689/3.385 = 0.4991$ where $3.385 = \sqrt{36/\pi}$

21f. 0.374, using $\sqrt{64/\pi}$ in the denominator

23a. weighted mean center = (2.733, 3.4)

aggregate travel distance to the weighted mean center: 13.494

aggregate travel distance to   (2,3): 13.482
                                     (3,3): 13.641
                                     (2,4): 13.797
                                     (3,4): 13.861

The median center is likely in the direction of (2,3), since it minimizes the aggregate travel distance.

23b. standard distance = 2.829

23c. relative standard distance = 2.829/3.656 = 0.774

CHAPTER 3

1b. $\binom{3}{2}0.3^{2}0.7^{1} = 0.189$

1c. $\binom{3}{2}0.3^{2}0.7^{1} + \binom{3}{3}0.3^{3}0.7^{0} = 0.189 + 0.027 = 0.216$

1d. $0.8^{3} = 0.512$

3. $1 - pr(0) - pr(1) = 1 - 0.97^{10} - 10\ (0.03\ )(0.97^{9}) = 0.0345$

5. $0.9^{10} = 0.349$

$10!/(8!2!)0.9^{8}0.1^{2} = 0.194$

7. 0.080

9. 1/16

11a. 0.17   b. 0.097   c. 5.882 e. 0.926

13a. 0.135 b. 0.247 c. 0.982

15a. 0.762   b. mean = 23/30; probability = 0.811

17. Poisson probability: 0.1494

Binomial with n = 30:   0.1413

n = 60:   0.1455

n = 120: 0.1475

21a. (0.9^4) (0.1)

21b. $1/(0.1) = 10$

23. Mean: $1/0.18$

Variance: $0.82/0.18^2$

25. 0.1383

26. 0.1539

27. probability of one earthquake = 0.2388. Probability of at least one earthquake = 0.2835

29. 0.3012

CHAPTER 4

1a. $z = (120-96)/32 = 0.75$ Answer: 0.2266

1b. $z = -6/32$; area = $0.5 - 0.4247 = 0.0753$

$z = 4/32$; area = $0.5 - 0.4483 = 0.0517$

Answer: $0.0753 + 0.0517 = 0.2270$

1c. $z = 1.28$ Answer = $90 + 1.28(32) = 96 + 40.96 = 136.96$

5. 0.0685 0.5061

7. 0.186, 0.164, 0.189, 0.189

9. $z = -150/110 = -1.36$; Answer: $0.5 - 0.4139 = 0.0869$
$z = 70/110$; Answer: $0.5 - 0.2389 = 0.2611$
$z = 1.28$; Answer: $250 + 1.28(110) = 390.8$

11. Mean = 22; $pr(X > 25) = 0.25$; $pr(X < 18) = 0.166$

CHAPTER 5

1. $n = 4 (1.96/0.2)^2 = 384$

3. $z = 2.53$, which is greater than the critical value of 1.96; reject null hypothesis

5. $z = 1.43$; Fail to reject null hypothesis, based on critical value of 1.645. $p = 0.0764$.

7. a.   pooled variance = $\{14.3(19) + 12(15)\} / 34 = 13.3$

$$t = (4.1-3.1) / \sqrt{(13.3/20 + 13.3/16) = 0.818} = 0.818$$

df$= 20 + 16 - 2 = 34$

critical value about 2.03

Fail to reject null hypothesis

p-value about 0.21

b.   unequal variances:

$$t = (4.1-3.1) / \sqrt{(14.3/20 + 12/16)} = 0.826$$

$df = 15$

critical $t = 2.13$

p-value about 0.21

9. $t = (6.4-4.2) / (4.4/\sqrt{17}) = 2.062$

$t_{crit} = 1.746$ (one-sided, $\alpha = 0.05$)

Reject H_0

p-value $= 0.029$

13. $t = 3.27$. Reject null hypothesis, using critical value of 1.96.

15. 90% confidence: 326.75, plus or minus 117.96

95% confidence: 326.75, plus or minus 143.40

17a. $t = 1.547$, based upon pooled std. dev. of 3.54. Fail to reject null hypothesis. $p = 0.1328$.

17b. $t = 1.55$. Fail to reject null hypothesis, using t-table to find critical value with 15 df.

19a. $z = 1.8974$. Since this is less than the two-sided critical value of $z = 1.96$, fail to reject null hypothesis.

19b. 0.56, plus or minus 0.0615. Note that this interval includes the value $p = 0.05$.

CHAPTER 6

1.

Boston		
	Mean	Std. Dev.
SAT	2.96	1.3
SUN	3.95	0.97
MON	3.26	1.00
TUE	2.38	0.92
WED	3.02	1.22
THU	3.53	1.07
FRI	3.23	1.15
OVERALL	3.21	1.11

ANOVA					
	SS	df	M.S.	F	sig.
Between	8.77	6	1.46	1.21	0.323
Within	42.2	35	1.2		
Total	50.97	41			

Fail to reject H_0

Levene's statistic: 0.143; p-value $= 0.989$ implies assumption of homoscedasticity is not rejected

Pittsburgh

ANOVA					
	SS	df	M.S.	F	sig.
Between	5.80	6	0.97	0.72	0.64
Within	47.13	35	1.36		
Total	52.93	41			

Fail to reject H_0

3. Assumptions:

    a.   Independent observations in each column

    b.   Homoscedasticity – equal variances in each column

    c.   Normality – data in each column come from a normal distribution

ANOVA is relatively robust with respect to assumptions (b) and (c) … this means that deviations from the assumptions will not drastically alter the results.

The Kruskal-Wallis test has more difficulty rejecting false hypotheses, in comparison with ANOVA.

5. (a)   Within sum of squares: $5(11.08^2 + 16.69^2 + 6.31^2) = 2205.7$

  (b)   Total sum of squares: $17(11.85^2) = 2387.3$

        Between sum of squares $= 2387.2 - 2205.7 = 181.5$

        Between mean square $= 181.5/2 = 90.75$

        Within mean square $= 2205.7/15 = 147.04$

        $F = 90.75/147.05 = 0.62$

        Critical value of F with 2,15 df and $\alpha = 0.05$ is 3.68.

        Fail to reject null hypothesis

  (c)   Kruskal-Wallis

        Sums of ranks in each column $= 64, 59.5$, and 47.5, respectively.

        $H = 12/(18*19) \{64^2/6 + 59.5^2/6 + 47.5^2/6) - 3(19) = 0.85$

        Adjusted value $= 0.85 / \{1 - (8-2)/(18^3 - 18) = 0.85/0.999 = 0.85$

        Critical value of chi-square with 2 df and $\alpha = 0.05$ is 5.99

        Fail to reject null hypothesis

7. Confidence intervals associated with a priori contrasts are narrower, and are therefore more powerful in rejecting null hypotheses that contrasts are equal to zero. This is because no adjustment has to be made for looking at the data after the analysis has been completed.

9. Total SS $= 24726.56$; Within SS $= 21329.37$; Between SS $= 3397.19$

    $F = 2.15$; critical value of F has 2 and 27 df, and is equal to 3.36.

    Fail to reject null hypothesis

11. Total SS $= 481.28$;   Within SS $= 425.70$; Between SS $= 55.58$

Within df = 4f; Between df = 3
Within Mean Square = 9.675
Between Mean Square = 18.5267
$F = 1.915$

13. Within SS = 46.10 df = 37; Mean square = 1.246

Between SS = 10.06; df = 2; Mean square = 5.03

$F = 4.037$

Critical value of F with 2 and 37 df = 3.25

Reject null hypothesis

ANSWERS FOR CHAPTER 9

1. slope, b: = −0.605.

$e^b = 0.546$
$p = 0.02$

Every extra km reduces odds of visiting the park by approximately 1/2 (0.546).

3. Multicollinearity implies correlation among the independent variables. It may be detected by examining a correlation matrix of the independent variables, or by checking 'Collinearity diagnostics' when running a regression in SPSS. It can be addressed by omitting a variable if it is highly correlated with other independent variables.

5. negative ... The high values of y are associated with low values of x.

ANSWERS FOR CHAPTER 10

1. Expected values

9.36 8.87 15.27 9.36 11.82 12.32

9.64 9.13 15.73 9.64 12.18 12.68

Observed chi-square = sum of {(observed − expected)^2/expected} for all 10 cells = 4.81. This is less than the critical value of 11.07 from the table on p. 221 with 5 df, and so the null hypothesis is not rejected.

3a. mean observed nearest neighbor distance $= 1 + 1 + 2 + 1 + 1 + 1 + 1 + 2.5$
$= 9.5 / 7 = 1.36$ km.

Mean nearest neighbor distance expected in a random pattern: $1/ \{2\sqrt{7 / 40}\} = 1.2$ km.

3b. $z = 0.68$, so do not reject the null hypothesis of randomness

3c. There are 81 points, implying a mean of 1 point per cell.

Chi-square $=$ sum of $[\{\text{pts. in a cell} - \text{mean}\}^2 / \text{mean}]$, summed over all cells.
$= 27 \, (0-1)^2 + 27(2-1)^2 = 54$.

$z = (54-80)/ \sqrt{160} = -2.06$; since this is less than -1.96, we conclude that the pattern is not random − because the z-score is negative, we conclude the pattern is uniform/dispersed (since the variance of points per cell is much lower than expected)

5 mean points per cell $= 1$

Chi-square $= 13(0-1)^2 + 5(2-1)^2 + 4(3-1)^2 = 34$.

Variance $=$ VMR $= 34/29$.

Chi-square falls between the critical values of 17.7 and 42.6 using the chi-square table with 29 df, so do not reject H_0

APPENDIX A: STATISTICAL TABLES

Table A.1 Random digits

Each location in the table is equally likely to be the digit 0, 1, 2, 3, 4, 5, 6, 7, 8, or 9.

39203	59841	91168	32021	82081	60164	3738S	52925	91004	71887
39965	79079	97829	95836	26651	12495	68275	20281	73978	07258
17752	87652	07004	95860	89325	56997	70904	91993	13209	50274
04284	63927	07533	60557	41339	16728	96512	11116	92345	04612
03440	97786	37416	24541	36408	63936	36480	87028	05094	95318
07466	12899	31434	06525	81175	38234	24468	30891	89620	50129
83343	72721	52695	36309	67961	73792	63300	89222	10618	24229
03745	48015	85373	77206	76214	85412	83510	73998	13500	65084
27975	70407	56983	07913	38682	89173	40739	40168	95705	46872
54284	28109	48080	80215	85753	64411	27938	56201	16005	49409
79521	93795	56291	03839	16098	44436	22678	37566	45822	26879
17817	48797	59971	28104	68171	05068	98190	33721	13991	73487
56213	82716	77356	91791	31267	19598	25159	28785	57736	72346
75194	03658	65212	50828	73031	12498	30153	80522	30866	05307
44549	28479	49939	43539	66337	61547	25104	27361	27060	17720
11543	45735	21121	46119	96548	48237	30815	01082	00715	18213
27327	47369	72686	74153	67849	91820	22255	91564	28009	19796
65332	83444	40231	84229	48713	46748	54693	63440	03439	97497
45214	30409	35466	73494	39421	86061	88928	55676	68453	66827
77929	36175	61017	71350	93393	32687	29040	74575	45306	22552
54366	88887	16301	19105	51147	31217	41907	42982	64904	63597
08535	65466	48869	58315	23905	24696	66332	22822	37808	78375
36947	67802	81864	59051	52076	34284	06530	51015	39540	61780
28323	33789	56413	16652	28571	53781	63579	42659	53203	29708
16748	41349	75175	66405	75745	33003	32043	01747	49361	61584

(Continued)

Table A.1 (Continued)

Each location in the table is equally likely to be the digit 0, 1, 2, 3, 4, 5, 6, 7, 8, or 9.

33178	69744	11252	49458	86585	85536	92257	24864	48761	31924
26466	93243	88962	31547	05650	29480	92795	39219	22342	60169
36535	14197	72029	40094	61100	17633	38541	08250	04353	13417
66835	93340	09121	97179	24446	47809	87930	83677	46036	07924
09357	02826	35480	92998	35244	39454	50956	36244	31511	40640
07296	75285	29833	78926	48012	97299	56635	57142	00203	77302
01106	48819	40679	96311	90666	91712	16907	65802	94408	76429
15742	99837	87999	36431	96530	84598	62879	82602	57911	18505
16523	51356	37907	65491	39889	49415	97503	09430	39471	12136
03536	42548	50478	54022	18614	03129	68513	08643	91870	93123
73445	35057	97928	83183	57729	35701	70757	28092	97686	90810
52017	99654	63051	87131	S7755	29329	52001	24808	54075	48002
63724	57039	06679	46472	92762	75952	54470	88720	57702	61299
16675	01990	38803	84706	24066	41937	26551	58381	04810	35915
01377	36919	49327	24518	61098	25962	04427	33234	04480	02438
49752	61849	05823	84198	18174	74419	10322	95196	47893	77825
40734	81595	96763	68282	34155	29452	94005	23972	66115	40478
64213	91973	62604	00789	21825	25568	00981	89250	24446	86013
24505	41214	03031	34756	31600	84374	36871	83645	80482	22081
34248	31337	78109	49077	10187	84757	45754	51435	52726	24296
60229	06451	61294	53777	17640	85533	10178	23212	02002	08264
36712	16560	35055	99750	53169	58659	37377	53580	16829	10472
94150	42762	54989	58564	12434	81297	36197	84099	55629	03717
36402	94992	51794	59245	87178	84460	58370	34416	75064	07568
15853	95261	90876	66395	72788	66605	08718	96740	45414	81015
84807	71928	78331	51465	39259	63729	32989	80330	57238	98955
98408	62427	04782	69732	83461	01420	68618	11575	24972	14040
61825	69602	11652	56412	22210	03517	40796	29470	49044	10343
39883	29540	45090	05811	62559	50967	66031	48501	05426	82446
68403	57420	50632	05400	81552	91661	37190	95155	26634	01135
58917	60176	48503	14559	18274	45809	09748	19716	15081	84704
72565	19292	16976	41309	04164	94000	19939	55374	26109	58722
58272	12730	89732	49176	14281	57181	02887	84072	91832	97489
92754	47117	98296	74972	38940	45352	58711	43014	95376	57402
34520	96779	25092	96327	05785	76439	10332	07534	79067	27126
18388	17135	08468	31149	82568	96509	32335	65895	64362	01431
06578	34257	67618	62744	93422	89236	53124	85750	98015	00038

Each location in the table is equally likely to be the digit 0, 1, 2, 3, 4, 5, 6, 7, 8, or 9.

67183	75783	54437	58890	02256	53920	61369	65913	65478	62319
26942	92564	92010	95670	75547	20940	06219	28040	10050	05974
06345	01152	49596	02064	85321	59627	28489	88186	74006	18320
24221	12108	16037	99857	73773	42506	60530	96317	29918	16918
83975	61251	82471	06941	48817	76078	68930	39693	87372	09600
86232	01398	50258	22868	71052	10127	48729	67613	59400	65886
04912	01051	33687	03296	17112	23843	16796	22332	91570	47197
15455	88237	91026	36454	18765	97891	11022	98774	00321	10386
88430	09861	45098	66176	59598	98527	11059	31626	10798	50313
48849	11583	63654	55670	89474	75232	14186	52377	19129	67166
33659	59617	40920	30295	07463	79923	83393	77120	38862	75503
60198	41729	19897	04805	09351	76734	10333	87776	36947	88618
55868	53145	66232	52007	81206	89543	66226	45709	37114	78075
22011	71396	95174	43043	68304	56773	83931	43631	50995	68130
90301	54934	08008	00565	67790	84760	82229	64147	28031	11609
07586	90936	21021	54066	87281	63574	41155	01740	29025	19909
09973	76136	87904	54419	34370	75071	56201	16768	61934	12083
59750	42528	19864	31595	72097	17005	24682	43560	74423	59197
74492	19327	17812	63897	65708	07709	13817	95943	07909	75504
69042	57646	38606	30549	34351	21432	50312	10566	43842	70046
16054	32268	29828	73413	53819	39324	13581	71841	94894	64223
17930	78622	70578	23048	73730	73507	69602	77174	32593	45565
46812	93896	65639	73905	45396	71653	01490	33674	16888	53434
04590	07459	04096	15216	56633	69845	85550	15141	56349	56117
99618	63788	86396	37564	12962	96090	70358	23378	63441	36828
34545	32273	45427	30693	49369	27427	28362	17307	45092	08302
04337	00565	27718	67942	19284	69126	51649	03469	88009	41916
73810	70135	72055	90111	71202	08210	76424	66364	63081	37784
60555	94102	39146	67795	05985	43280	97202	35613	25369	47959
58261	16861	39080	22820	46555	32213	38440	32662	48259	61197
98765	65802	44467	03358	38894	34290	31107	25519	26585	34852
39157	58231	30710	09394	04012	49122	26283	34946	23590	25663
08143	91252	23181	51183	52102	85298	52008	48688	86779	21722
66806	72352	64500	89120	13493	85813	93999	12558	24852	04575
08289	82806	36490	96421	81718	63075	54178	39209	03050	47089
12989	31280	71466	72234	26922	04753	61943	86149	26938	53736
44154	63471	30657	62298	56461	48879	54108	97126	43219	95349
63788	18000	10049	49041	28807	64190	39753	17397	48026	76947

Table A.2 Normal distribution

The tabled entries represent the proportion p of the total area under the curve that is in the tail of the normal curve, to the right of the indicated value of z. (Example: 0.0694 or 6.94% of the area is to the right of $z = 1.48$. This is found by using the $z = 1.4$ row, and the 0.08 column, of the table.) If the value of z is negative, the tabled entry corresponding to the absolute value of z represents the area less than z. (Example: 0.3015 or 30.15% of the area is to the left of $z = -0.52$ and this is found by using $z = +0.52$ in the table.)

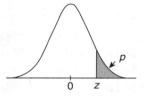

| z | Second decimal place of z | | | | | | | | | |
	.00	.01	.02	.03	.04	.05	.06	.07	.08	.09
0.0	.5000	.4960	.4920	.4880	.4840	.4801	.4761	.4721	.4681	.4641
0.1	.4602	.4562	.4522	.4483	.4443	.4404	.4364	.4325	.4286	.4247
0.2	.4207	.4168	.4129	.4090	.4052	.4013	.3974	.3936	.3897	.3859
0.3	.3821	.3783	.3745	.3707	.3669	.3632	.3594	.3557	.3520	.3483
0.4	.3446	.3409	.3372	.3336	.3300	.3264	.3228	.3192	.3156	.3121
0.5	.3085	.3050	.3015	.2981	.2946	.2912	.2877	.2843	.2810	.2776
0.6	.2743	.2709	.2676	.2643	.2611	.2578	.2546	.2514	.2483	.2451
0.7	.2420	.2389	.2358	.2327	.2297	.2266	.2236	.2206	.2177	.2148
0.8	.2119	.2090	.2061	.2033	.2005	.1977	.1949	.1922	.1894	.1867
0.9	.1841	.1814	.1788	.1762	.1736	.1711	.1685	.1660	.1635	.1611
1.0	.1587	.1562	.1539	.1515	.1492	.1469	.1446	.1423	.1401	.1379
1.1	.1357	.1335	.1314	.1292	.1271	.1251	.1230	.1210	.1190	.1170
1.2	.1151	.1131	.1112	.1093	.1075	.1056	.1038	.1020	.1003	.0985
1.3	.0968	.0951	.0934	.0918	.0901	.0885	.0869	.0853	.0838	.0823
1.4	.0808	.0793	.0778	.0764	.0749	.0735	.0721	.0708	.0694	.0681
1.5	.0668	.0655	.0643	.0630	.0618	.0606	.0594	.0582	.0571	.0559
1.6	0548	.0537	.0526	.0516	.0505	.0495	.0485	.0475	.0465	.0455
1.7	.0446	.0436	.0427	.0418	.0409	.0401	.0392	.0384	.0375	.0367
1.8	.0359	.0351	.0344	.0336	.0329	.0322	.0314	.0307	.0301	.0294
1.9	.0287	.0281	.0274	.0268	.0262	.0256	.0250	.0244	.0239	.0233
2.0	.0228	.0222	.0217	.0212	.0207	.0202	.0197	.0192	.0188	.0183
2.1	.0179	.0174	.0170	.0166	.0162	.0158	.0154	.0150	.0146	.0143
2.2	.0139	.0136	.0132	.0129	.0125	.0122	.0119	.0116	.0113	.0110
2.3	.0107	.0104	.0102	.0099	.0096	.0094	.0091	.0089	.0087	.0084
2.4	.0082	.0080	.0078	.0075	.0073	.0071	.0069	.0068	.0066	.0064
2.5	.0062	.0060	.0059	.0057	.0055	.0054	.0052	.0051	.0049	.0048
2.6	.0047	.0045	.0044	.0043	.0041	.0040	.0039	.0038	.0037	.0036
2.7	.0035	.0034	.0033	.0032	.0031	.0030	.0029	.0028	.0027	.0026
2.8	.0026	.0025	.0024	.0023	.0023	.0022	.0021	.0021	.0020	.0019
2.9	.0019	.0018	.0018	.0017	.0016	.0016	.0015	.0015	.0014	.0014
3.0	.0013	.0013	.0013	.0012	.0012	.0011	.0011	.0011	.0010	.0010

Adapted with rounding from Table II of Fisher and Yates (1974).

Table A.3 Student's *t*-distribution

For various degrees of freedom (df), the tabled entries represent the critical values of *t* above which a specified proportion *p* of the *t* distribution falls. (Example: for df = 9, a *t* of 2.262 is surpassed by .025 or 2.5% of the total distribution.)

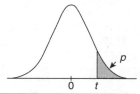

df	| *p* (one-tailed probabilities)				
	.10	.05	.025	.01	.005
1	3.078	6.314	12.706	31.821	63.657
2	1.886	2.920	4.303	6.965	9.925
3	1.638	2.353	3.182	4.541	5.841
4	1.533	2.132	2.776	3.747	4.604
5	1.476	2.015	2.571	3.365	4.032
6	1.440	1.943	2.447	3.143	3.707
7	1.415	1.895	2.365	2.998	3.499
8	1.397	1.860	2.306	2.896	3.355
9	1.383	1.833	2.262	2.821	3.250
10	1.372	1.812	2.228	2.764	3.169
11	1.363	1.796	2.201	2.718	3.106
12	1.356	1.782	2.179	2.681	3.055
13	1.350	1.771	2.160	2.650	3.012
14	1.345	1.761	2.145	2.624	2.977
15	1.341	1.753	2.131	2.602	2.947
16	1.337	1.746	2.120	2.583	2.921
17	1.333	1.740	2.110	2.567	2.898
18	1.330	1.734	2.101	2.552	2.878
19	1.328	1.729	2.093	2.539	2.861
20	1.325	1.725	2.086	2.528	2.845
21	1.323	1.721	2.080	2.518	2.831
22	1.321	1.717	2.074	2.508	2.819
23	1.319	1.714	2.069	2.500	2.807
24	1.318	1.711	2.064	2.492	2.797
25	1.316	1.708	2.060	2.485	2.787
26	1.315	1.706	2.056	2.479	2.779
27	1.314	1.703	2.052	2.473	2.771
28	1.313	1.701	2.048	2.467	2.763
29	1.311	1.699	2.045	2.462	2.756

(Continued)

Table A.3   (Continued)

df	p (one-tailed probabilities)				
	.10	.05	.025	.01	.005
30	1.310	1.697	2.042	2.457	2.750
40	1.303	1.684	2.021	2.423	2.704
60	1.296	1.671	2.000	2.390	2.660
120	1.289	1.658	1.980	2.358	2.617
∞	1.282	1.645	1.960	2.326	2.576

Adapted from Table III of Fisher and Yates (1974).

Table A.4 Cumulative distribution of student's t-distribution

t/v	Degrees of freedom (df)									
	1	2	3	4	5	6	7	8	9	10
0.0	0.50000	0.50000	0.50000	0.50000	0.50000	0.50000	0.50000	0.50000	0.50000	0.50000
0.1	0.53173	0.53527	0.53667	0.53742	0.53788	0.53820	0.53843	0.53860	0.53873	0.53884
0.2	0.56283	0.57002	0.57286	0.57438	0.57532	0.57596	0.57642	0.57676	0.57704	0.57726
0.3	0.56283	0.57002	0.57286	0.57438	0.57532	0.57596	0.57642	0.57676	0.57704	0.57726
0.4	0.62112	0.63608	0.64203	0.64520	0.64716	0.64850	0.64946	0.65019	0.65076	0.65122
0.5	0.64758	0.66667	0.67428	0.67834	0.68085	0.68256	0.68380	0.68473	0.68546	0.68605
0.6	0.67202	0.69529	0.70460	0.70958	0.71267	0.71477	0.71629	0.71745	0.71835	0.71907
0.7	0.69440	0.72181	0.73284	0.73875	0.74243	0.74493	0.74674	0.74811	0.74919	0.75006
0.8	0.71478	0.74618	0.75890	0.76574	0.76999	0.77289	0.77500	0.77659	0.77784	0.77885
0.9	0.73326	0.76845	0.78277	0.79050	0.79531	0.79860	0.80099	0.80280	0.80422	0.80536
1.0	0.75000	0.78868	0.80450	0.81305	0.81839	0.82204	0.82469	0.82670	0.82828	0.82955
1.1	0.76515	0.80698	0.82416	0.83346	0.83927	0.84325	0.84614	0.84834	0.85006	0.85145
1.2	0.77886	0.82349	0.84187	0.85182	0.85805	0.86232	0.86541	0.86777	0.86961	0.87110
1.3	0.79129	0.83838	0.85777	0.86827	0.87485	0.87935	0.88262	0.88510	0.88705	0.88862
1.4	0.80257	0.85177	0.87200	0.88295	0.88980	0.89448	0.89788	0.90046	0.90249	0.90412
1.5	0.81283	0.86380	0.88471	0.89600	0.90305	0.90786	0.91135	0.91400	0.91608	0.91775
1.6	0.82219	0.87464	0.89605	0.90758	0.91475	0.91964	0.92318	0.92587	0.92797	0.92966
1.7	0.83075	0.88439	0.90615	0.91782	0.92506	0.92998	0.93354	0.93622	0.93833	0.94002
1.8	0.83859	0.89317	0.91516	0.92688	0.93412	0.93902	0.94256	0.94522	0.94731	0.94897
1.9	0.84579	0.90109	0.92318	0.93488	0.94207	0.94691	0.95040	0.95302	0.95506	0.95669
2.0	0.85242	0.90825	0.93034	0.94194	0.94903	0.95379	0.95719	0.95974	0.96172	0.96331
2.1	0.85854	0.91473	0.93672	0.94817	0.95512	0.95976	0.96306	0.96553	0.96744	0.96896
2.2	0.86420	0.92060	0.94241	0.95367	0.96045	0.96495	0.96813	0.97050	0.97233	0.97378
2.3	0.86945	0.92593	0.94751	0.95853	0.96511	0.96945	0.97250	0.97476	0.97650	0.97787

(Continued)

Table A.4 (Continued)

| | Degrees of freedom (df) | | | | | | | | | |
t/v	1	2	3	4	5	6	7	8	9	10
2.4	0.87433	0.93077	0.95206	0.96282	0.96919	0.97335	0.97627	0.97841	0.98005	0.98134
2.5	0.87888	0.93519	0.95615	0.96662	0.97275	0.97674	0.97950	0.98153	0.98307	0.98428
2.6	0.88313	0.93923	0.95981	0.96998	0.97587	0.97967	0.98229	0.98419	0.98563	0.98675
2.7	0.88709	0.94292	0.96311	0.97295	0.97861	0.98221	0.98468	0.98646	0.98780	0.98884
2.8	0.89081	0.94630	0.96607	0.97559	0.98100	0.98442	0.98674	0.98840	0.98964	0.99060
2.9	0.89430	0.94941	0.96875	0.97794	0.98310	0.98633	0.98851	0.99005	0.99120	0.99208
3.0	0.89758	0.95227	0.97116	0.98003	0.98495	0.98800	0.99003	0.99146	0.99252	0.99333
3.1	0.90067	0.95490	0.97335	0.98189	0.98657	0.98944	0.99134	0.99267	0.99364	0.99437
3.2	0.90359	0.95733	0.97533	0.98355	0.98800	0.99070	0.99247	0.99369	0.99459	0.99525
3.3	0.90634	0.95958	0.97713	0.98503	0.98926	0.99180	0.99344	0.99457	0.99539	0.99599
3.4	0.90895	0.96166	0.97877	0.98636	0.99037	0.99275	0.99428	0.99532	0.99606	0.99661
3.5	0.91141	0.96358	0.98026	0.98755	0.99136	0.99359	0.99500	0.99596	0.99664	0.99714
3.6	0.91376	0.96538	0.98162	0.98862	0.99223	0.99432	0.99563	0.99651	0.99713	0.99758
3.7	0.91598	0.96705	0.98286	0.98958	0.99300	0.99496	0.99617	0.99698	0.99754	0.99795
3,8	0.91809	0.96860	0.98400	0.99045	0.99369	0.99552	0.99664	0.99738	0.99789	0.99826
3.9	0.92010	0.97005	0.98504	0.99123	0.99430	0.99601	0.99705	0.99773	0.99819	0.99852
4.0	0.92202	0.97141	0.98600	0.99193	0.99484	0.99644	0.99741	0.99803	0.99845	0.99874
4.2	0.92560	0.97386	0.98768	0.99315	0.99575	0.99716	0.99798	0.99850	0.99885	0.99909
4.4	0.92887	0.97602	0.98912	0.99415	0.99649	0.99772	0.99842	0.99886	0.99914	0.99933
4.6	0.93186	0.97792	0.99034	0.99498	0.99708	0.99815	0.99876	0.99912	0.99936	0.99951
4.8	0.93462	0.97962	0.99140	0.99568	0.99756	0.99850	0.99902	0.99932	0.99951	0.99964
5.0	0.93717	0.98113	0.99230	0.99625	0.99795	0.99877	0.99922	0.99947	0.99963	0.99973
5.2	0.93952	0.98248	0.99309	0.99674	0.99827	0.99899	0.99937	0.99959	0.99972	0.99980

Degrees of freedom (df)

t/v	1	2	3	4	5	6	7	8	9	10
5.4	0.94171	0.98369	0.99378	0.99715	0.99853	0.99917	0.99950	0.99968	0.99978	0.99985
5.6	0.94375	0.98478	0.99437	0.99750	0.99875	0.99931	0.99959	0.99975	0.99983	0.99989
5.8	0.94565	0.98577	0.99490	0.99780	0.99893	0.99942	0.99967	0.99980	0.99987	0.99991
6.0	0.94743	0.98666	0.99536	0.99806	0.99908	0.99952	0.99973	0.99984	0.99990	0.99993
6.2	0.94910	0.98748	0.99577	0.99828	0.99920	0.99959	0.99978	0.99987	0.99992	0.99995
6.4	0.95066	0.98822	0.99614	0.99847	0.99931	0.99966	0.99982	0.99990	0.99994	0.99996
6.6	0.95214	0.98890	0.99646	0.99863	0.99940	0.99971	0.99985	0.99992	0.99995	0.99997
6.8	0.95352	0.98953	0.99675	0.99878	0.99948	0.99975	0.99987	0.99993	0.99996	0.99998
7.0	0.95483	0.99010	0.99701	0.99890	0.99954	0.99979	0.99990	0.99994	0.99997	0.99998
7.2	0.95607	0.99063	0.99724	0.99901	0.99960	0.99982	0.99991	0.99995	0.99997	0.99999
7.4	0.95724	0.99111	0.99745	0.99911	0.99964	0.99984	0.99993	0.99996	0.99998	0.99999
7.6	0.95836	0.99156	0.99764	0.99920	0.99969	0.99986	0.99994	0.99997	0.99998	0.99999
7.8	0.95941	0.99198	0.99781	0.99927	0.99972	0.99988	0.99995	0.99997	0.99999	0.99999
8.0	0.96042	0.99237	0.99796	0.99934	0.99975	0.99990	0.99996	0.99998	0.99999	0.99999

Note: Entries in the table give the probability of a value less than the specified value of t, for given degrees of freedom. For example, when df = 4, 79.05% of t-values will be less than $t = 0.9$.
Source: Pearson and Hartley (1966) (by permission).

Degrees of freedom (df)

Δv	11	12	13	14	15	16	17	18	19	20
0.0	0.50000	0.50000	0.50000	0.50000	0.50000	0.50000	0.50000	0.50000	0.50000	0.50000
0.1	0.53893	0.53900	0.53907	0.53912	0.53917	0.53921	0.53924	0.53928	0.53930	0.53933
0.2	0.57744	0.57759	0.57771	0.57782	0.57792	0.57800	0.57807	0.57814	0.57820	0.57825
0.3	0.61511	0.61534	0.61554	0.61571	0.61585	0.61598	0.61609	0.61619	0.61628	0.61636
0.4	0.65159	0.65191	0.65217	0.65240	0.65260	0.65278	0.65293	0.65307	0.65319	0.65330
0.5	0.68654	0.68694	0.68728	0.68758	0.68783	0.68806	0.68826	0.68843	0.68859	0.68873
0.6	0.71967	0.72017	0.72059	0.72095	0.72127	0.72155	0.72179	0.72201	0.72220	0.72238
0.7	0.75077	0.75136	0.75187	0.75230	0.75268	0.75301	0.75330	0.75356	0.75380	0.75400
0.8	0.77968	0.78037	0.78096	0.78146	0.78190	0.78229	0.78263	0.78293	0.78320	0.78344
0.9	0.80630	0.80709	0.80776	0.80883	0.80883	0.80927	0.80965	0.81000	0.81031	0.81058
1.0	0.83060	0.83148	0.83222	0.83286	0.83341	0.83390	0.83433	0.83472	0.83506	0.83537
1.1	0.85259	0.85355	0.85436	0.85506	0.85566	0.85620	0.85667	0.85709	0.85746	0.85780
1.2	0.87233	0.87335	0.87422	0.87497	0.87562	0.87620	0.87670	0.87715	0.87756	0.87792
1.3	0.88991	0.89099	0.89191	0.89270	0.89339	0.89399	0.89452	0.89500	0.89542	0.89581
1.4	0.90546	0.90658	0.90754	0.90836	0.90907	0.90970	0.91025	0.91074	0.91118	0.91158
1.5	0.91912	0.92027	0.92125	0.92209	0.92282	0.92346	0.92402	0.92452	0.92498	0.92538
1.6	0.93105	0.93221	0.93320	0.93404	0.93478	0.93542	0.93599	0.93650	0.93695	0.93736
1.7	0.94140	0.94256	0.94354	0.94439	0.94512	0.94576	0.94632	0.94683	0.94728	0.94768
1.8	0.95034	0.95148	0.95245	0.95328	0.95400	0.95463	0.95518	0.95568	0.95612	0.95652
1.9	0.95802	0.95914	0.96008	0.96089	0.96158	0.96220	0.96273	0.96321	0.96364	0.96403
2.0	0.96460	0.96567	0.96658	0.96736	0.96803	0.96861	0.96913	0.96959	0.97000	0.97037
2.1	0.97020	0.97123	0.97209	0.97283	0.97347	0.97403	0.97452	0.97495	0.97534	0.97569
2.2	0.97496	0.97593	0.97675	0.97745	0.97805	0.97858	0.97904	0.97945	0.97981	0.98014

Degrees of freedom (df)

v	11	12	13	14	15	16	17	18	19	20
2.3	0.86945	0.92593	0.94751	0.95853	0.96511	0.96945	0.97250	0.97476	0.97650	0.97787
2.4	0.98238	0.98324	0.98396	0.98457	0.98509	0.98554	0.98594	0.98629	0.98660	0.98688
2.5	0.98525	0.98604	0.98671	0.98727	0.98775	0.98816	0.98853	0.98885	0.98913	0.98938
2.6	0.98765	0.98839	0.98900	0.98951	0.98995	0.99033	0.99066	0.99095	0.99121	0.99144
2.7	0.98967	0.99035	0.99090	0.99137	0.99177	0.99211	0.99241	0.99267	0.99290	0.99311
2.8	0.99136	0.99198	0.99249	0.99291	0.99327	0.99358	0.99385	0.99408	0.99429	0.99447
2.9	0.99278	0.99334	0.99380	0.99418	0.99450	0.99478	0.99502	0.99523	0.99541	0.99557
3.0	0.99396	0.99447	0.99488	0.99522	0.99551	0.99576	0.99597	0.99616	0.99632	0.99646
3.1	0.99495	0.99541	0.99578	0.99608	0.99634	0.99656	0.99675	0.99691	0.99705	0.99718
3.2	0.99577	0.99618	0.99652	0.99679	0.99702	0.99721	0.99738	0.99752	0.99764	0.99775
3.3	0.99646	0.99683	0.99713	0.99737	0.99757	0.99774	0.99789	0.99801	0.99812	0.99821
3.4	0.99703	0.99737	0.99763	0.99784	0.99802	0.99817	0.99830	0.99840	0.99850	0.99858
3.5	0.99751	0.99781	0.99804	0.99823	0.99839	0.99852	0.99863	0.99872	0.99880	0.99887
3.6	0.99791	0.99818	0.99838	0.99855	0.99869	0.99880	0.99890	0.99898	0.99905	0.99911
3.7	0.99825	0.99848	0.99867	0.99881	0.99893	0.99903	0.99911	0.99918	0.99924	0.99929
3.8	0.99853	0.99874	0.99890	0.99902	0.99913	0.99921	0.99928	0.99934	0.99939	0.99944
3.9	0.99876	0.99895	0.99909	0.99920	0.99929	0.99936	0.99942	0.99948	0.99952	0.99956
4.0	0.99896	0.99912	0.99924	0.99934	0.99942	0.99948	0.99954	0.99958	0.99962	0.99965
4.2	0.99926	0.99938	0.99948	0.99955	0.99961	0.99966	0.99970	0.99973	0.99976	0.99978
4.4	0.99947	0.99957	0.99964	0.99970	0.99974	0.99978	0.99980	0.99983	0.99988	0.99986
4.6	0.99962	0.99969	0.99975	0.99979	0.99983	0.99985	0.99987	0.99989	0.99990	0.99991
4.8	0.99972	0.99978	0.99983	0.99986	0.99988	0.99990	0.99992	0.99993	0.99994	0.99995
5.0	0.99980	0.99985	0.99988	0.99990	0.99992	0.99993	0.99995	0.99995	0.99996	0.99997
5.2	0.99985	0.99989	0.99992	0.99993	0.99995	0.99996	0.99996	0.99997	0.99997	0.99998

(Continued)

(Continued)

					Degrees of freedom (df)					
/v	11	12	13	14	15	16	17	18	19	20
5.4	0.99989	0.99992	0.99994	0.99995	0.99996	0.99997	1.99998	0.99998	0.99998	0.99999
5.6	0.99992	0.99994	0.99996	0.99997	0.99997	0.99998	1.99998	0.99990	0.99999	0.99999
5.8	0.99994	0.99996	0.99997	0.99998	0.99998	0.99999	0.99999	0.99999	0.99999	0.99999
6.0	0.99995	0.99997	0.99998	0.99998	0.99999	0.99999	0.99999	0.99999		
6.2	0.99997	0.99998	0.99998	0.99999	0.99999	0.99999				
6.4	0.99997	0.99998	0.99999	0.99999	0.99999					
6.6	0.99998	0.99999	0.99999	0.99999						
6.8	0.99998	0.99999	0.99999							
7.0	0.99999	0.99999								

Table A.5 F-distribution

For various pairs of degrees of freedom $v1$, $v2$, the tabled entries represent the critical values of F above which a proportion p of the distribution falls. (Example: for df = $v1$, $v2$ = 4,16 an F = 2.33 has 10% of the area to the right of it.) Tables are provided for values of p equal to .10, .05, .01.

p = .10 values

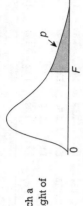

Degrees of freedom for denominator $v2$ — Degrees of freedom for numerator $v1$

$v2$	1	2	3	4	5	6	7	8	9	10	12	15	20	30	40	60	120	∞
1	39.86	49.50	53.59	55.83	57.24	58.20	58.91	59.14	59.86	60.19	60.71	61.22	61.71	62.26	62.53	62.79	63.06	63.33
2	8.53	9.00	9.16	9.24	9.29	9.33	9.35	9.37	9.38	9.39	9.41	9.42	9.41	9.46	9.47	9.47	9.48	9.49
3	5.54	5.46	5.39	5.34	5.31	5.28	5.27	5.25	5.24	5.23	5.22	5.20	5.18	5.17	5.16	5.15	5.14	5.13
4	4.54	4.32	4.19	4.11	4.05	4.01	3.98	3.95	3.94	3.92	3.90	3.87	3.84	3.82	3.80	3.79	3.78	3.76
5	4.06	3.78	3.62	3.52	3.45	3.40	3.37	3.34	3.32	3.30	3.27	3.24	3.21	3.17	3.16	3.14	3.12	3.10
6	3.78	3.46	3.29	3.18	3.11	3.05	3.01	2.98	2.96	2.94	2.90	2.87	2.84	2.80	2.78	2.76	2.74	2.72
7	3.59	3.26	3.07	2.96	2.88	2.83	2.78	2.75	2.72	2.70	2.67	2.63	2.59	2.56	2.54	2.51	2.49	2.47
8	3.46	3.11	2.92	2.81	2.73	2.67	2.62	2.59	2.56	2.54	2.50	2.46	2.42	2.38	2.36	2.34	2.32	2.29
9	3.36	3.01	2.81	2.69	2.61	2.55	2.51	2.47	2.44	2.42	2.38	2.34	2.30	2.25	2.23	2.21	2.18	2.16
10	3.29	2.92	2.73	2.61	2.52	2.46	2.41	2.38	2.35	2.32	2.28	2.24	2.20	2.16	2.13	2.11	2.08	2.06
11	3.23	2.86	2.66	2.54	2.45	2.39	2.34	2.30	2.27	2.25	2.21	2.17	2.12	2.08	2.05	2.03	2.00	1.97
12	3.18	2.81	2.61	2.48	2.39	2.33	2.28	2.24	2.21	2.19	2.15	2.10	2.06	2.01	1.99	1.96	1.93	1.90
13	3.14	2.76	2.56	2.43	2.35	2.28	2.23	2.20	2.16	2.14	2.10	2.05	2.01	1.96	1.93	1.90	1.88	1.85
14	3.10	2.73	2.52	2.39	2.31	2.24	2.19	2.15	2.12	2.10	2.05	2.01	1.96	1.91	1.89	1.86	1.83	1.80
15	3.07	2.70	2.49	2.36	2.27	2.21	2.16	2.12	2.09	2.06	2.02	1.97	1.92	1.87	1.85	1.82	1.79	1.76
16	3.05	2.67	2.46	2.33	2.24	2.18	2.13	2.09	2.06	2.03	1.99	1.94	1.89	1.84	1.81	1.78	1.75	1.72
17	3.03	2.64	2.44	2.31	2.22	2.15	2.10	2.06	2.03	2.00	1.96	1.91	1.86	1.81	1.78	1.75	1.72	1.69
18	3.01	2.62	2.42	2.29	2.20	2.13	2.08	2.04	2.00	1.98	1.93	1.89	1.84	1.78	1.75	1.72	1.69	1.66
19	2.99	2.61	2.40	2.27	2.18	2.11	2.06	2.02	1.98	1.96	1.91	1.86	1.81	1.76	1.73	1.70	1.67	1.63
20	2.97	2.59	2.38	2.25	2.16	2.09	2.04	2.00	1.96	1.94	1.89	1.84	1.79	1.74	1.71	1.68	1.64	1.61
21	2.96	2.57	2.36	2.23	2.14	2.08	2.02	1.98	1.95	1.92	1.87	1.83	1.78	1.72	1.69	1.66	1.62	1.59
22	2.95	2.56	2.35	2.22	2.13	2.06	2.01	1.97	1.93	1.90	1.86	1.81	1.76	1.70	1.67	1.64	1.60	1.57
23	2.94	2.55	2.34	2.21	2.11	2.05	1.99	1.95	1.92	1.89	1.84	1.80	1.74	1.69	1.66	1.62	1.59	1.55
24	2.93	2.54	2.33	2.19	2.10	2.04	1.98	1.94	1.91	1.88	1.83	1.78	1.73	1.67	1.64	1.61	1.57	1.53
30	2.88	2.49	2.28	2.14	2.05	1.98	1.93	1.88	1.85	1.82	1.77	1.72	1.67	1.61	1.57	1.54	1.50	1.46
40	2.84	2.44	2.23	2.09	2.00	1.93	1.87	1.83	1.79	1.76	1.71	1.66	1.61	1.54	1.51	1.47	1.42	1.38
60	2.79	2.39	2.18	2.04	1.95	1.87	1.82	1.77	1.74	1.71	1.66	1.60	1.54	1.48	1.44	1.40	1.35	1.29
120	2.75	2.35	2.13	1.99	1.90	1.82	1.77	1.72	1.68	1.65	1.60	1.55	1.48	1.41	1.37	1.32	1.26	1.19
∞	2.71	2.30	2.08	1.94	1.85	1.77	1.72	1.67	1.63	1.60	1.55	1.49	1.42	1.34	1.30	1.24	1.17	1.00

(Continued)

Table A.5 (Continued)

$P = 0.05$ values

Degrees of freedom for denominator $v2$ / Degrees of freedom for numerator $v1$

$v2$	1	2	3	4	5	6	7	8	9	10	12	15	20	30	40	60	120	∞
1	161.4	199.5	215.7	224.6	230.2	234.0	236.8	238.9	240.5	241.9	243.9	245.9	248.0	250.1	251.1	252.2	253.3	254.3
2	18.51	19.00	19.16	19.25	19.30	19.33	19.35	19.37	19.38	19.40	19.41	19.43	19.45	19.46	19.47	19.48	19.49	19.50
3	10.13	9.55	9.28	9.12	9.01	8.94	8.89	8.85	8.81	8.79	8.74	8.70	8.66	8.62	8.59	8.57	8.55	8.53
4	7.71	6.94	6.59	6.39	6.26	6.16	6.09	6.04	6.00	5.96	5.91	5.86	5.80	5.75	5.72	5.69	5.66	5.63
5	6.61	5.79	5.41	5.19	5.05	4.95	4.88	4.82	4.77	4.74	4.68	4.62	4.56	4.50	4.46	4.43	4.40	4.36
6	5.99	5.14	4.76	4.53	4.39	4.28	4.21	4.15	4.10	4.06	4.00	3.94	3.87	3.81	3.77	3.74	3.70	3.67
7	5.59	4.74	4.35	4.12	3.97	3.87	3.79	3.73	3.68	3.64	3.57	3.51	3.44	3.38	3.34	3.30	3.27	3.23
8	5.32	4.46	4.07	3.84	3.69	3.58	3.50	3.44	3.39	3.35	3.28	3.22	3.15	3.08	3.04	3.01	2.97	2.93
9	5.12	4.26	3.86	3.63	3.48	3.37	3.29	3.23	3.18	3.14	3.07	3.01	2.94	2.86	2.83	2.79	2.75	2.71
10	4.96	4.10	3.71	3.48	3.33	3.22	3.14	3.07	3.02	2.98	2.91	2.85	2.77	2.70	2.66	2.62	2.58	2.54
11	4.84	3.98	3.59	3.36	3.20	3.09	3.01	2.95	2.90	2.85	2.79	2.72	2.65	2.57	2.53	2.49	2.45	2.40
12	4.75	3.89	3.49	3.26	3.11	3.00	2.91	2.85	2.80	2.75	2.69	2.62	2.54	2.47	2.43	2.38	2.34	2.30
13	4.67	3.81	3.41	3.18	3.03	2.92	2.83	2.77	2.71	2.67	2.60	2.53	2.46	2.38	2.34	2.30	2.25	2.21
14	4.60	3.74	3.34	3.11	2.96	2.85	2.76	2.70	2.65	2.60	2.53	2.46	2.39	2.31	2.27	2.22	2.18	2.13
15	4.54	3.68	3.29	3.06	2.90	2.79	2.71	2.64	2.59	2.54	2.48	2.40	2.33	2.25	2.20	2.16	2.11	2.07
16	4.49	3.63	3.24	3.01	2.85	2.74	2.66	2.59	2.54	2.49	2.42	2.35	2.28	2.19	2.15	2.11	2.06	2.01
17	4.45	3.59	3.20	2.96	2.81	2.70	2.61	2.55	2.49	2.45	2.38	2.31	2.23	2.15	2.10	2.06	2.01	1.96
18	4.41	3.55	3.16	2.93	2.77	2.66	2.58	2.51	2.46	2.41	2.34	2.27	2.19	2.11	2.06	2.02	1.97	1.92
19	4.38	3.52	3.13	2.90	2.74	2.63	2.54	2.48	2.42	2.38	2.31	2.23	2.16	2.07	2.03	1.98	1.93	1.88
20	4.35	3.49	3.10	2.87	2.71	2.60	2.51	2.45	2.39	2.35	2.28	2.20	2.12	2.04	1.99	1.95	1.90	1.84
21	4.32	3.47	3.07	2.84	2.68	2.57	2.49	2.42	2.37	2.32	2.25	2.18	2.10	2.01	1.96	1.92	1.87	1.81
22	4.30	3.44	3.05	2.82	2.66	2.55	2.46	2.40	2.34	2.30	2.23	2.15	2.07	1.98	1.94	1.89	1.84	1.78
23	4.28	3.42	3.03	2.80	2.64	2.53	2.44	2.37	2.32	2.27	2.20	2.13	2.05	1.96	1.91	1.86	1.81	1.76
24	4.26	3.40	3.01	2.78	2.62	2.51	2.42	2.36	2.30	2.25	2.18	2.11	2.03	1.94	1.89	1.84	1.79	1.73
30	4.17	3.32	2.92	2.69	2.53	2.42	2.33	2.27	2.21	2.16	2.09	2.01	1.93	1.84	1.79	1.74	1.68	1.62
40	4.08	3.23	2.84	2.61	2.45	2.34	2.25	2.18	2.12	2.08	2.00	1.92	1.84	1.74	1.69	1.64	1.58	1.51
60	4.00	3.15	2.76	2.53	2.37	2.25	2.17	2.10	2.04	1.99	1.92	1.84	1.75	1.65	1.59	1.53	1.47	1.39
120	3.92	3.07	2.68	2.45	2.29	2.17	2.09	2.02	1.96	1.91	1.83	1.75	1.66	1.55	1.50	1.43	1.35	1.25
∞	3.84	3.00	2.60	2.37	2.21	2.10	2.01	1.94	1.88	1.83	1.75	1.67	1.57	1.46	1.39	1.32	1.22	1.00

P = 0.01 values

Degrees of freedom for denominator

Degrees of freedom for numerator v1

v2	1	2	3	4	5	6	7	8	9	10	12	15	20	30	40	60	120	∞
1	4052	4999.5	5403	5625	5764	5859	5928	5981	6022	6056	6106	6157	6209	6261	6287	6313	6339	6366
2	93.50	99.00	99.17	99.25	99.30	99.33	99.36	99.37	99.39	99.40	99.42	99.43	99.45	99.47	99.47	99.48	99.49	99.50
3	34.12	30.82	29.46	28.71	28.24	27.91	27.67	27.49	27.35	27.23	27.05	26.87	26.69	26.50	26.41	26.32	26.22	26.13
4	21.20	18.00	16.69	15.98	15.52	15.21	14.98	14.80	14.66	14.55	14.37	14.20	14.02	13.84	13.75	13.65	13.56	13.6
5	16.26	13.27	12.06	11.39	10.97	10.67	10.46	10.29	10.16	10.05	9.89	9.72	9.55	9.38	9.29	9.20	9.11	9.02
6	13.75	10.92	9.78	9.15	8.75	8.47	8.26	8.10	7.98	7.87	7.72	7.56	7.40	7.23	7.14	7.06	6.97	6.88
7	12.25	9.55	8.45	7.85	7.46	7.19	6.99	6.84	6.72	6.62	6.47	6.31	6.16	5.99	5.91	5.82	5.74	5.65
8	11.26	8.65	7.59	7.01	6.63	6.37	6.18	6.03	5.91	5.81	5.67	5.52	5.36	5.20	5.12	5.03	4.95	4.86
9	10.56	8.02	6.99	6.42	6.06	5.80	5.61	5.47	5.35	5.26	5.11	4.96	4.81	4.65	4.57	4.48	4.40	4.31
10	10.04	7.56	6.55	5.99	5.64	5.39	5.20	5.06	4.94	4.85	4.71	4.56	4.41	4.25	4.17	4.08	4.00	3.91
11	9.65	7.21	6.22	5.67	5.32	5.07	4.89	4.74	4.63	4.54	4.40	4.25	4.10	3.94	3.86	3.78	3.69	3.60
12	9.33	6.93	5.95	5.41	5.06	4.82	4.64	4.50	4.39	4.30	4.16	4.01	3.86	3.70	3.62	3.54	3.45	3.36
13	9.07	6.70	5.74	5.21	4.86	4.62	4.44	4.30	4.19	4.10	3.96	3.82	3.66	3.51	3.43	3.34	3.25	3.17
14	5.86	6.51	5.56	5.04	4.69	4.46	4.28	4.14	4.03	3.94	3.80	3.66	3.51	3.35	3.27	3.18	3.09	3.00
15	8.68	6.36	5.42	4.89	4.56	4.32	4.14	4.00	3.89	3.80	3.67	3.52	3.37	3.21	3.13	3.05	2.96	2.87
16	8.53	6.23	5.29	4.77	4.44	4.20	4.03	3.89	3.78	3.69	3.55	3.41	3.26	3.10	3.02	2.93	2.84	2.75
17	8.40	6.11	5.18	4.67	4.34	4.10	3.93	3.79	3.68	3.59	3.46	3.31	3.16	3.00	2.92	2.83	2.75	2.65
18	8.29	6.01	5.09	4.58	4.25	4.01	3.84	3.71	3.60	3.51	3.37	3.23	3.08	2.92	2.84	2.75	2.66	2.57
19	8.18	5.93	5.01	4.50	4.17	3.94	3.77	3.63	3.52	3.43	3.30	3.15	3.00	2.84	2.76	2.67	2.58	2.49
20	8.10	5.85	4.94	4.43	4.10	3.87	3.70	3.56	3.46	3.37	3.23	3.09	2.94	2.78	2.69	2.61	2.52	2.42
21	8.02	5.78	4.87	4.37	4.04	3.81	3.64	3.51	3.40	3.31	3.17	3.03	2.88	2.72	2.64	2.55	2.46	2.36
22	7.95	5.72	4.82	4.31	3.99	3.76	3.59	3.45	3.35	3.26	3.12	2.98	2.83	2.67	2.58	2.50	2.40	2.31
23	7.88	5.66	4.76	4.26	3.94	3.71	3.54	3.41	3.30	3.21	3.07	2.93	2.78	2.62	2.54	2.45	2.35	2.26
24	7.82	5.61	4.72	4.22	3.90	3.67	3.50	3.36	3.26	3.17	3.03	2.89	2.74	2.58	2.49	2.40	2.31	2.21
30	7.56	5.39	4.51	4.02	3.70	3.47	3.30	3.17	3.07	2.98	2.84	2.70	2.55	2.39	2.30	2.21	2.11	2.01
40	7.31	5.18	4.31	3.83	3.51	3.29	3.12	2.99	2.89	2.80	2.66	2.52	2.37	2.20	2.11	2.02	1.92	1.80
60	7.08	4.98	4.13	3.65	3.34	3.12	2.95	2.82	2.72	2.63	2.50	2.35	2.20	2.03	1.94	1.84	1.73	1.60
120	6.85	4.79	3.95	3.48	3.17	2.96	2.79	2.66	2.56	2.47	2.34	2.19	2.03	1.86	1.76	1.66	1.53	1.38
∞	6.63	4.61	3.78	3.32	3.02	2.80	2.64	2.51	2.41	2.32	2.18	2.04	1.88	1.70	1.59	1.47	1.32	1.00

Adapted from Table 18 of Pearson and Hartley (1966).

Table A. 6 χ^2 distribution

For various degrees of freedom (df), the tabled entries
represent the values of χ^2 above which a proportion p of the
distribution falls. (Example: for df = 5, χ^2 = 11.070 is
exceeded by p = .05 or 5% of the distribution.)

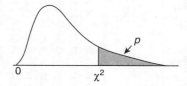

df	.99	.95	.90	.10	.05	.01	.001
				p			
1	.03157	.00393	.0158	2.706	3.841	6.635	10.827
2	.0201	.103	.211	4.605	5.991	9.210	13.815
3	.115	.352	.584	6.251	7.815	11.345	16.266
4	.297	.711	1.064	7.779	9.488	13.277	18.467
5	.554	1.145	1.610	9.236	11.070	15.086	20.515
6	.872	1.635	2.204	10.645	12.592	16.812	22.457
7	1.239	2.167	2.833	12.017	14.067	18.475	24.322
8	1.646	2.733	3.490	13.362	15.507	20.090	26.125
9	2.088	3.325	4.168	14.684	16.919	21.666	27.877
10	2.558	3.940	4.865	15.987	18.307	23.209	29.588
11	3.053	4.575	5.578	17.275	19.675	24.725	31.264
12	3.571	5.226	6.304	18.549	21.026	26.217	32.909
13	4.107	5.892	7.042	19.812	22.362	27.688	34.528
14	4.660	6.571	7.790	21.064	23.685	29.141	36.123
15	5.229	7.261	8.547	22.307	24.996	30.578	37.697
16	5.812	7.962	9.312	23.542	26.296	32.000	39.252
17	6.408	8.672	10.085	24.769	27.587	33.409	40.790
18	7.015	9.390	10.865	25.989	28.869	34.805	42.312
19	7.633	10.117	11.651	27.204	30.144	36.191	43.820
20	8.260	10.851	12.443	28.412	31.410	37.566	45.315
21	8.897	11.591	13.240	29.615	32.671	38.932	46.797
22	9.542	12.338	14.041	30.813	33.924	40.289	48.268
23	10.196	13.091	14.848	32.007	35.172	41.638	49.728
24	10.856	13.848	15.659	33.196	36.415	42.980	51.179
25	11.524	14.611	16.473	34.382	37.652	44.314	52.620
26	12.198	15.379	17.292	35.563	38.885	45.642	54.052
27	12.879	16.151	18.114	36.741	40.113	46.963	55.476
28	13.565	16.928	18.939	37.916	41.337	48.278	56.893
29	14.256	17.708	19.768	39.087	42.557	49.588	58.302
30	14.953	18.493	20.599	40.256	43.773	50.892	59.703

Adapted from Table IV of Fisher and Yates (1974).

APPENDIX B: MATHEMATICAL CONVENTIONS AND NOTATION

The amount of mathematical notation used in this book is actually quite small, but, nevertheless, it is useful to review some basic notation and mathematical conventions.

B.1 MATHEMATICAL CONVENTIONS

By the term 'mathematical conventions', we are not referring here to the gatherings of mathematicians at conferences, but rather to the standards that are used in the writing and use of mathematical material. The primary conventions we are concerned with are those regarding parentheses and the ordering of mathematical operations. In a mathematical expression, one performs operations in the following order, arranged from operations performed first to those performed last:

1. factorials (the factorial of an integer m is the product of the integers from 1 to m, and is further defined below)

2. powers and roots

3. multiplication and division

4. addition and subtraction.

Thus the expression:

$$3 + 10/5^2 \qquad \text{(B.1)}$$

is evaluated by first squaring 5, then finding $10/25 = 0.4$, and then adding 3 to find the result of 3.4. One does not simply go from left to right; if you did, you would incorrectly add 10 to 3, then divide by 5 to get 2.6, and then square 2.6, for a final (incorrect) answer of 6.76.

If there is more than one operation in any of the four categories above, one carries out those particular operations from left to right. Thus, to evaluate:

$$3 + 10/5 + 6 \times 7 \qquad\qquad\qquad (B.2)$$

one would do the division first, and the multiplication second, yielding

$$3 + 2 + 42 = 47. \qquad\qquad\qquad (B.3)$$

Although it would be unusual to see it written this way,

$$6 / 3 / 3 \qquad\qquad\qquad (B.4)$$

is equal to 2/3, since 6/3 would be carried out first.

Operations within parentheses are always performed before those that are not within parentheses, and those within nested parentheses are dealt with by performing the operations within the innermost set of parentheses first. So, for example:

$$3 \times ((5 + 3)^2 / 2) + 4 = 3 \times (8^2 / 2) + 4 = 3 \times 32 + 4 = 100. \qquad (B.5)$$

Although these basic principles are taught before the high-school years, it is not uncommon to need a little review! It is important to realize too that it is not just students of statistics that need brushing up – software developers and decision-makers sometimes do not abide by these conventions. For example, new variables that were created within the geographic information system (GIS) *ArcView* 3.1 were created by simply carrying out operations from left to right. Although parentheses were recognized, the fundamental order of operations, as outlined above, was not. This leads to visions of planners and others all over the world making decisions based upon inaccurate information!

Suppose we have data on the proportion of people commuting by train (variable 1), the number of people who commute by bus (variable 2), and the total number of commuters (variable 3) for a number of census tracts in our database. Thinking that *ArcView* will surely use the standard order of mathematical operations, we compute a new variable reflecting the proportion of people who commute by bus or train (variable 4) via:

$$\text{Var. } 4 = \text{Var. } 1 + \text{Var. } 2 / \text{Var. } 3 \qquad\qquad (B.6)$$

ArcView will provide us with a column of answers where:

$$\text{Var. } 4 = (\text{Var. } 1 + \text{Var. } 2) / \text{Var. } 3 \qquad\qquad (B.7)$$

when in fact what we wanted was:

$$\text{Var. } 4 = \text{Var. } 1 + (\text{Var. } 2 / \text{Var. } 3) \qquad\qquad (B.8)$$

One way of ensuring that problems like this do not arise is to use extra sets of parentheses, as in the last equation (and in fact to obtain the desired variable within *ArcView*, they *must* be used).

B.2 MATHEMATICAL NOTATION

The mathematical notation used most often in this book is the summation notation. The Greek letter 'Σ' is used as a shorthand way of indicating that a sum is to be taken.

For example,

$$\sum_{i=1}^{i=n} x_i \tag{B.9}$$

denotes that the sum of n observations is to be taken; the expression is equivalent to:

$$x_1 + x_2 + \ldots + x_n \tag{B.10}$$

The '$i = 1$' under the symbol refers to where the sum of terms begins, and the '$i = n$' refers to where it terminates. Thus:

$$\sum_{i=3}^{i=5} x_i = x_3 + x_4 + x_5 \tag{B.11}$$

implies that we are to sum only the third, fourth, and fifth observations. There are a number of rules that govern the use of this notation. These may be summarized as follows, where a is a constant, n is the number of observations, and x and y are variables:

$$\sum_{i=1}^{i=n} a = na$$

$$\sum_{i=1}^{i=n} ax_i = a \sum_{i=1}^{i=n} x_i \tag{B.12}$$

$$\sum_{i=1}^{i=n} (x_i + y_i) = \sum_{i=1}^{i=n} x_i + \sum_{i=1}^{i=n} y_i$$

The first states that summing a constant n times yields a result of an. Thus:

$$\sum_{i=1}^{i=3} 4 = 4 + 4 + 4 = 4 \times 3 = 12. \tag{B.13}$$

The second rule in B.12 indicates that constants may be taken outside of the summation sign. So, for example:

$$\sum_{i=1}^{i=3} 3x_i = 3 \sum_{i=1}^{i=3} x_i = 3(x_1 + x_2 + x_3). \tag{B.14}$$

The third rule implies that the order of addition does not matter when sums of sums are being taken.

Other conventions include:

$$\sum_{i=1}^{i=n} x_i y_i = x_1 y_1 + x_2 y_2 + \ldots + x_n y_n$$

$$\sum_{i=1}^{i=n} x_i^2 = x_1^2 + x_2^2 + \ldots + x_n^2 \qquad (B.15)$$

$$\left(\sum_{i=1}^{i=n} x_i \right)^2 = (x_1 + x_2 + \ldots + x_n)^2$$

Shorthand versions of the summation notation leave out the upper limit of the summation, and sometimes the lower limit as well. This is done in those situations where *all* of the terms, and not just some subset of them, are to be summed. The following are all equivalent:

$$\sum_{i=1}^{i=n} x_i = \sum_{i}^{n} x_i = \sum_{i} x_i = \sum x_i \qquad (B.16)$$

It should also be recognized that the letter 'i' is used in this notation simply as an indicator (to indicate which observations or terms to sum); we could just as easily use any other letter:

$$\sum_{i=1}^{i=n} x_i = \sum_{k=1}^{k=n} x_k . \qquad (B.17)$$

In each case, we find the sum by adding up all of the n observations. In fact, we often have use for more than one summation indicator. Double summations are required when we want to denote the sum of all of the observations in a table. A table of commuting flows such as the one in Table B.1 indicates the origins and destinations of individuals. The value of any cell is denoted x_{ij} and this refers to the number of commuters from origin i who commute to destination j. The number of commuters going to destination j from all origins is $\sum_{i=1}^{i=n} x_{ij}$ (where there are n transportation zones), and the number of commuters leaving origin i for all destinations is $\sum_{j=1}^{j=n} x_{ij}$. The total number of commuters is designated by the double summation, $\sum_{i=1}^{n} \sum_{j=1}^{n} x_{ij}$. Using the data in Table B.1, for example, we find that:

$$\sum_{i} x_{i2} = 160, \ \sum_{j} x_{1j} = 220, \ \text{and} \ \sum_{i} \sum_{j} x_{ij} = 500$$

Whereas the summation notation refers to the addition of terms, the product notation applies to the multiplication of terms. It is denoted by the upper-case Greek letter pi (Π), and is used in the same way as the summation notation. For example,

$$\prod_{i=1}^{n}(x_i + y_i) = (x_1 + y_1)(x_2 + y_2) \dots (x_n + y_n) \qquad \text{(B.18)}$$

The *factorial* of a positive integer, n, is equal to the product of the first n integers. Surprisingly perhaps, factorials are denoted by an exclamation point! Thus:

$$5! = 5 \times 4 \times 3 \times 2 \times 1 = 120 \qquad \text{(B.19)}$$

Note that we could express factorials in terms of the product notation:

$$n! = \prod_{i=1}^{n} i. \qquad \text{(B.20)}$$

There is also a convention that $0! = 1$; factorials are not defined for negative integers or for nonintegers.

Factorials arise in the calculation *of combinations.* Combinations refer to the number of possible outcomes that particular probability experiments may have (see Section 3.3). Specifically, the number of ways that r items may be chosen from a group of n items is denoted by $\binom{n}{r}$ and is equal to:

$$\binom{n}{r} = \frac{n!}{r!(n-r)!} \qquad \text{(B.21)}$$

For example,

$$\binom{5}{2} = \frac{5!}{2!3!} = \frac{120}{2 * 6} = 10 \qquad \text{(B.22)}$$

What does this mean? If, for example, we group income into $n = 5$ categories, then there are ten ways to choose two of them. If we label the five categories (a) through (e), then the ten possible combinations of two income categories are *ab ac ad ae bc bd be cd ce de.*

B.2.1 More Examples

$$6! = 6 \times 5 \times 4 \times 3 \times 2 \times 1 = 720 \qquad \text{(B.23)}$$

$$\prod_{i=1}^{i=4} i^2 = 1^2 2^2 3^2 4^2 = 576 \qquad \text{(B.24)}$$

$$34 + (26 / 13) \times 12 = 58 \qquad \text{(B.25)}$$

Now let $a = 3$, and let the values of a set $(n = 3)$ of x and y values be $x_1 = 4, x_2 = 5, x_3 = 6, y_1 = 7, y_2 = 8, y_3 = 9$. Then:

$$\sum_{2} ax_i = 3(4 + 5 + 6) = 45$$

$$\sum_{i=1}^{2} x_i y_i = (4)(7) + (5)(8) = 68$$

$$\sum x_i^3 = 4^3 + 5^3 + 6^3 = 405$$

$$\bar{x} = \frac{\sum x_i}{n} = \frac{4 + 5 + 6}{3} = 5$$

$$s_y^2 = \frac{\sum(y_i - \bar{y})^2}{n - 1} = \frac{(7 - 8)^2 + (8 - 8)^2 + (9 - 8)^2}{3 - 1} = 1$$

$$\prod_{i=1}^{n} x_i = (4)(5)(6) = 120$$

$$\prod_{i=1}^{n} y_i = (7)(8)(9) = 504$$

$$\prod_{i=1}^{n} (x_i + y_i) = (4 + 7)(5 + 8)(6 + 9) = 2145$$

(B.26)

Note that the sum of products does not equal the product of sums:

$$\sum x_i y_i = (4 \times 7) + (5 \times 8) + (6 \times 9) = 122$$
$$\neq \sum x_i \sum y_i = (4 + 5 + 6) \times (7 + 8 + 9) = 360.$$

(B.27)

Table B.1   **Hypothetical commuting data**

	Destination		
Origin	1	2	3
1	130	40	50
2	20	100	10
3	30	20	100

APPENDIX C: REVIEW AND EXTENSION OF SOME PROBABILITY THEORY

A discrete random variable, X, has a probability distribution (sometimes called a probability mass function) denoted by $pr(X = x) = pr(x)$, where x is the value taken by X. A continuous random variable has a probability distribution (also called a probability density function or pdf) denoted by $f(x)$. The likelihood of getting a specific value x is zero, since the distribution is continuous. The likelihood of getting a value within a range, $a < x < b$ is equal to the area under the curve, $f(x)$, that lies between a and b. For those familiar with calculus, this area is given as the integral of $f(x)$ from a to b:

$$\Pr(a < X < b) = \int_a^b f(x)dx \tag{C.1}$$

For those not familiar with calculus, the integral sign may be thought of as similar to the summation sign; the only difference is that, with a continuous random variable, we have an infinite number of values to sum over. Since the probability of getting a value between minus and plus infinity is equal to one, the total area under the curve $f(x)$ must equal one:

$$\int_{-\infty}^{+\infty} f(x)dx = 1 \tag{C.2}$$

Cumulative distribution functions are denoted with an upper-case 'F', and they tell us the likelihood that the random variable will be less than or equal to a particular value. For a discrete random variable, the probability of obtaining a value less than or equal to a is:

$$F(a) = \sum_{x \le a} pr(X = x) \tag{C.3}$$

For a continuous random variable, we have:

$$F(a) = pr(X \le a) = \int_{+\infty}^{a} f(x) \tag{C.4}$$

C.1 EXPECTED VALUES

The expected value of a random variable, $E[X]$, is also known as the theoretical mean and is denoted by μ. The expected value is given as the weighted average of the possible values the random variable can take on, where the weights are the likelihoods of obtaining those values. For a discrete random variable,

$$E[X] = \mu = \sum_x \frac{x\{pr(X = x)\}}{\sum_x pr(X = x)} = \sum_x x\{pr(X = x)\} \tag{C.5}$$

For a continuous random variable,

$$E[X] = \mu = \frac{\int\limits_{-\infty}^{+\infty} xf(x)dx}{\int\limits_{-\infty}^{+\infty} f(x)dx} = \int\limits_{-\infty}^{+\infty} x\, f(x)dx \tag{C.6}$$

As an example, consider the experiment that consists of tossing a die. What is the expected value of the die? This is equivalent to asking what the expected average is of a large number of tosses. Using Equation C.5,

$$\sum_{x=1}^{6} 1\left(\frac{1}{6}\right) + 2\left(\frac{1}{6}\right) + 3\left(\frac{1}{6}\right) + \dots 6\left(\frac{1}{6}\right) = 3.5 \tag{C.7}$$

What is the expected value of a uniform random variable – that is, a random variable that has equally likely outcomes over the range (a, b)? Such a random variable has a probability density function given by:

$$f(x) = \frac{1}{b - a} \tag{C.8}$$

The expected value is then:

$$E[X] = \mu = \int_a^b \frac{x}{b - a} dx = \frac{1}{b - a}\left(\frac{b^2 - a^2}{2}\right) = \frac{a + b}{2} \tag{C.9}$$

The expected value of any function of a random variable, $g(x)$, is a weighted average of the values of $g(x)$, where the weights are again the likelihoods of obtaining the values of $g(x)$. Thus for discrete and continuous variables we have:

$$\left. \begin{aligned} E[g(X)] &= \sum_x g(x)pr(X = x) \\ E[g(x)] &= \int\limits_{-\infty}^{+\infty} g(x)f(x)dx \end{aligned} \right\} \tag{C.10}$$

for discrete and continuous random variables, respectively. Useful rules for working with expected values are: (i) the expected value of a constant is simply equal to the constant; (ii) the expected value of a constant times a random variable is equal to the constant times the expected value of the random variable; and (iii) the expected value of a sum is equal to the sum of the expected values. These rules are summarized below:

$$
\left.\begin{array}{ll}
E[a] = a & \text{(i)} \\
E[bX] = bE[X] & \text{(ii)} \\
E[a + bX] = a + bE[X] & \text{(i, ii, and iii)}
\end{array}\right\} \tag{C.11}
$$

C.2 VARIANCE OF A RANDOM VARIABLE

The variance of a random variable, $\sigma^2 = V[X]$, is the expected value of the squared deviation of an observation from the mean:

$$
V[X] = \sigma^2 = E[(X - E[X])^2] = E[(X - \mu)^2] \tag{C.12}
$$

Using the rules for expected values above,

$$
\begin{aligned}
V[X] &= E[X^2 - 2X\mu + \mu^2] = E[X^2] - 2\mu E[X] + \mu^2 \\
&= E[X^2] - \mu^2
\end{aligned} \tag{C.13}
$$

To illustrate, let us return to the experiment involving the toss of a die. The variance of the random variable X in this case is equal to $E[X^2] - 3.5^2$. The expected value of X^2 is found using Equation C.10:

$$
E[X^2] = 1^2\left(\frac{1}{6}\right) + 2^2\left(\frac{1}{6}\right) + \ldots + 6^2\left(\frac{1}{6}\right) = 15.17 \tag{C.14}
$$

The variance is therefore equal to $15.17 - 3.5^2 = 2.92$. To illustrate the derivation of the variance using a continuous variable, let us continue with the example of a uniform random variable. We have:

$$
E[X^2] = \int_a^b x^2 \frac{1}{b-a} = \frac{b^3 - a^3}{3(b-a)} \tag{C.15}
$$

Then:

$$
V[X] = \frac{b^3 - a^3}{3(b-a)} - \frac{(a+b)^2}{4} = \frac{b-a}{12} \tag{C.16}
$$

C.3 COVARIANCE OF RANDOM VARIABLES

How do two variables co-vary? Is there a tendency for one variable to exhibit high values when the other does? Or are the variables independent? The covariance of two random variables, X and Y, is defined as the expected value of the product of the two deviations from their respective means:

$$\text{Cov}[X,Y] = E[(X - \mu_X)(Y - \mu_Y)] \qquad (C.17)$$

This may be rewritten in the form:

$$
\begin{aligned}
\text{Cov}[X,Y] &= E[(X - \mu_X)(Y - \mu_Y)] \\
&= E[XY] - \mu_Y E[X] - \mu_X E[Y] + \mu_X \mu_Y \\
&= E[XY] - \mu_X \mu_Y - \mu_X \mu_Y + \mu_X \mu_Y \\
&= E[XY] - \mu_X \mu_y
\end{aligned}
\qquad (C.18)
$$

To find the observed covariance for a set of data, we find the average value of the product of deviations:

$$\text{Cov}[x,y] = \sum_{i=1}^{n} \frac{(x_i - \bar{x})(y_i - \bar{y})}{n} \qquad (C.19)$$

The theoretical correlation coefficient is the standardized covariance:

$$\rho = \frac{\text{Cov}[X,Y]}{\sigma_x \sigma_y} \qquad (C.20)$$

and the observed correlation coefficient is found by standardizing the observed covariance (C.19):

$$r = \frac{\sum_i (x_i - \bar{x})(y_i - \bar{y})}{n s_x s_y} \qquad (C.21)$$

BIBLIOGRAPHY

Ackerman, W.V. 1998. Socioeconomic correlates of increasing crime rates in smaller communities. *Professional Geographer* 50: 372–87.

Akella, M.R., Bang, C., Beutner, R., Delmelle, E.M., Wilson, G., Blatt, A., Batta, R., and Rogerson, P. 2003. Evaluating the reliability of automated collision notification systems. *Accident Analysis and Prevention* 35: 349–60.

Allen, J.P., and Turner, E. 1996. Spatial patterns of immigrant assimilation. *Professional Geographer* 48: 140–55.

Andrews, D.F., and Herzberg, A.M. 1985. *Data: a collection of problems from many fields for the student and research worker.* New York: Springer.

Anselin, L. 1995. Local indicators of spatial association – LISA. *Geographical Analysis* 27: 93–115.

Anselin, L. 2005. Exploring spatial data with GeoDa: a workbook. Center for Spatially Integrated Social Science at http://www.csiss.org/clearinghouse/GeoDa/geodaworkbook.pdf

Anselin, Luc, Syabri, I. and Kho, Y. (2006). GeoDa: An introduction to spatial data analysis. *Geographical Analysis* 38 (1): 5-22.

Bachi, R. 1963. Standard distance measures and related methods for spatial analysis. *Papers of the Regional Science Association* 10: 83–132.

Bai, S.–B., Wang, J., Lu, G.–N., Zhou, P.–G, Hou, S.–S., and Xu, S.–N. 2010. GIS-based logistic regression for landslide susceptibility mapping of the Zhongxian segment in the Three Gorges area, China. *Geomorphology* 115 (1–2): 23–31.

Bailey, A., and Gatrell, A. 1995. *Interactive spatial data analysis.* Harlow: Longman (published in the USA by Wiley).

Beckmann, P. 1971. *A history of pi.* New York: St. Martin's Press.

Besag, J., and Newell, J. 1991. The detection of clusters in rare diseases. *Journal of the Royal Statistical Society Series A* 154: 143–55.

Bostrom, N. 2001. Cars in the next lane really do go faster. *Plus* (November). Available at www.pass.maths.org.uk/issue17/features/traffic/index.html

Brunsdon, C., Fotheringham, A.S., and Charlton, M. 1996. Geographically weighted regression: a method for exploring spatial nonstationarity. *Geographical Analysis* 28: 281–98.

Brunsdon, C., Fotheringham, A.S., and Charlton, M. 1999. Some notes on parametric significance tests for geographically weighted regression. *Journal of Regional Science* 39: 497–524.

Casetti, E. 1972. Generating models by the expansion method: applications to geographic research. *Geographical Analysis* 4: 81–91.

Chun, Y., and Griffith, D.A. 2013. *Spatial statistics and geostatistics: theory and applications for geographic information science and technology.* Thousand Oaks, CA: Sage Publications.

City of Milwaukee. http://assessments.milwaukee.gov/mainsales.html (accessed December 2013).

Clark, P., and Evans, F.C. 1954. Distance to nearest neighbor as a measure of spatial relationships in populations. *Ecology* 35: 445–53.

Clarke, A.E., and Holly, B. 1996. The organization of production in high technology industries: an empirical assessment. *Professional Geographer* 48: 127–39.

Cliff, A., and Ord, J.K. 1975. The comparison of means when samples consist of spatially autocorrelated observations. *Environment and Planning A* 7: 725–34.

Clifford, P., and Richardson, S. 1985. Testing the association between two spatial processes. *Statistics and Decisions* Supplement No. 2: 155–60.

Cohen, J., 1995. *How many people can the earth support?* New York: W.W. Norton and Co.

Cornish, S.L. 1997. Strategies for the acquisition of market intelligence and implications for the transferability of information inputs. *Annals of the Association of American Geographers* 87: 451–70.

Cressie, N. 1993. *Statistical analysis of spatial data.* New York: Wiley.

Cressie, N. 2011. *Statistics for spatio-temporal data.* New York: Wiley.

Cringoli, G., Taddei, R., Rinaldi, L., Veneziano, V., Musella, V., Cascone,, C., Sibilio, G., and Malone, J.B. 2004. Use of remote sensing and geographical information systems to identify environmental features that influence the distribution of paramphistomosis in sheep from the southern Italian Apennines. *Veterinary Parasitology* 122 (1): 15–26.

Curtiss, J., and McIntosh, R. 1950. The interrelations of certain analytic and synthetic phytosociological characters. *Ecology* 31: 434–55.

Dagel, K.C. 1997. Defining drought in marginal areas: the role of perception. *Professional Geographer* 49: 192–202.

Davies Withers, S. 2002. Quantitative methods: Bayesian inference, Bayesian thinking. *Progress in Human Geography* 26: 553–66.

Dawson, C.B., and Riggs, T.D. 2004. Highway relativity. *College Mathematics Journal* 35: 246–50.

deSmith, M.J., Goodchild, M.F., and Longley, P.A. 2009. *Geospatial analysis: a comprehensive guide to principles, techniques, and software tools.* Leicester, UK: Troubador; full text available at http://www. spatialanalysisonline.com/HTML

Dunhill, A.M. 2011. Using remote sensing and a geographic information system to quantify rock exposure area in England and Wales: Implications for paleodiversity studies. *Geology* 39: 111–14.

Duque, J.C., Church, R.L., and Middleton, R.S. 2011. The max *p*-regions problem. *Geographical Analysis* 43: 104–126.

Duque, J.C., Anselin, L., and Rey, S.J. 2012. The *p*-regions problem. *Journal of Regional Science* 52: 397–419.

Easterlin, R. 1980. *Birth and fortune: the impact of numbers on personal welfare.* New York: Basic Books.

Eilon, S., Watson-Gandy, C.D.T. and Christofides, N. 1971. *Distribution management: mathematical modeling and practical analysis.* London: Griffin.

Fan, C. 1996. Economic opportunities and internal migration: a case study of Guangdong Province, China. *Professional Geographer* 48: 28–45.

Fischer, M., and Getis, A. 2009. *Handbook of applied spatial analysis: software tools, methods, and applications.* New York: Springer.

Fisher, R.A., and Yates, F. 1974. *Statistical tables for biological, agricultural, and medical research*, 6th edition. London: Longman.

Fotheringham, A.S., and Rogerson, P. 1993. GIS and spatial analytical problems. *International Journal of Geographical Information Systems* 7: 3–19.

Fotheringham, A.S., and Rogerson, P. 2008. *Handbook of spatial analysis*. London: Sage.

Fotheringham, A.S., and Wong, D. 1991. The modifiable area unit problem in multivariate statistical analysis. *Environment and Planning A* 23: 1025–44.

Fotheringham, A.S., Charlton, M.E., and Brunsdon, C. 1998. Geographically weighted regression: a natural evolution of the expansion method for spatial data analysis. *Environment and Planning A* 30: 1905–27.

Fotheringham, A.S., Brunsdon, C., and Charlton, M.E. 2000. *Quantitative geography: perspectives on spatial data analysis*. London: Sage.

Fotheringham, A.S., Brunsdon, C., and Charlton, M.E. 2002. *Geographically weighted regression: the analysis of spatially varying relationships*. New York: Wiley.

Gardner, M. 1976. On the fabric of inductive logic and some probability paradoxes. *Scientific American*, March, 119-24.

Gao, X., Asami, Y., and Ching, C.–J.F. 2006. An empirical evaluation of spatial regression models. *Computers and Geosciences* 32: 1040–51.

Gehlke, C., and Biehl, K. 1934. Certain effects of grouping upon the size of the correlation coefficient in census tract material. *Journal of the American Statistical Association* 29: 169–70.

Geoda. Spatial statistical software available at http://geodacenter.asu.edu/projects/opengeoda

Getis, A., and Ord, J. 1992. The analysis of spatial association by use of distance statistics. *Geographical Analysis* 24: 189–206.

Gott, R. 1993. Implications of the Copernican principle for our future prospects. *Nature* 363: 315–19.

Griffith, D.A. 1978. A spatially adjusted ANOVA model. *Geographical Analysis* 10: 296–301.

Griffith, D.A. 1987. *Spatial autocorrelation: a primer*. Washington, DC: Association of American Geographers.

Griffith, D.A. 1996. Computational simplifications for space–time forecasting within GIS: the neighbourhood spatial forecasting model. In *Spatial analysis: modelling in a GIS environment*. Eds. P. Longley and M. Batty. pp. 247–60. Cambridge: Geoinformation International (distributed by Wiley).

Griffith, D.A., Doyle, P.G., Wheeler, D.C., and Johnson, D.L. 1998. A tale of two swaths: urban childhood blood-lead levels across Syracuse, New York. *Annals of the Association of American Geographers* 88: 640–65.

Hadi, A.S., and Ling, R.F. 1998. Some cautionary notes on the use of principal components regression. *American Statistician* 52(1): 15–19.

Haining, R. 1990a. *Spatial data analysis in the social and environmental sciences*. Cambridge: Cambridge University Press.

Haining, R. 1990b. The use of added variable plots in regression modelling with spatial data. *Professional Geographer* 42: 336–45.

Haining, R. 2003. *Spatial data analysis: theory and practice*. Cambridge: Cambridge University Press.

Hemmasi, M., and Prorok, C. 2002. Women's migration and quality of life in Turkey. *Geoforum* 33: 399–411.

Hendryx, M., Fedorko, E. and Anesetti-Rothermel, A. 2010. A geographical information system-based analysis of cancer mortality and population exposure to coal mining activities in West Virginia, United States of America. *Geospatial Health*, 4: 243–56.

IJmker, J., Stauch, G., Hartmann, K., Diekmann, B., Dietze, E., Opitz, S., Wunnemann, B., and Lehmkuhl, F. 2012. Environmental conditions in the Donggi Cona lake catchment, NE Tibetan Plateau, based on factor analysis of geochemical data. *Journal of Asian Earth Sciences* 44: 176–188.

Johnson, B.W., and McCulloch, R.E. 1987. Added variable plots in linear regression. *Technometrics* 29: 427–33.

Jones III, J.P., and Casetti, E. 1992. *Applications of the expansion method*. London: Routledge.

Karlin, S., and Taylor, H.M. 1975. *A first course in stochastic processes*. New York: Academic Press.

Keylock, C.J., and Dorling, D. 2004. What kind of quantitative methods for what kind of geography? *Area* 36: 358–66.

Ligmann-Zielinska, A., and Jankowski, P. 2012 Impact of proximity-adjusted preferences on rank-order stability in geographical multicriteria decision analysis. *Journal of Geographical Systems* 14: 167–87.

Liu, C.W., Lin, K.H., and Kuo, Y.M. 2003. Application of factor analysis in the assessment of ground water quality in a blackfoot disease area in Taiwan. *Science of the Total Environment* 313(1–3): 77–89.

Lloyd, C., and Shuttleworth, I. 2005. Analyzing commuting using local regression techniques. *Environment and Planning A* 37: 81–103.

Longley, P., Brooks, S.M., McDonnell, R., and Macmillan, B. 1998. *Geocomputation: a primer*. Chichester: Wiley.

MacDonald, G.M., Szeicz, J.M., Claricoates, J., and Dale, K.A. 1998. Response of the central Canadian treeline to recent climatic changes. *Annals of the Association of American Geographers* 88: 183–208.

Mallows, C. 1998. The zeroth problem. *American Statistician* 52(1): 1–9.

Mardia, K.V., and Jupp, P. 1999. *Directional statistics*. New York: Wiley.

McGrew, J.C.J., and Monroe, C.B. 2000. *Introduction to statistical problem solving in geography*. Boston, MA: McGraw-Hill.

Meehl, P. 1990. Why summaries of research on psychological theories are often uninterpretable. *Psychological Reports* 66: 195–244. (Monograph Supplement 1–V66.)

Moran, P. A. P. 1948. The interpretation of statistical maps. *Journal of the Royal Statistical Society Series B* 10: 245–51.

Moran, P. A. P. 1950. Notes on continuous stochastic phenomena. *Biometrika* 37: 17–23.

Murtagh, F. 1985. A survey of algorithms for contiguity-constrained clustering and related problems. *The Computer Journal* 28: 82–8.

Myers, D., Lee, S.W., and Choi, S.S. 1997. Constraints of housing age and migration on residential mobility. *Professional Geographer* 49: 14–28.

Nelson, P.W. 1997. Migration, sources of income, and community change in the Pacific Northwest. *Professional Geographer* 49: 418–30.

O'Loughlin, J., Ward, M.D., Lofdahl, C.L., Cohen, J.S., Brown, D.S., Reilly, D., Gleditsch, K.S., and Shin, M. 1998. The diffusion of democracy, 1946–1994. *Annals of the Association of American Geographers* 88: 545–74.

O'Reilly, K., and Webster, G.R. 1998. A sociodemographic and partisan analysis of voting in three anti-gay rights referenda in Oregon. *Professional Geographer* 50: 498–515.

Ord, J., and Getis, A. 1995. Local spatial autocorrelation statistics: distributional issues and an application. *Geographical Analysis* 27: 286-306.

Ormrod, R.K., and Cole, D.B. 1996. The vote on Colorado's Amendment Two. *Professional Geographer* 48: 14–27.

Oxford English Dictionary, online at www.oed.com/view/Entry/189322?

Pearson, E.S., and Hartley, H.O. (eds.) (1966) *Biometrika tables for statisticians*, vol. 1. Cambridge: Cambridge University Press.

Plane, D., and Rogerson, P. 1991. Tracking the baby boom, the baby bust, and the echo generations: how age composition regulates US migration. *Professional Geographer* 43: 416–39.

Plane, D., and Rogerson, P. 1994. *The geographical analysis of population: with applications to planning and business.* New York: Wiley.

Poon, J.P.H., Eldredge, B., and Yeung, D. 2004. Rank size distribution of international financial centers. *International Regional Science Review* 27: 411–30.

Redelmeier, D.A., and Tibshirani, R.J. 2000. Are those other drivers really going faster? *Chance* 13(3): 8–14. Available at http://people.brandeis.edu/~moshep/Projects/DoTheyReallyMoveFaster/133.redelmeier.pdf

Rey, S., and Montouri, B. 1999. US regional income convergence. *Regional Studies* 33: 146–56.

Robinson, W. 1950. Ecological correlation and the behavior of individuals. *American Sociological Review* 15: 351–57.

Rogers, A. 1975. *Matrix population models.* Thousand Oaks, CA: Sage.

Rogerson, P. 1987. Changes in U.S. national mobility levels. *Professional Geographer* 39: 344–51.

Rogerson, P., and Plane, D. 1998. The dynamics of neighborhood composition. *Environment and Planning A* 30: 1461–72.

Rogerson, P., Weng, R., and Lin, G. 1993. The spatial separation between parents and their adult children. *Annals of the Association of American Geographers* 83: 656–71.

Sachs, L. 1984. *Applied statistics: a handbook of techniques.* New York: Springer-Verlag.

Sasson, C., Cudnik, M.T., Nassel, A., Semple, H., Magid, D.J., Sayre, M., Keseg, D., Haukoos, J.S., and Warden, C.R. 2012. Identifying high-risk geographic areas for cardiac arrest using three methods for cluster analysis. *Academic Emergency Medicine* 19(2): 139–46.

Scheffé, H. 1959. *The analysis of variance.* New York: Wiley.

Shearmur, R., Apparicio, P., Lizion, P., and Polese, M. 2007. Space, time, and local employment growth: an application of spatial regression analysis. *Growth and Change* 38: 696–722.

Slocum, T. 1990. The use of quantitative methods in major geographical journals, 1956–1986. *Professional Geographer* 42: 84–94.

Staeheli, L., and Clarke, S.E. 2003. The new politics of citizenship: structuring participation by household, work and identity. *Urban Geography* 24: 103–26.

Standing, L., Sproule, R., and Khouzam, N. 1991. Empirical statistics: IV. Illustrating Meehl's sixth law of soft psychology: everything correlates with everything. *Psychological Reports* 69: 123–6.

Stouffer, S. 1940. Intervening opportunities: a theory relating mobility and distance. *American Sociological Review* 5: 845–67.

Takhteyev, Y., Gruzd, A., and Wellman, B. 2012. Geography of Twitter networks. *Social Networks* 34 (1): 73–81.

Tu, J. 2011. Spatially varying relationships between land use and water quality across an urbanization gradient explored by geographically weighted regression. *Applied Geography* 31: 376-92.

Tukey, J. W. 1972. *Some graphic and semigraphic displays.* In *Statistical papers in honor of George W. Snedecor.* Ed. T.A. Bancroft. Ames, IA: Iowa State University Press.

U.S. Bureau of the Census. 2011. *Mean center of population for the United States: 1790 to 2010.* Available at http://www.census.gov/geo/reference/pdfs/cenpop2010/centerpop_mean2010.pdf

Velleman, P.F., and Hoaglin, D.G., 1981. *Applications, basics, and computing of exploratory data analysis.* Belmont, CA: Wadsworth.

Waller, L.A., and Gotway, C.A. 2004. *Applied statistics for public health data.* New York: Wiley.

Weisberg, S. 1985. *Applied linear regression.* New York: Wiley.

Williams, K.R.S., and Parker, K.C. 1997. Trends in interdiurnal temperature variation for the central United States, 1945–1985. *Professional Geographer* 49: 342–55.

Wyllie, D.S., and Smith, G.C. 1996. Effects of extroversion on the routine spatial behavior of middle adolescents. *Professional Geographer* 48: 166–80.

INDEX

Pages in italics denote figures, pages in **bold** denote tables.